Excursions in the Land of Statistical Physics

Selected Reviews *by* Michael E. Fisher

Excursions in the Land of Statistical Physics

Selected Reviews *by* Michael E. Fisher

World Scientific

NEW JERSEY · LONDON · SINGAPORE · BEIJING · SHANGHAI · HONG KONG · TAIPEI · CHENNAI · TOKYO

Published by

World Scientific Publishing Co. Pte. Ltd.
5 Toh Tuck Link, Singapore 596224
USA office: 27 Warren Street, Suite 401-402, Hackensack, NJ 07601
UK office: 57 Shelton Street, Covent Garden, London WC2H 9HE

Library of Congress Cataloging-in-Publication Data
Names: Fisher, Michael E. (Michael Ellis), author.
Title: Excursions in the land of statistical physics / Michael E. Fisher, The University of Maryland, College Park, USA.
Description: Hackensack, NJ : World Scientific, [2016] | Includes bibliographical references.
Identifiers: LCCN 2016030573| ISBN 9789813144897 (hardcover ; alk. paper) |
  ISBN 9813144890 (hardcover ; alk. paper) | ISBN 9789813144903 (softcover ; alk. paper) |
  ISBN 9813144904 (softcover ; alk. paper)
Subjects: LCSH: Statistical physics.
Classification: LCC QC174.8 .F58 2016 | DDC 530.15/95--dc23
LC record available at https://lccn.loc.gov/2016030573

British Library Cataloguing-in-Publication Data
A catalogue record for this book is available from the British Library.

Cover photo credit: BBVA Foundation – Royal Spanish Physical Society (RSEF)

The author/editor and publisher gratefully acknowledge the permission granted to reproduce the copyright material in this book.

While every effort has been made to contact the publishers of reprinted papers prior to publication, we have not been successful in some cases. Where we could not contact the publishers, we have acknowledged the source of the material. Proper credit will be accorded to these publications in future editions of this work after permission is granted.

Copyright © 2017 by World Scientific Publishing Co. Pte. Ltd.

All rights reserved. This book, or parts thereof, may not be reproduced in any form or by any means, electronic or mechanical, including photocopying, recording or any information storage and retrieval system now known or to be invented, without written permission from the publisher.

For photocopying of material in this volume, please pay a copying fee through the Copyright Clearance Center, Inc., 222 Rosewood Drive, Danvers, MA 01923, USA. In this case permission to photocopy is not required from the publisher.

Preface

The invitation to assemble this book was a surprise. Undoubtedly a primary source for the invitation was the conference to honor Chen Ning Yang whom I first met personally on 19 November 1963 when Freeman Dyson invited me to speak at the Institute for Advanced Study in Princeton. Before my talk Freeman took me for lunch with the Director, J. Robert Oppenheimer; also joining the lunch was C. N. Yang and his guest, T. T. Wu, whose work I also knew.

In accepting the invitation to publish a collection of my review articles I was moved by the fact that I had always talked about my findings in a general way once I had been intrigued by a field. And, if I had actually contributed to an area of science, I was often moved to express what I had learned in its simplest terms and with lots of figures. Indeed, since my earliest days I had often framed my thoughts in terms of two- or three-dimensional images. It was also encouraging that my propensity for writing reviews — with crucial references carefully listed — had been recognized formally, in 1983, by the U.S. National Academy of Sciences through their James Murray Luck Award for excellence in scientific reviewing.

Given my interests in the Ising model, which had been inspired at King's College, London, by my postdoctoral advisor, Cyril Domb, I already knew of C. N. Yang via his important work evaluating the spontaneous magnetization of the two-dimensional Ising model — not to mention his 1957 Nobel Prize. By then, he had also published (in 1954) his to-be-famous work with Robert Mills that eventually led to the Conference: "*60 Years of Yang–Mills Gauge Field Theories: C. N. Yang's Contributions to Physics.*" My invited role at this Conference was to review Yang's work in Statistical Physics — which task I was delighted to undertake! Indeed, the resulting review, "*Statistical Physics in the Oeuvre of Chen Ning Yang,*" is the last (i.e., tenth) article reproduced in this volume.

On the other hand, the first article, written in 1991, is by Cyril Domb. It describes the first 16 years of my academic life — up to my move to Cornell University in 1966. Domb briefly mentions my early research work on ultra-high speed electronic analogue computation; that led to a book published jointly with my Ph.D. advisor, Donald M. MacKay, in 1962. However, my later studies were largely based on Domb's inspiring suggestions, many already featured in his inaugural lecture: *"Statistical Physics and its Problems."* It is my hope that this introductory article will provide readers with an overview of many of the topics later taken up and reviewed in greater depth.

The next article, *"The Theory of Condensation and the Critical Point,"* was based on a talk given in March 1965 at a Conference celebrating the 100th Anniversary of the University of Kentucky. The three other speakers were Lars Onsager (renowned for his 1944 explicit solution of the square-lattice Ising model in zero field), C. N. Yang, and Mark Kac (who had hosted my 1963–64 visit to the United States).

The analysis presented confirmed the idea of Joseph Mayer that an analytic singularity arises in thermodynamic functions at a first-order condensation point, such as the transition between liquid and gas; but a hard-to-detect *essential singularity* was predicted [as later confirmed rigorously by S. N. Isakov in *Commun. Math. Phys.* **95** (1984) 427–443]. However, the first-order boundary in, say, the pressure-temperature plane ends at a *critical point*: at this point were defined various basic *critical exponents*: α, β, γ and δ and relations between the exponents were justified. An exactly soluble one-dimensional model exhibiting the exponent relations was advanced.

The third article, *"The States of Matter — A Theoretical Perspective,"* was presented in Houston in 1979 at a Welch Foundation Conference on Chemical Research. Incidentally, it is graced by a photograph of the author at a stage when his beard was still (mostly!) black; and, in Figure 1, it contains an excellent photo of Lars Onsager — to whom the presentation was dedicated. In the remaining 36 figures are shown many phase diagrams, including for *tricritical* and *bicritical points*. Also defined are *Lifshitz points*, *protocriticality*, *Kosterlitz–Thouless points*, etc.

Figure 3 of Review 3, presents the *Plane of Theory* as a function of the *dimensionality*, d, and the *symmetry number*, n. The correlation exponents ν and η are introduced and the ideas of *scaling* and *conformal covariance* are explained briefly. A final feature of this section — true to Welch Foundation traditions — is a Discussion to which the principle contributors (apart from the author) are Paul A. Fleury (Bell Laboratories), Benjamin Widom (Cornell University), and Henry Eyring (University of Utah).

The next review, Number 4, starts with the claim that: "The first lecture in a conference should stimulate rather than strain, should place a subject in perspective rather than delve into detail, should respect the history but survey recent happenings, and, ideally, look to the future." The article *"Walks, Walls, Wetting and Melting"* attempts to do that; as regards the basics it reexamines what can be learned from random walkers on a lattice — in large part in $d = 1$ or 2 dimensions (although in the last section $d > 2$ is considered for lipid membranes).

Specifically, *vicious walkers* and their collective *reunions* are introduced and analyzed. This leads to insights in to a host of interactions between interfaces, between interfaces and rigid walls, and, effectively, between rigid walls. Phase transitions can arise in one-dimensional systems which, in turn, yield descriptions of wetting transitions to form an adsorbed phase, of the melting of surface phases, and of transitions in $p \times 1$ commensurate adsorbed phases.

In November 1985 Herman Feshbach of M.I.T. organized a Symposium at the American Academy of Arts and Sciences in honor of the Centennial of the birth Niels Bohr. He asked me to present a talk about condensed matter physics and its open problems with attention to the relevance of Bohr's ideas regarding the foundations of quantum mechanics (about which he had held strong views). Thinking about that led to the title: *"Condensed Matter Physics: Does Quantum Mechanics Matter?"* for the fifth article. Of course, the question was not meant facetiously. Indeed, it turns out that for many of the issues addressed in current condensed matter physics, quantum mechanics plays little role.

Thus, as explained in the section, *"Planck's Constant is Worth Only One Extra Dimension!,"* even if necessary, quantum mechanics is often not the main theoretical difficulty. So while for the *"Stability of Matter"* it is essential, for *"Polymeric Matter"* it is fundamentally irrelevant! And we also know that for the behavior near critical points at finite-temperature, e.g., in ferro- or antiferromagnets, quantum mechanics plays no role. Likewise in understanding *"Modulated Matter"* with long periods, as in, e.g., Ag_3Mg

and Ag$_3$Zn, the so-called ANNNI model plays a useful role, but quantum mechanics may be neglected. Similar remarks apply to *"Quasicrystals."* By contrast the opposite applies for the *"Quantum Hall Effect"!*

Josiah Willard Gibbs was, as I emphasized at Yale in May 1989: *"the outstanding theoretical physicist born and bred in the United States."* So, in *"Phases and Phase Diagrams: Gibbs's Legacy Today,"* it was a privilege, in Review No. 6, to present recent developments in his fundamental formulation of thermodynamics. To understand what may and what may not appear in a phase diagram, *convexity* proves a crucial concept. Critical end-points are tricky, especially as regards Gibbs's important "phase rule." Bicritical points even introduce singularities in the phase boundaries themselves.

Then, *anomalous first-order transitions*, were considered: in such a transition (illustrated in Figures 17 and 19 of Article 6) the pressure, for example, jumps discontinuously as a function of the volume (rather than *vice-versa* as in a standard gas–liquid transition). Such transitions (which must be *immovable*) have never been seen in real systems! Never-the-less, as explained, they may arise in models with *many-body forces*; so their non-observation set limits on many-body interactions. Finally, *random thermodynamic systems* were addressed briefly.

The seventh review selected is short and sweet: *"How to Simulate Fluid Criticality: The Simplest Ionic Model has Ising Behavior but the Proof is Not So Obvious!"* — only five pages long including references. But to make up for the brevity one should note that Reference 16 — added later — lists 14 articles, subsequently published, that are crucial to verifying theoretically the detailed critical behavior of the basic, hard-sphere ionic models.

Furthermore, it is worth noting that there were earlier, plausible, experimentally-based suggestions as to a different, in fact *classical*, critical behavior that, however, were simply contradicted by later experiments. See, on the one hand, M. E. Fisher, *J. Stat. Phys.* **75** (1994) 1–36, but, on the other hand, note studies by W. Schröer summarized in reference [2], *Adv. Chem. Phys.* **110** (2001) 1; and, already earlier, by S. Wiegand *et al.*, *J. Chem. Phys.* **106** (1997) 2777, and M. Kleemeier *et al.*, *J. Chem. Phys.* **110** (1999) 3085.

A complete change of topic, to BioPhysics, enters in review No. 8, coauthored with A. B. "Tolya" Kolomeisky, and entitled: *"Molecular Motors: A Theorist's Perspective."* Statistical mechanics still plays a crucial role: motor proteins, specifically, kinesins, myosins, and dyneins, are enzyme molecules which convert chemical energy in to mechanical work. Existing experimental techniques enable studies of *individual molecules*! Continuum "ratchet" models have been used but discrete stochastic models account for many aspects: the load- and [ATP]-dependence of motor velocities, behavior under stall, etc. Such models provide a flexible approach to many cellular mechanisms and processes — including [in later work: Y. Zhang and M. E. Fisher, *Phys. Rev. E* **82** (2010) 011923] the interactions of many motors.

Renormalization Group Theory (or RGT) — the conceptual child of my long-time Cornell colleague, Kenneth G. Wilson, who alas died prematurely in June 2013 — is the topic of review No. 9. This version was partially based on a talk given at a Memorial Symposium held at Cornell in November of that year; hence the full title: *"Renormalization Group Theory, the Epsilon Expansion, and Ken Wilson as I knew him."* But a principle aim was to provide a historical and conceptual account of RGT, especially as regards its roots in the theory of critical phenomena.

Kadanoff's scaling picture played an important role in Ken Wilson's account; but in many ways a crucial aspect was the conception of an unbounded space of Hamiltonians and the idea of flows in this space. Within that space Ken's "vision" of a variety of fixed points plays a vital role for the concepts of *universality* and *scaling*. The origin of the *epsilon expansion* (in powers of the deviation of the spatial dimensionality from $d = 4$) is related briefly.

In conclusion, it is my hope that a wide range of non-experts and, perhaps some experts also, will enjoy reading the variety of reviews collected in this volume.

Michael E. Fisher
19 June 2016

Contents

Preface	v
1. Michael Fisher at King's College London *Cyril Domb*	1
2. The Theory of Condensation and the Critical Point *Michael E. Fisher*	25
3. The States of Matter — A Theoretical Perspective *Michael E. Fisher*	59
4. Walks, Walls, Wetting, and Melting *Michael E. Fisher*	143
5. Condensed Matter Physics: Does Quantum Mechanics Matter? *Michael E. Fisher*	207
6. Phases and Phase Diagrams: Gibbs's Legacy Today *Michael E. Fisher*	255
7. How to Simulate Fluid Criticality: The Simplest Ionic Model has Ising Behavior but the Proof is not so Obvious! *Michael E. Fisher*	291
8. Molecular Motors: A Theorist's Perspective *Anatoly B. Kolomeisky and Michael E. Fisher*	297

9. Renormalization Group Theory, the Epsilon Expansion and
   Ken Wilson as I knew Him
   Michael E. Fisher   323

10. Statistical Physics in the Oeuvre of Chen Ning Yang
    Michael E. Fisher   367

1

Michael Fisher at King's College London*

Cyril Domb

Bar-Ilan University, Ramat-Gan 52100, Israel

MICHAEL Fisher spent the first 16 years of his academic life in the Physics Department of King's College, London, starting as an undergraduate and ending as a full professor. A survey is undertaken of his activities and achievements during the various periods of this phase of his career.

1. Introduction

The Physics Department at King's College can be proud that Michael Fisher spent the first 16 years of his academic career in the department. Starting as an undergraduate in 1948, he graduated in 1951, and left from 1951–1953 to perform his National Service as a teacher in the Air Force. In 1953 he came back as a graduate student, and also served as a tutor, demonstrator and part-time lecturer. His Ph.D. thesis was presented in 1957, and he held a Senior Research Fellowship awarded by the Government from 1956–1958. In January 1958 the College appointed him to a lectureship in Theoretical Physics, in 1962 he was appointed by London University to a Readership (equivalent to Associate Professorship in the USA) which had become vacant; in 1964 he obtained a personal promotion to the rank of Full Professor. In the summer of 1966 he left King's to take up an appointment at Cornell.

The first surprising thing about Michael Fisher's academic studies is that, despite his unusual mathematical ability, he did not go to Cambridge.

---
*Reprinted from *Physica A* **177**, 1–21 (1991), with permission from Elsevier.

There are strong pressures in British Schools to cream off the top mathematical talent and direct it to Cambridge where the teaching is indeed first-rate. But unusual circumstances combined to stop Michael from following the conventional path to Cambridge; he attributes his choice of King's amongst the Colleges of London University to the warm reception given to him when he applied for a place. His subsequent career demonstrates that he did not suffer by going to King's.

The King's College Physics Department has a distinguished history. James Clark Maxwell was professor there from 1860–1865, and three Nobel Prizes were won by its professors between 1917 and 1947 (Barkla, Richardson and Appleton). In 1947 Appleton moved to Edinburgh, and John Randall was appointed to succeed him as head of department. Randall had been responsible (with H. A. Boot) for the invention of the strapped magnetron, usually regarded as the most significant radar development of World War 2. On arriving at King's he took a crucial decision to leave his previous research field of solid state physics and move into biophysics. This decision played a key role in one of the most important discoveries of the 20th century, the unravelling of the structure of DNA, and the part played by the King's College physics department has been described in detail elsewhere.[1] The whole of this outstanding work took place during Michael's period at King's so the atmosphere in the department was exciting and stimulating.

The second major King's appointment in 1947 was Charles Coulson to the Chair of Theoretical Physics. Coulson, who was one of the founders of Theoretical Chemistry, was also a superb lecturer and teacher. Coulson left King's for Oxford in 1952, and was succeeded by Christopher Longuet-Higgins, who stayed only two years before leaving for Cambridge. When I was appointed to succeed Longuet-Higgins in 1954 Randall expressed two hopes, that I would produce good research, and that I would last longer than my predecessors. It is gratifying to record that at least I fulfilled the second aspiration — I stayed 27 years.

2. Graduate Student 1953–1956

Michael re-joined King's College in 1953 to study for a Ph.D. in analogue computing under the guidance of Donald MacKay. Michael regards Donald as his first "teacher" who gave him great encouragement as an undergraduate. One of the pioneers of information theory in Britain, MacKay was later appointed to a Chair in this subject established at the new residential university of Keele, Staffordshire. A persuasive speaker, and clear thinker, he was quickly impressed by Michael's unusual abilities, and was widely reported to have said that Michael would be a professor before the age of 30. He was not far wrong.

Early in Michael's student career an event of great consequence took place in his life. The student physics society (the Maxwell Society) was organizing a fresher's meeting, and Michael attempted to enroll a new student with a Spanish background. Sorrel Castillejo, who had just transferred to King's from Regent St. Polytechnic (where she had gained a degree in mathematics and physics). Sorrel's father had been a professor in Spain, and after the Spanish civil war he was forced to flee to Europe and subsequently to England. Michael was interested in Spanish music, and, on a later, less public occasion, initiated a discussion with Sorrel on the topic. This triggered a friendship which blossomed and ripened into marriage in 1954. Anyone who has enjoyed the Fishers' generous hospitality will testify how well the two are matched. In addition, the genes which represent ability in Physics have been passed onto the next generation, and there are now two additional Fishers well established, and showing great promise in research. (There are also good physics genes on the Castillejo side; Sorrel's brother Leonardo is a distinguished particle physicist who became Professor of Physics at University College, London.)

When I arrived at King's in October 1954, my first graduate student was John Zucker to whom I gave a problem related to the solidified inert gases. John had joined the College in 1950, and remembers the vociferous manner in which Michael argued points with Longuet-Higgins at theoretical physics seminars, showing little of the deference that one might expect from a new graduate student to his professor. I soon found that I shared many mathematical interests with Michael; I had just published a paper on iterative processes for which Michael showed particular enthusiasm because of his current activity with computational methods. There was an unusual type of iterative process in two variables which I had been turning over in my mind, a particular example of which had been used by Gauss. It had led Gauss to a functional equation, and I wondered whether one could not examine this type of process generally with a view to other applications. I put the suggestion to Michael, who responded positively, and we collaborated to publish a paper in Proc. Cambridge Philos. Soc. on the topic.[2]

Theoretical physicists were low on the priority list as regards room accommodation in the department. Most of the biophysics and other groups consisted of experimentalists who needed space for their laboratory work. Theoreticians who need only pen and paper (in the pre-computer days) were assigned to cubbyholes in which the old King's College specialized. A large windowless room in the bowels of the College was allocated to theoretical graduate students, and this had within it a smaller windowless room with space for two students. John Zucker relates that he was asked by Michael to share the inner room with him as he was the only one who did not smoke at all.

This accommodation had fruitful consequences for Zucker. One of the problems with which he was concerned was anharmonicity in solidified gases, and I had suggested that since one was here concerned essentially with short range effects, a type of approach taking near neighbours into account should be reasonable. For the more massive gases it was possible to use perturbation theory, but for lighter gases like deuterium, hydrogen and helium it was clear that the anharmonic terms could no longer be treated as a perturbation. John spent several months hunting unsuccessfully for a suitable method; by chance he mentioned the problem to Michael who immediately told him of a paper by Coulson and McWeeny[3] which effectively solved the problem and led eventually to his Ph.D.

A second problem with which I had been concerned was the Debye Θ value of solid helium whose reduced value is well over twice those of the heavier inert gas solids. Excellent measurements were available from Oxford, and I had suggested an empirical approximation[4] which led to a non-linear differential equation. Experimental values inserted into this differential equation gave consistent results, and I suggested to Zucker that he might try to solve this equation ab initio. Zucker found a number of exact solutions for unrealistic potentials, and tried numerical integration for the physically significant potentials, starting from these exact solutions. But the equations were non-linear and unstable, and it was not clear that anything meaningful was resulting.

One day, John mentioned the problem to Michael, and they spent a little while discussing it. *The next morning* Michael came to King's with a complete paper written on this particular equation and with some very pertinent comments on a related general class of non-linear differential equations. In fact he showed that my empirical approximation had been ill-advised, and the equation which I had proposed possessed only trivial solutions of no physical consequence!

There is an anti-climactic end to this story. The paper written overnight in 1957 took 4 years to get into print. Rumour has it that the manuscript was mislaid in the editorial offices of Proc. Cambridge Philos. Soc. and the paper appeared only in 1961.[5]

In 1955 I had delivered my inaugural lecture entitled "Statistical Physics and its Problems".[6] In it I had endeavoured to describe in terms intelligible to a general audience the challenge presented by discontinuous behaviour in physical systems, the importance of the concept of long-range order in explaining these discontinuities, and the difference between crystals, liquids and glasses. I had given some indication of the fascinating mathematical problems which arise in the statistics of order–disorder transitions and the Ising model, together with a brief reference to Onsager's tour de force exact solution, and the configurational problems of counting polygons on lattices. Michael told me afterwards that he was impressed by the variety of problems awaiting solution, and that if an opportunity arose he would be happy to work in the field. I had by then convinced him (with the help of Hartree) that the future of analogue computing looked pretty bleak.

In the summer of 1955 I visited Israel and met Aaron Katchalsky who had recently established a polymer division at the Weizmann Institute. He showed great interest in the branch of statistical mechanics in which I was working, and outlined to me the problem of the conformation of polyelectrolyte molecules in solution which ought to fit in this category. Primitive attempts had been made to tackle the problem, but he felt that a more sophisticated attack was warranted. When I returned I told Michael of this discussion, and we decided to use it as the basis for an application to the Department of Scientific and Industrial Research for a Senior Fellowhip. Our bid was successful, and Michael was guaranteed at least two years in which to start research in this area of statistical physics.

Michael's Ph.D. thesis presented in 1957 formed the basis for a book "Analogue computing at ultra-high speed — an experimental and theoretical study" published jointly with MacKay in 1962. Altogether his work in computing led him to the publication of a dozen papers.

Before moving to the next stage in his career, mention should be made of Michael's unusual natural talent for graphics, which all who have heard his lectures or collaborated with him must have surely noted. It was manifested at King's in his beautifully designed posters advertising the programme of the Maxwell Society, which organized public lectures for staff and students in the Physics Department. These are clearly remembered by all of his student contemporaries. Another skill from which his colleagues and friends have benefited is playing the flamenco guitar. Michael's interest in flamenco music is serious, and he has published a significant number of scholarly articles on the subject.

3. Research Fellow 1956–1958

Martin Sykes, my last student at Oxford, had shown a remarkable aptitude for counting configurations on lattices which provide the basis of the series expansion approach to critical behaviour. Together he and I had made considerable progress in unravelling the critical properties of a variety of two and three dimensional models, and had come to the conclusion, for example, that these properties were determined essentially by dimension and not by lattice structure.[7] When Sykes had completed his Ph.D. I arranged for him to spend a year in Neel's laboratory in Grenoble. Néel had been trying to account for the experimental results of Weiss and Forrer on nickel[8] and had formulated a theory which required interactions with hundreds of neighbours. Sykes and I demonstrated[9] that if one used a correct statistical theory, the experimental results could be well fitted by a model using only nearest neighbour interactions.

In 1957 I was able to secure an ICI Fellowship in London University for Sykes, and he joined our group at King's College. We were very keen to improve our configuration counting techniques so as to extend our series expansions (it was before the computer era) and we had devised a method which made use of returns to the origin in random walks with no immediate reversals. Michael showed great interest in our results, and I suggested to him that it would be interesting to investigate the general properties of such walks. I outlined a few ideas, which Michael took up with enthusiasm, and as a result we published a definitive paper[10] entitled "Random walks with restricted reversals".

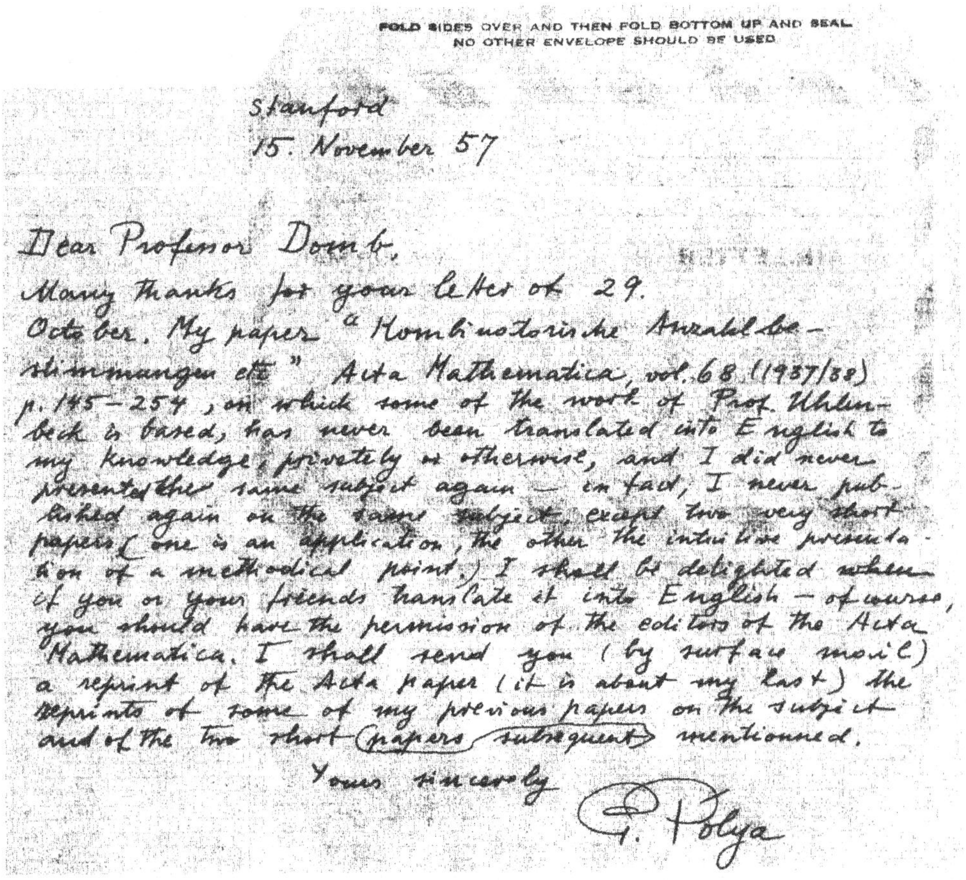

> FOLD SIDES OVER AND THEN FOLD BOTTOM UP AND SEAL
> NO OTHER ENVELOPE SHOULD BE USED
>
> Stanford
> 15. November 57
>
> Dear Professor Domb,
> Many thanks for your letter of 29. October. My paper "Kombinatorische Anzahlbestimmungen etc." Acta Mathematica, vol. 68 (1937/38) p. 145-254, on which some of the work of Prof. Uhlenbeck is based, has never been translated into English to my knowledge privately or otherwise, and I did never presented the same subject again — in fact, I never published again on the same subject except two very short papers (one is an application, the other the intuitive presentation of a methodical point.) I shall be delighted when if you or your friends translate it into English — of course, you should have the permission of the editors of the Acta Mathematica. I shall send you (by surface mail) a reprint of the Acta paper (it is about my last) the reprints of some of my previous papers on the subject and of the two short (papers subsequent) mentioned.
> Yours sincerely
> G. Polya

Figure 1. Letter received by the author from the distinguished mathematician George Polya giving permission for a classic paper in German to be translated into English. The proposal for such a translation came from Michael Fisher.

More generally Sykes and I were trying to develop a systematic approach to the enumeration of graphs on lattices[11] and were making use of the graph theoretical ideas of Uhlenbeck and his collaborators[12] based on the classic paper of Polya.[13] This paper was in German, and Michael proposed that I write to Polya to ask if we could translate the paper into English. My letter evoked a positive response (Figure 1), but we were unable to mobilize the necessary finances for the translation. However, Michael's initiative secured for me a personal reprint, the last available, of this outstanding work.

At that time, there were three members of staff in theoretical physics at King's, but none of them worked in statistical physics. The Reader, Lewis Elton, as interested in nuclear physics, the Lecturer, F. Booth, in biophysics, and the Assistant Lecturer, Michael Bernal, in computing. Hence, in my plans to establish a research group in statistical physics I depended solely on the support of Fisher and Sykes. We organized two sets of colloquia, the first for our small theoretical group, and the second with wider scope, for the research workers of the King's Physics Department and, hopefully, of other Colleges of London University.

```
                KING'S COLLEGE LONDON
                DEPARTMENT OF PHYSICS
                       SEMINARS

THURSDAY 23rd MAY.
DR. F. PIRANI            The Clock Paradox Controversy.


THURSDAY 30th MAY.
PROFESSOR C. DOMB.       Configuration Counting in Crystal
                         Statistics.


THURSDAY 13th JUNE.
DR. P.W. HIGGS.          What is Field Theory?


THURSDAY 20th JUNE.
MR. J. ZUCKER and DR. M.J.M. BERNAL.
                         The Equation of State of the Rare Gases
                         and Henkel's Anharmonic Approximation.


THURSDAY 27th JUNE.
DR. M.E. FISHER.         The Mathematics of Random Walks.



The seminars are held in the Physics Department (Room 21 c) on
Thursdays at 2.15 p.m. and are preceded by coffee at 2.05 p.m.
```

Figure 2. A typical programme of theoretical physics seminars at King's College (Summer 1957).

It is interesting to look at typical programmes at this period which I found in my files (Figures 2 and 3). Peter Higgs (Figure 2) was a contemporary, and good friend, of Michael Fisher who had been a graduate student of Christopher Longuet-Higgins. His thesis was concerned with molecular vibrations, but he had developed a strong interest in field theory which he now wished to pursue. Michael emphasized to me that Peter was very able, and felt that it was worth making an effort to keep him in our group. But I could not see how we could possibly secure a grant to accommodate a field theorist, and Peter went off to Edinburgh to work with Kemmer. During the past decade his work has received wide recognition, and Michael's early assessment of his ability has clearly been vindicated.

In regard to the wider colloquium (Figure 3), the following observations are relevant. Aaron Klug was later awarded a Noble prize for his work on the structure of viruses, Brian Pippard became Cavendish professor and John Hammersley was then a lonely pioneer of Monte Carlo methods which afterwards became so widespread; the lecture on "Winkles, Twinkles and Crinkles" dealt with wave structure and was given by *Michael* Longuet-Higgins, the oceanographer, and younger brother of Christopher.

The problem on polyelectrolytes to which Michael had been assigned turned out to be tough, but he was able to draw meaningful conclusions by the use of high temperature expansions.[14] The two years of his research fellowship were coming to a close; it might have been possible to apply for renewal, but the Lectureship in Theoretical Physics at King's became vacant (Booth moved to another institution) and Michael applied for it. There were several other applicants.

When the committee met to decide who to appoint, several of the members pointed out that the College did not normally appoint *initially* to a tenured position (a Lectureship carries tenure, an Assistant Lectureship does not). They proposed that Michael's first appointment should be as Assistant Lecturer, and perhaps after a couple of years he could be promoted. I countered with the argument that if one did not make unconventional gestures to people of unusual ability one was liable to lose them. It was gratifying that my view received the backing of John Randall, Head of the Department, and Peter Noble, the Principal, and Michael was appointed Lecturer from January 1st, 1958. It was an appointment that rapidly produced excellent dividends.

Michaelmas Term 1956-7

PHYSICS COLLOQUIA AND SEMINARS

The letter following the title of the lecture indicates the time and place of the lecture - see the 'key' below.

WEDNESDAY 17th OCTOBER.
DR.J.G.POWLES.   Nuclear Magnetic Resonance in Solids.   (B).

THURSDAY 25th OCTOBER.
DR.G.E.BACON.   Neutron Diffraction.   (K).

THURSDAY 1st NOVEMBER.
MR.J.M.HAMMERSLEY.   Monte Carlo Methods.   (K).

THURSDAY 8th NOVEMBER.
DR.L.R.B.ELTON.   A High Energy approximation to Scattering.   (K).

WEDNESDAY 14th NOVEMBER.
DR.A.B.PIPPARD.   The Simple Laws of Superconductivity.   (B).

THURSDAY 22nd NOVEMBER.
DR.A.KLUG.   The Structure of Viruses.   (K).

THURSDAY 29th NOVEMBER.
DR.M.S.LONGUET-HIGGINS. Winkles, Twinkles and Crinkles.   (K).

THURSDAY 6th DECEMBER.
DR.A.J.B.ROBERTSON.   Problems of Surface Action.   (K).

WEDNESDAY 12th DECEMBER.
DR.E.P.WOHLFART.   Some Recent Developments in the Theory   (B).
                                          of Ferromagnetism.

KEY.

Those lectures marked (B) are <u>Intercollegiate Colloquia on the **Physics of the Solid State**</u> and are held in the Physics Department of Birkbeck College on Wednesdays at 4.15.p.m.

Those marked (K) are held in the Physics Department of King's College (Room 21C) on Thursdays at 2.15.p.m.

Figure 3. A typical programme of University of London physics seminars and colloquia (Winter 1956–1957).

4. Lecturer 1958–1962

The day that Michael Fisher took up his new appointment I left King's for a seven month sabbatical with Elliott Montroll at Maryland. I had first met Elliott at Oxford a few years previously when we found that we had many interests in common. He had given me an open invitation for any bright students to spend time working in his department. I had recommended Ren Potts (of the Potts model) with whom he had collaborated in first rate exact work on the effect of impurities on the lattice vibration spectrum. When he requested further students, I asked if I might come myself, and he responded positively. However, I took a rest from statistical mechanics, since Elliott was then very absorbed in the theory of disordered lattices.

About mid-March I received a cable from Fisher and Sykes (Figure 4) saying that they had rigorous proofs that the self-avoiding walk limit on the simple quadratic lattice was greater than 2.52 and less than 2.70. The story is as follows. Onsager had observed privately to Temperley[15] that the boundary tension of the Ising model was given correctly by considering only a simple class of "forward moving" walks. Temperley then suggested that since this class of walks is by definition self-avoiding, it might also yield the correct SAW limit which would then be $\sqrt{2}+1 \simeq 2.414$. Michael immediately realized that this could not be correct. He defined a set of more

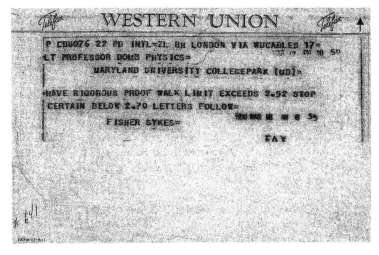

Figure 4. Cable received by the author in March 1958 during a sabbatical leave at the University of Maryland informing him of the refutation of certain current conjectures regarding the SAW limit for the simple quadratic lattice. Cable sent by Michael Fisher and Martin Sykes.

sophisticated forward moving walks which are also self-avoiding but have a higher limit of 2.4384. In fact. he introduced a hierarchy of such walks with steadily increasing limits, the 10th term of which yields a value 2.5767.

The value of e ($\simeq$2.718) had also been put forward speculatively by Lehman and Weiss[16] as a possible exact SAW limit for this lattice, and this proved a little more difficult to scotch. Some time previously, I had suggested to Sykes that he enumerate SAW's, and use the same type of analysis for them as we had used for the Ising model. By suitably manipulating the data an upper bound was derived which just excluded this speculation.

Fisher and Sykes published a comprehensive paper[17] which outlined clearly the configurational relationship between the Ising and SAW problems. Onsager's observation showed that there must be a remarkable cancellation of higher order graphs in the boundary tension enumerations. As far as I am aware this observation has still not received a proper explanation. The value of 2.639, which Fisher and Sykes gave as a final estimate of the SAW limit, differs only marginally from the best estimates today.

Soon after I returned to King's Michael produced another idea with far reaching implications. Sykes and I had become convinced that the susceptibility exponent γ for the Ising model was 7/4 for all two dimensional lattices. Michael felt that this could be derived by summing the correlations. Kauffman and Onsager[18] had shown that at T_c the correlations decayed as $r^{-1/4}$; Onsager in his original paper[19] had pointed out that away from T_c the correlations decayed as $\exp[-r(T - T_c)]$. If these two results were combined to give a correlation behaviour of the form

$$\Gamma(r, T) \sim r^{-1/4} \exp[-r(T - T_c)], \qquad (1)$$

the desired value of 7/4 would result. For an antiferromagnet the behaviour of the initial susceptibility would parallel that of the energy, and it would therefore have an infinite slope at T_c.[20]

I thought that these results were sufficiently important to write to Onsager about them. In his reply (Figure 5) Onsager expressed satisfaction with the result for the antiferromagnet, but felt that the ferromagnet needed a more detailed analysis to take proper account of the spectrum of eigenvalues. Whilst we were waiting for Onsager's reply, Michael had reached the same conclusion independently, and had modified his treatment appropriately.[#]

[#]The calculation is reproduced in Ref. 53.

DEPARTMENT OF CHEMISTRY
YALE UNIVERSITY
NEW HAVEN, CONNECTICUT

STERLING CHEMISTRY LABORATORY
225 PROSPECT STREET

18 February 1959

Dr. C. Domb
King's College, University of London
Strand, W.C. 2, London, England

Dear Dr. Domb:

Fisher's result for the anti-ferromagnetic case looks reasonable; but his analysis for the ferromagnetic case is not convincing.

My conjecture (I. 104) is of course related to (I. 108) as well as to (I. 83a); cf. also the discussion of (102). (Consider the Fourier coefficient of $e^{-2\gamma' 2r}$). The conjecture agrees with a separate exact computation for the diagonal direction. Incidentally, if (104) is right it suggests that a difference equation related to (108) will describe the most persistent part of the propagation of order, leading to

$$\omega_{\ell,m} \sim k^{-1/2} e^{-2(H^*-H)k}$$

in Fisher's notation, rather than the formula he uses. Just how this connects with results at and below the critical point is a question that seems to call for quite profound analysis. An eigenwert problem which crops up when you compute long range order the hard way suggests that terms of higher degree in $e^{-2\gamma'}$ crowd in and become important near the critical point.

As regards Σ_a there is small need for a better approximation than the one used in (III. 72), which is a special case of the exact formula for correlation along a main diagonal given by single determinants of the form (45), whose elements are the Fourier coefficients of the simple generating function

$$((ss' - e^{i\omega})/(ss'^{-1} - e^{-i\omega}))^{1/2} \equiv \sum a_n e^{in\omega}$$

$$\Delta = \begin{vmatrix} a_0 & a_1 \\ a_{-1} & a_0 \\ & & \ddots \end{vmatrix}, \quad s = \sinh 2H, \quad s' = \sinh' 2H$$

N.B. At T_c, $a_r = \frac{2}{\pi} \frac{1}{1-2r}$
and the value is then given by III.72 and my letter.

Figure 5. Letter sent to the author by Lars Onsager reacting to Michael Fisher's calculation of initial susceptibility of the Ising model in two dimensions. (The N.B. footnote was inserted by Michael Fisher.)

DEPARTMENT OF CHEMISTRY
YALE UNIVERSITY
NEW HAVEN, CONNECTICUT

STERLING CHEMISTRY LABORATORY
225 PROSPECT STREET

Dr. C. Domb   Page 2   18 February 1959

This result can be derived by way of a simple commutation rule from an application of the infinitesimal star-triangle transformation, which was quite helpful in the original solution of the Ising problem.

Still, as regards Δ_k in general, the best I can do so far is to compute Δ_∞ and its minors (below T_c) from the general theory of recurrent determinants, which I hope to publish before long.

That approach yields the degree of order and higher odd correlations in the ordered state without complete solution of the associated eigenwert problem but the latter may perhaps prove useful in dealing with the segments of the origin recurrent determinants, if the appropriate projection operators have tractable Fourier series in terms of the elliptic angles.

                                        Yours sincerely,

                                        Lars Onsager

LO:mrb

P.S. A generalization of (II.58) pertinent to the computation of $w_{\ell m}$ involves the angles $i m' \varphi_k + \ell'(k\pi/n) + \frac{1}{2}\delta_k'$

$m' = 0, 1, \ldots, m-1, \ell m$
$\ell' = 0, 1, \ldots (\ell-1)$ or ℓ.

This gives me a strong feeling that (I.104) is right.

                                        L.O.

Figure 5.   (*Continued*).

The critical behaviour predicted for an antiferromagnet meant that the susceptibility must have a maximum *above* T_c, and not at T_c as had always been assumed previously from mean field theories. Sykes examined the series expansions and found that there was indeed such a maximum. The important conclusions were published in the first volume of the *new* "Physical Review Letters" journal.[21]

Sykes and Fisher went on to provide estimates of the antiferromagnetic susceptibility for common two and three dimensional lattices which they presented in papers submitted to Physical Review. The papers were rejected by the referee who claimed that they were too long and detailed! I suggested that they try Physica, and I was gratified to hear very quickly that the papers were accepted.[22] To this day they remain classic references on the antiferromagnetic susceptibility of the Ising model.

Another topic pursued by Michael at this time was the transformation of Ising models. He showed how the star–triangle and decoration transformations could be generalized to yield exact results for a fascinating variety of different models.[23] He pursued these ideas further to calculate the exact solution of a particular antiferromagnetic model *in a non-zero field*.[24] The model differed markedly from the standard Ising model, but the exact results provided valuable insights on a number of key points.

It was encouraging at this stage that our research group was attracting graduate students of ability. The physics syllabus at King's allowed students to take a theoretical option in their final year and we did our best to make the option stimulating and inspiring. Virtually all who reached the required standard in the theoretical option decided to continue in research, although not all stayed at King's. Of those who did, many subsequently obtained faculty appointments in British and other Universities, and among those present during Michael's period at King's were David Burley, Basil Hiley, John Essam, Cyril Isenberg, David Gaunt, Geoff Joyce, John Stephenson, Norris Dalton, Dave Wood, Jill Bonner, F. T. Hioe, P. G. Watson and Douglas Hunter. There were many different fruitful avenues to pursue in our research programme, and there was little chance of getting in one another's way.

During my sabbatical in Maryland I became aware of the potential importance of percolation processes in solid state physics. Although Broadbent and Hammersley[25] had introduced the statistical problem in 1957 their ideas had not penetrated into the world of physics. When I returned to

King's, I introduced the basic ideas to our research group, and showed how series expansions could be derived at low and high densities. We were then presented with a new field to which all the techniques of the Ising model could be applied. Michael took up the challenge with alacrity, and quickly set to work with his graduate student John Essam deriving closed form solutions.[26] He contributed significantly to the development of our ideas in this field and introduced the important concept of the *covering lattice*[27] which serves as a link between site and bond percolation processes. His student, John Essam, became a leading world authority on the subject.

In 1961 another vista opened before our eyes with the publication of George Baker's paper applying the Padé approximant to Ising model series.[28] His results firstly confirmed our conjectures relating to high temperature critical exponents. Secondly, and of more importance, he was able to derive information on critical behaviour from low temperature series in which the singularity of physical interest is usually masked by other spurious singularities. It was clear that there was ample scope for further investigating this remarkable new technique, and applying it to new expansions which we were in the process of deriving. Michael Fisher's earlier experience with computation was particularly useful in dealing with the technical problems posed in the calculation of Padé approximants, and he soon gained wide experience in their application. He subsequently became a leading expositor at conferences devoted to the Padé approximant, and the innovator of *partial differential approximants*[29] which tackle the question of generalizing the technique to functions of several variables.

George Baker came to spend a year's leave with our group, and thereby initiated a research collaboration which bore fruit for many subsequent years. Another visitor who established a cordial relationship with our group was Donald Betts, and for several years we enjoyed a close association with the University of Alberta. Michael Fisher was instrumental in bringing John Nagle to spend a year with us, who introduced us to Onsager's ideas on the ice-problem, and showed how to develop and use series expansions for ferro-electric models. Again, we established a valuable continuing association.

From time to time new methods were devised for deriving Onsager's exact solution, and in 1960 Hurst and Green showed how the result could be obtained by the use of Pfaffians.[30] Michael Fisher, who had been involved previously with Temperley in a discussion of the problem of fitting monomers and dimers on a lattice,[31] realized that this technique enabled an important exact result to be calculated for the dimer problem. The same idea occurred independently to Temperley, and they published a joint paper giving the result.[32] Michael subsequently published the details of his calculation,[33] and then found out that a third researcher, P.W. Kasteleyn, had made the same discovery independently.[34]

In the spring of 1962 Lewis Elton decided to leave King's College to join the new University of Surrey, and the Readership in Theoretical Physics fell vacant and Michael applied for it. This time the Appointments Committee showed no hesitation. Without even bothering to convene formally, they agreed that Michael Fisher was the best candidate. He was duly appointed at the beginning of the new academic year in 1962.

5. Reader 1962–1964

In the Spring of 1962 I had a further communication from Elliott Montroll. He was organizing a Conference on "Irreversible Thermodynamics and the Statistical Mechanics of Phase Transitions" at Brown University, Providence, Rhode Island, in honour of the 30th anniversary of Onsager's paper on irreversible thermodynamics, and the 20th anniversary of the publication of his formula for the partition function of the Ising model. (An honorary degree was conferred on Onsager during the conference.) Elliott was interested in the names of bright young people who might be participants. I strongly recommended Michael Fisher, and in due course an invitation was sent to him. When he learned that Piet Kasteleyn was also participating, and was to speak on the exact solution of the dimer problem, Michael decided to concentrate on the Ising model, and to present a comprehensive survey of exact and conjectured data relating to it.[35]

Although he had visited Brookhaven in the summer of 1959, I believe that this was the first conference that Michael attented in the USA. I learned from Elliott that he had an immediate impact. Among the many contacts which Michael made were Mark Kac and George Uhlenbeck, who had recently moved to the newly established Rockefeller Institute in New York. Michael received an invitation from Mark Kac to spend the academic year 1963–1964 at Rockefeller, which he subsequently accepted. This proved to be the first major step towards his move to the USA.

The papers presented at the Conference were published in a special issue of J. Math. Phys. (Vol. 4, No. 2) of which Elliott was then editor.

Back at King's, Michael pursued a number of fruitful ideas. He showed that the close relationship between the specific heat and susceptibility of an antiferromagnet is not confined to the Ising model.[36] With Jill Bonner he explored the properties of the linear anisotropic Ising–Heisenberg chain by numerical methods (the paper which they published subsequently[37] became a citation classic). He responded to Mel Green's proposal for correlations in fluids showing that it did not accord with Onsager.[38] But the paper which was to have the most far reaching consequences was written with Essam, and carried the formalistic title "Padé approximant studies of the lattice gas and Ising ferromagnet below the critical point".[39] Most of the paper was concerned with numerical estimates of low temperature critical exponents, but the final section introduced the heuristic droplet model which led to the exponent relation

$$\alpha' + 2\beta + \gamma' = 2. \qquad (2)$$

This was the first suggestion that there might be relations between critical exponents.

The paper sparked off Rushbrooke's classic note[40] introducing his thermodynamic inequality

$$\alpha' + 2\beta + \gamma' \geq 2. \qquad (3)$$

Rushbrooke has told me that for some time he had been turning over in his mind the possibility of applying the thermodynamic relation for $C_H - C_M$ to behaviour near the critical point, but he had been troubled about the manipulation of exponents on the low temperature side of T_c because of the spontaneous magnetization. Essam and Fisher calculations clarified the picture for him, and led to his publication, and this in turn promoted the discovery of a number of other important thermodynamic inequalities.[41]

Michael Fisher developed his ideas on the droplet model and presented them in a lecture given at the Centennial Conference on Phase Transformation held at the University of Kentucky in 1965. This was eventually published in the short lived journal "Physics",[42] and contained the innovating suggestion that the free energy has an essential singularity at the phase boundary.

Finally, with David Gaunt, he pursued an investigation which seemed to me to be totally divorced from reality — the Ising model and self-avoiding walks in space dimensions of four and above.[43] Expansions were derived for critical parameters as a function of dimension. Of course they were then unaware of the concept of a critical dimension at which the exponents attain their mean field values, but I believe that the expansion in ϵ ($=4-d$)[44] which subsequently played such a central role in the renormalization group development, owes its origin to this earlier work.

6. Leave of Absence in the USA 1963–1964

When Michael left for Rockefeller in 1963 we felt that considerable theoretical progress had been made in unravelling the critical behaviour of the Ising and Heisenberg models, but contact with experiment had been minimal. During his stay in the USA Michael met with experimentalists, presented them with the challenge of the new theoretical work, and stimulated them to carry out experiments which could test the theory. He studied the Soviet literature and came across the seminal work of Voronel on the specific heat of argon which he fitted to the King's College lattice gas data.[45] A talk given at a Gordon Conference stimulated Jim Kouvel to suggest that useful information was contained in old experimental data, but it needed to be extracted properly.[46] At the end of his stay he delivered a series of lectures entitled "The nature of critical points" at Boulder, Colorado,[47] which, whilst basically theoretical, paid serious attention to experimental results. These lectures played an important part in putting critical phenomena on the map.

However, the most significant consequence of his stay at Rockefeller was the publication of the review "Correlation functions and the critical region of simple fluids".[48] In it Michael classified the status of the classic work of Ornstein and Zernike on fluids, and related it to the ideas on correlations which he had been developing for the Ising model. (He gives credit to Uhlenbeck for directing him to the relevant literature, and for critically reading the manuscript.) Michael outlined the modifications to Ornstein and Zernike which the modern theories necessitated, introduced the critical exponents ν and η and related them to the susceptibility exponent γ by the formula which has subsequently always been associated with his name. Taking Fourier transforms he showed how the classical theories of critical scattering must be modified to incorporate modern ideas. Finally he gave numerical estimates of ν, η in three dimensions ($\nu \approx 0.644$, $\eta \approx 0.060$) based on the Ising model work which he was currently pursuing with his student R. J. Burford. This paper can legitimately be described as a landmark.

From the letters which Michael wrote to me at that time it is clear that he found the research atmosphere in the USA very stimulating, and he was beginning to attract high level offers of appointment. I tried to paint as rosy a picture as I could of future prospects in England, but he replied:

"I must admit that statistical mechanics (and other fields of my special interests) do seem to me relatively undervalued in England (where are the F.R.S.'s for example?)... For some reason most of my scientific contacts outside King's do seem to have developed with people abroad (here and in Holland). Perhaps it is my fault that I have never approached ... senior people but making such contacts seems much easier over here".

However, I was gratified that Michael decided to return for the beginning of the 1964–1965 academic year, and that King's College recognized his greatly enhanced status by promoting him to a personal chair.

7. Professor 1964–1966

Michael brought back with him the interesting news of an international conference to be held in April 1965 at the Bureau of Standards in Washington devoted solely to critical phenomena. Mel Green was the host, and Michael was a member of the organizing committee. The aim of the conference was to bring together leading theoreticians and experimentalists in the field, so that each group became informed of what the other was doing, to assess the current situation, and formulate the major problems to be tackled in the future.

The conference was an outstanding success and undoubtedly served as a catalyst for the remarkable progress during the years which followed. Michael's contribution was devoted to the theory of correlations, and its implication for critical scattering of neutrons, the new technique which was then emerging as a powerful experimental tool. He was particularly concerned to promote a search for experimental evidence of the elusive exponent η; it might be small in three dimensions, but he was sure that it was non-zero. Incidentally, it was at this conference that Michael proposed the notation for critical exponents which has been accepted universally.

When we returned to London I suggested to Michael that the time was ripe for a review of theoretical developments in critical phenomena. My own earlier review[49] was now out of date. I was serving as Chairman of the Editorial Board of "Reports on Progress in Physics" which had a well earned reputation for first class up to date reviews, and I was convinced that Michael could do a first rate job. We would commission an independent experimental review and the two would be published together. After a good deal of persuasion Michael reluctantly agreed; on several occasions during the next year he told me that he deeply regretted this agreement since it restricted his research activities. He was also particularly annoyed that when the review appeared in 1967[50] it was full of misprints, and needed a subsequent "correction sheet". But the review had a very substantial impact and soon achieved the status of a citation classic. I marked my personal reprint "Not to be taken from this room".

A year or two earlier Michael had made the acquaintance of David Ruelle in Paris, and had been impressed by the rigorous approach which he was developing towards the fundamentals of statistical mechanics. Michael himself had been thinking along similar lines in response to the stimulus of Uhlenbeck and van Kampen. At Michael's suggestion I invited him over to give some lectures at King's. He and Michael collaborated in at least one publication, which supplemented other papers which Michael had published in the same area.[51]

There were several fruitful avenues along which Michael was directing graduate students. Having been born in the West Indies himself, Michael welcomed the opportunity to supervise a West Indian student, A. E. Ferdinand, with whom he explored exact two-dimensional solutions for finite systems. The work involved exacting calculations with elliptic functions, and led to important conclusions relating to finite size and surface effects.[52] With R. J. Burford he was completing the calculations for the definitive paper to be published on correlations in the Ising model;[53] and with David Gaunt he was looking at the nature of the transition in a hard square lattice gas.[54]

But there are two papers of this period to which I should like to refer in more detail since they seem to provide an excellent characterization of Michael's versatility. In 1963 whilst Michael was at Rockefeller, I spent several months on sabbatical at The Weizmann Institute in Israel working on SAW's. On the basis of John Martin's Computer enumerations,[55] I had become convinced[56] that the mean squared length exponent θ (often called ν subsequently) ($\langle R_N^2 \rangle \sim N^{2\theta}$) is 3/4 in two dimensions and 3/5 in three dimensions. (In those days we were biased in the direction of rational exponent values.) But what about the shape of the walk? How did it deviate from Gaussian? After careful numerical analysis with Gillis and Wilmers,[57] we suggested that the shape was asymptotically $\exp[-(r/r_0)^\delta]$ with δ equal to 4 in two dimensions and 2.5 in three dimensions. When I told these results to Michael he made a plausible assumption that the walk had a limiting shape, and after Fourier transforming, derived the exponent relation[58]

$$\theta + 1/\delta = 1. \tag{4}$$

This was satisfied by our estimates in two and three dimensions, as well as by Gaussian walks. We were convinced that it must be correct, as in fact was confirmed subsequently with the emergence of the scaling theories.

The second paper "On hearing the shape of a drum" was stimulated by some work of Marc Kac[59] who posed the question "Is it possible to determine the shape of a membrane from its spectrum of characteristic vibrations?" Using continuum methods Kac derived a number of significant relations from the spectrum, but had a general heuristic feeling that the shape was not uniquely determined by the spectrum.

Michael Fisher re-formulated the problem for a discrete model, with the membrane replaced by point masses at the vertices of a lattice, and having a harmonic interaction between nearest neighbours. He related the problem to random walks on the lattice, and readily derived the same relations that Kac had derived for the continuum model. He then considered the same problem for finite linear graphs, and showed that for all linear graphs of 6 or fewer points the spectrum was unique. But he was able to produce a counterexample of two *distinct* graphs of much higher order with identical spectra. He was led to this counterexample by returns to the origin in random walks which have been mentioned earlier. The paper was published in the first volume of the new Journal of Combinatorial Theory.[60]

In December 1965 Michael Fisher informed the King's College authorities of his intention to resign from his appointment at the end of the academic year. Of the tempting offers from the USA it is particularly fortunate that he chose to go to Cornell where he met up with Ken Wilson. The renormalization group approach which was a direct consequence of this meeting, and for which Wilson subsequently received a Nobel prize, was a direct consequence of this decision. In recognition of his outstanding achievements at King's College, Michael was later elected a Fellow of the College.

Acknowledgements

I am deeply grateful to John Zucker and Peter Higgs for valuable information relating to the early years at King's.

References

1. J. D. Watson, *The Double Helix* (Atheneum Press, 1968).
2. C. Domb and M. E. Fisher, *Proc. Cambridge Philos. Soc.* **52**, 652 (1956).
3. C. A. Coulson and R. McWeeny, *Proc. Cambridge Philos. Soc.* **44**, 413 (1948).
4. C. Domb and J. S. Dugdale, *Prog. Low. Temp. Phys.* **2**, 338 (1957).
5. M. E. Fisher and I. J. Zucker, *Proc. Cambridge Philos. Soc.* **57**, 107 (1961).
6. C. Domb, *Sci. Prog.* **403**, 402 (1955).
7. C. Domb and M. F. Sykes, *Proc. R. Soc. A* **235**, 247 (1956).
8. P. Weiss and R. Forrer, *Ann. Phys.* **5**, 153 (1926).
9. C. Domb and M. F. Sykes, *Proc. R. Soc. A* **240**, 214 (1957).
10. C. Domb and M. E. Fisher, *Proc. Cambridge Philos. Soc.* **54**, 48 (1958).
11. C. Domb and M. F. Sykes, *Philos. Mag.* **2**, 733 (1957).
12. E.g. R. J. Riddel and G. E. Uhlenbeck, *J. Chem. Phys.* **21**, 2056 (1953).
13. G. Polya, *Acta Math.* **68**, 145 (1937).
14. M. E. Fisher, *J. Chem. Phys.* **28**, 756 (1958).
15. H. N. V. Temperley, *Proc. Cambridge Philos. Soc.* **48**, 683 (1952); *Phys. Rev.* **103**, 1 (1958).
16. R. S. Lehman and G. H. Weiss, *J. Soc. Indust. Appl. Math.* **6**, 257 (1958).
17. M. E. Fisher and M. F. Sykes, *Phys. Rev.* **114**, 45 (1959).
18. B. Kauffman and L. Onsager, *Phys. Rev.* **76**, 1244 (1949).
19. L. Onsager, *Phys. Rev.* **65**, 117 (1944).
20. M. E. Fisher, *Physica* **25**, 521 (1959).
21. M. F. Sykes and M. E. Fisher, *Phys. Rev. Lett.* **1**, 321 (1958).
22. M. E. Fisher and M. F. Sykes, *Physica* **28**, 919, 939 (1962).
23. M. E. Fisher, *Phys. Rev.* **113**, 969 (1959).
24. M. E. Fisher, *Proc. R. Soc. A* **254**, 66 (1960); **256**, 502 (1960).
25. S. R. Broadbent and J. M. Hammersley, *Proc. Cambridge Philos. Soc.* **53**, 629 (1957).
26. M. E. Fisher and J. W. Essam, *J. Math. Phys.* **2** 609 (1961).
27. M. E. Fisher, *Proc. IBM Symp. on Combinatorial Problems, March 16–18, 1964*, 179 (1966).
28. G. A. Baker, *Phys. Rev.* **124**, 768 (1961).
29. M. E. Fisher, *Physica B* **86–88**, 590 (1977).
30. C. A. Hurst and H. S. Green, *J. Chem. Phys.* **33**, 1059 (1960).
31. M. E. Fisher and H. N. V. Temperley, *Rev. Mod. Phys.* **32**, 1029 (1960).
32. H. N. V. Temperley and M. E. Fisher, *Philos. Mag.* **6**, 1061 (1961).
33. M. E. Fisher, *Phys. Rev.* **124**, 1664 (1961).
34. P. W. Kasteleyn, *J. Math. Phys.* **4**, 287 (1963).
35. M. E. Fisher, *J. Math. Phys.* **4**, 278 (1963).
36. M. E. Fisher, *Philos. Mag.* **7**, 1731 (1962).
37. J. Bonner and M. E. Fisher, *Phys. Rev.* **135**, A 640 (1964).

38. M. E. Fisher, *Physica* **28**, 172 (1962).
39. J. W. Essam and M. E. Fisher, *J. Chem. Phys.* **38**, 802 (1963).
40. G. S. Rushbrooke, *J. Chem. Phys.* **39**, 842 (1963).
41. See, e.g., H. E. Stanley, *Phase Transitions and Critical Phenomena*, ch. 4 (Oxford Univ. Press, Oxford, 1971).
42. M. E. Fisher, *Physics* **3**, 255 (1967).
43. M. E. Fisher and D. S. Gaunt, *Phys. Rev.* **133**, A 224 (1964).
44. K. G. Wilson and M. E. Fisher, *Phys. Rev. Lett.* **28**, 240 (1972).
45. M. E. Fisher, *Phys. Rev.* **136**, A 1599 (1964).
46. J. S. Kouvel and M. E. Fisher, *Phys. Rev.* **136**, A 1626 (1964).
47. M. E. Fisher, *Lectures in Theoretical Physics*, Vol. VIIC, p. 1 (Univ. of Colorado Press, Boulder, CO, 1965).
48. M. E. Fisher, *J. Math. Phys.* **5**, 944 (1964).
49. C. Domb, *Adv. Phys.* **9**, 149, 264 (1960).
50. M. E. Fisher, *Rep. Prog. Phys.* **30**, 615 (1967).
51. M. E. Fisher and D. Ruelle, *J. Math. Phys.* **7**, 260 (1966); M. E. Fisher, *Arch. Rat. Mech. Anal.* **17**, 377 (1964); *J. Chem. Phys.* **42**, 3852 (1965).
52. M. E. Fisher and A. E. Ferdinand, *Phys. Rev. Lett.* **19**, 169 (1967).
53. M. E. Fisher and R. J. Burford, *Phys. Rev.* **156**, 583 (1967).
54. D. S. Gaunt and M. E. Fisher, *J. Chem. Phys.* **43**, 2840 (1965).
55. J. L. Martin, *Proc. Cambridge Philos. Soc.* **58**, 92 (1962).
56. C. Domb, *J. Chem. Phys.* **38**, 2957 (1963).
57. C. Domb, J. Gillis and G. Wilmers, *Proc. Phys. Soc.* **85**, 685 (1965).
58. M. E. Fisher, *J. Chem. Phys.* **44**, 616 (1966).
59. M. Kac, *Am. Math. Monthly* **73**, 1 (1966).
60. M. E. Fisher, *J. Combin. Theory* **1**, 105 (1966).

2

The Theory of Condensation and the Critical Point*

Michael E. Fisher

Baker Laboratory, Cornell University, Ithaca, New York 14850

THE droplet or cluster theory of condensation is reviewed critically and extended. It is shown to imply that the condensation point is marked by a singularity of the thermodynamic potential as conjectured by Mayer. The singularity turns out to be an essential singularity at which all derivatives of the thermodynamic variables remain finite. The theory also yields an understanding of the uniqueness of the critical point (in contrast to an extended critical region or Derby-hat type of behaviour) and leads to relations between the various critical point singularities.

A one-dimensional model is described with a Hamiltonian containing short-range many-body potentials, The exact solution of the model is sketched and shown to exhibit condensation and critical phenomena for suitable (fixed) potentials. The analysis confirms the conclusions of the cluster theory and thereby lends support to the validity of its underlying assumptions.

1. Introduction

In this lecture I will review some old ideas concerning the theory of condensation and will describe some new work and new conclusions that may also throw light on the nature of the critical region. To introduce our topic let us pick out some of the highlights in the historical development.

---

*This article is the text of a lecture given at the Centennial Conference on Phase Transformation held at the University of Kentucky, Lexington, Kentucky, 18th–20th March 1965. It is reproduced here

Onsager by his exact solution of the two-dimensional Ising model over twenty years ago[1] demonstrated clearly how a phase transition would come out of statistical mechanics alone if only one were clever enough to compute the partition function precisely: there is no need of any additional assumptions, special procedures or the like. This is the philosophy to which I will adhere throughout, namely, to describe the *equilibrium* properties of a physical system we need only calculate the partition function and use the formalism of statistical mehcanics.

In 1952 Yang and Lee,[2] showed how the discontinuities and related mathematical singularities which characterize the thermodynamic potentials of systems exhibiting phase transitions could "grow" from the completely smooth and analytic partition functions of finite systems. This occurs, of course, only when one proceeds to the "thermodynamic limit" in which the volume of the system becomes infinite while the intensive variables remain finite. The analysis of Lee and Yang is very general and does not therefore reveal the detailed nature of the singularities to be expected in any particular type of phase transition.

More recently Kac, Uhlenbeck and Hemmer,[3] have shown how the old ideas of van der Waals give some answers to the question of the nature of the condensation singularities, at least in the case of a system of particles interacting with very weak, long range attractive forces (in addition to the ever present strong short range repulsive forces). More precisely the van der Waals description of the condensation of a fluid becomes rigorously correct if the limit of infinitely long-range and infinitely weak attractive pair interaction forces is taken *after* the thermodynamic limit.[3,4] While it adds appreciably to our understanding of phase transitions, this conclusion is not entirely satisfying since the systems in which one sees condensation phenomena experimentally are characterized, in the main, by attractive forces of quite short range. One suspects (and indeed the soluble one-dimensional examples tell us) that the behaviour of such systems may be appreciably different.

---

from *Physics* **3**, 255–283 (1967), essentially in the same form as presented and later circulated privately except for the addition of Appendix B. The other three speakers at the Contennial Conference were Lars Onsager, Mark Kac and C. N. Yang. Attention should be drawn to the following more recent articles: (a) as regards the condensation point; J. S. Langer, *Ann. Phys. (N.Y.)* **41**, 108 (1967); (b) concerning critical point singularities, B. Widom, *J. Chem. Phys.* **43**, 3892, 3898 (1965); C. Domb and D. L. Hunter, *Proc. Phys. Soc.* **86**, 1147 (1965); L. P. Kadanoff, *Physics* **2**, 263 (1966); A. Z. Patashinskii and V. L. Pokrovskii, *Soviet Phys. JETP* **23**, 292 (1966); (c) in regard to related types of one-dimensional model: L. K. Runnels, *J. Chem. Phys.* **42**, 212 (1965); D. Poland and H. Scheraga, *J. Chem. Phys.* **45**, 1464 (1966); M. E. Fisher, *J. Chem. Phys.* **45**, 1469 (1966).

This observation forms the point of departure of the present discussion. We will endeavour to discover the type of analytic behaviour which occurs in the neighbourhood of a condensation point and will conclude that is probably quite different, and much more subtle, than the classical van der Waals description would suggest. I will also have something to say about the nature and uniqueness of the critical point which marks the limit of the condensation points. Our theme will thus be one of "hunt the singularity"!

2. Mayer's Conjecture

Experimentally one observes that a system of molecules interacting through short range repulsive forces (giving the molecules an essentially incompressible "core") and short range attractive forces, will undergo an abrupt transformation from a gaseous to a liquid state (or to a solid state at low enough temperatures) even at *very low (gaseous) densities*. It is this low density condensation phenomena occurring well below the critical temperature, which we shall consider. It is significant, furthermore, that the isothermal compressibility of the gas

$$K_T = \rho^{-1}(\partial \rho/\partial p)_T \tag{1}$$

is observed to *remain finite* as the condensation point $\rho = \rho_\sigma$ is approached (see Figure 1). The symbols p, ρ and T denote, as usual, the pressure, density and temperature.

If the system is allowed to go into a *non-equilibrium* state one can also find experimentally a metastable "continuation" of the gaseous isotherm into a region describing a supercooled vapour (see Figure 1). Although, when suitable experimental precautions are taken, this isotherm can be quite prolonged and reproducible it must be stressed that the system is no longer in a state of complete thermodynamic equilibrium. Furthermore it may be shown rigorously that such nonequilibrium metastable states *cannot* be found in a correct statistical mechanical calculation based on the total partition function for the system.[5] (If they *are* found the calculation has been incorrect!).

As is well known the van der Waals, and equivalent theories of condensation yield an isotherm with a "loop" (as shown in Figure 2) that continues analytically through the condensation region and even includes an "unstable" portion with negative compressibility. The correct

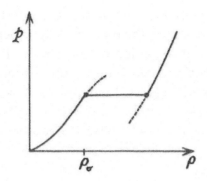

Figure 1. An isotherm illustrating condensation in a real system.

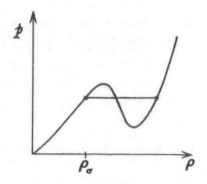

Figure 2. Condensation in a van der Waals system.

equilibrium isotherm must be found by a Maxwell, or better, Gibbs construction which yields the horizontal "two-phase" portion of the isotherm so "cutting off" the "metastable" and "unstable" portions of the loop. (This construction comes automatically out of the rigorous Kac, Uhlenbeck and Hemmer theory.[3,4])

In consequence the gaseous isotherm remains perfectly smooth and analytic as the condensation point is approached: there is *no singularity* (in the mathematical sense of a nonanalytic point) on the isotherm to "warn" of the onset of condensation. On the contrary the isotherm may be analytically continued to larger densities to yield the original looped isotherm. (The fact that the van der Waals type of theory describes a "metastable" continuation of the isotherm in this way is sometimes considered to be an advantage of the theory; in my opinion, however, the comments made above indicate that this is unjustified.)

The converse suggestion, namely, that the gaseous isotherm should exhibit some sort of mathematical singularity at the condensation point, which could thus be located even if the two-phase and liquid parts of the isotherm were unknown, was made by Mayer.[6] He based his arguments on the fundamental expansions

$$p/kT = \pi(z) = \sum_{l=1}^{\infty} b_l z^l, \tag{2}$$

$$\rho = z \frac{\partial \pi}{\partial z} = \sum_{l=1}^{\infty} l b_l z^l, \tag{3}$$

and on the related virial expansion

$$p/kT = \sum_{l=1}^{\infty} \beta_l \rho^l, \tag{4}$$

obtained by eliminating the activity z. Mayer tried to analyse the form of the cluster coefficients b_l for large l since one knows from the theory of analytic functions that this determines the nature of the singularities of the function. The radius of convergence of a power series such as (2) is determined by the behaviour of the coefficients for large l through

$$r_0 = \liminf_{l \to \infty} |b_l|^{-1/l}. \tag{5}$$

On the circle of convergence $|z| = r_0$, there must be at least one singularity z_0 of the function $\pi(z)$ and no other singularities in the complex plane can lie closer to the origin. Furthermore if the coefficients are of uniform sign for large l the dominant singularity z_0 lies on the positive real axis.

In its strong form we may express Mayer's conjecture by the assertion that a closest singularity $z = z_0$ of $\pi(z)$ will lie on the positive real axis and will occur at the condensation point $z = z_\sigma$. In this form the conjecture is quite possibly wrong in most realistic cases, since accumulating evidence suggests that the b_l will oscillate in sign for large l (essentially owing to the repulsive cores of the molecular interactions) so that the singularity z_0 determining the circle of convergence will lie on the *negative* real axis (i.e. in an unphysical region). I want to consider, however, the more general form of the conjecture which asserts merely that the function $\pi(z)$, defined by the power series (2) and its analytic continuation, has, *on the real axis*, a *nearest singularity* $z = z_1$ which occurs *at the condensation point* $z = z_\sigma$. We might then have $z_1 > |z_0|$, so that z_1 lies outside the circle of convergence of (2), but since the series may be analytically continued along the real axis up to z_1 this does not matter.

Of course, the van der Waals theory constitutes a counterexample to this conjecture since the nearest singularity on the real z axis occurs beyond the condensation point i.e. $z_1 > z_\sigma$. In fact the value of z_1 corresponds to the first point of infinite compressibility on the analytically continued isotherm (which is sometimes termed the "limit of metastability of the gaseous phase"). In a review of this question Katsura[7] has argued that this should be the general case. As we have pointed out, however, the van der Waals and similar theories cannot be taken as guides on this point owing to the unrealistic nature of the interactions they imply.

Another argument sometimes used suggests that the question of a singularity at $z = z_\sigma$ in the activity series is an artificial one since, it is asserted, the virial series might have *no* singularity at the corresponding condensation point $\rho = \rho_\sigma$ so that the singularity at z_σ could be of no physical significance. As an example of this the ideal Bose-Einstein gas is cited. In that case the activity series has a singularity at the Bose-Einstein condensation point (i.e. $z_1 = z_\sigma$) but the virial series is quite analytic at the corresponding density (and the isotherm can be continued on to higher densities). When one looks more closely, however, one finds that this example is also artificial because of the special shape of the isotherm near condensation. At fixed T this is given by

$$p = p_\sigma - A(\rho_\sigma - \rho)^2 + \cdots \tag{6}$$

which means that the isothermal compressibility becomes *infinite* as the condensation point is approached (see Figure 3).

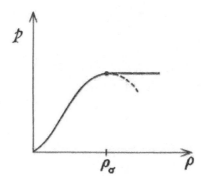

Figure 3. Condensation in an ideal Bose-Einstein fluid.

This feature is, of course, not found in real condensation processes. Conversely one may prove quite easily (see Appendix A) that if *at fixed temperature the pressure $p(\rho)$ is analytic in the density at $\rho = \rho_\sigma$ and if*

$$K_T(\rho_{\sigma-}) = \lim_{\rho \to \rho_{\sigma-}} K_T(\rho) < \infty \tag{7}$$

then the pressure $\bar{p}(z)$ is also analytic in the activity at $z = z_\sigma$. This shows that if the compressibility remains finite as a condensation point is approached, then a singularity in the activity series necessarily implies one in the virial series. Thus one cannot "escape" a singularity and hence it presumably has some physical significance.

3. Surface Tension and the Droplet Model

When we ask for the nature of the condensation singularities in a system with predominantly short range forces we are hampered because, apart from some examples I will describe later, there are no known model systems which are exactly soluble and exhibit condensation. It is true that we know rigorously that a two-dimensional Ising system with nearest neighbour forces undergoes condensation but since Onsager's solution is restricted, in "magnetic language", to zero field, that is in "fluid language" to $z = z_\sigma(T)$, one does not know what happens *as* the condensation point is approached. One may get a physical idea as to the "cause" of condensation, however, oy considering a real gas or, for that matter, a lattice gas or Ising model, at low densities and temperatures. Evidently most configurations of the system will consist of distributions of isolated molecules well separated from one another. There will also be present, however, clusters of two, three or more molecules bound together more-or-less tightly by the attractive forces but isolated, for the most part, from other clusters. Clusters of different sizes will be in mutual statistical equilibrium, associating and disassociating, but even fairly large clusters resembling "droplets" of the liquid phase will have some, generally rather small, chance of occurring.

Consider the potential energy E_l of such a not too small cluster of l molecules. This may be decomposed into a bulk term, determined by the binding energy per molecule in the liquid or condensed phase, say E_0, and a remainder W which is, evidently, associated with the loss of binding energy at the "surface" of the cluster. Thus

$$E_l = lE_0 + W, \tag{8}$$

where W is positive and may be taken as proportional to a "suitably defined" surface area s of the cluster, that is

$$W \approx ws. \tag{9}$$

For a lattice gas with nearest neighbour attractive forces the surface s may be defined unambiguously in terms of the numbers of "unsaturated bonds" surrounding the cluster (that is bonds between occupied and unoccupied sites) and the surface energy density w is similarly well defined. More generally, however, it is difficult to define s precisely in a way that is not somewhat arbitrary. Nevertheless for a large enough cluster (which is not too "drawn out" like a piece of seaweed for example!) the surface area is fairly well defined.

Now it is clear that the surface of a cluster gives it stability; because the surface energy is positive the cluster will, at low temperatures, tend to stay in compact configurations with relatively small surfaces. This tenlency to shrink will, speaking loosely, be opposed by the entropy of the cluster which might similarly be expected to be of the form

$$S_l = lS_0 + \omega s \tag{10}$$

where S_0 is the entropy per molecule in the bulk fluid. The surface entropy density ω is a measure of the number of different configurations of the cluster which have the same surface area. If now the temperature is lowered (or, what is essentially equivalent, if the activity is raised) the entropy will be less important and it becomes advantageous for clusters to combine to form droplets and for droplets to grow further by amalgamation thereby reducing the total surface area and hence lowering the total energy. Indeed if conditions are sufficiently favourable the droplets should continue to grow rapidly to macroscopic size. A macroscopic droplet represents, of course, the liquid phase and so its presence indicates that condensation has taken place!

This picture of the condensation of a gas was put forward apparently quite independently by Frenkel, Band and Bijl some twenty-seven years ago.[8–10] Bijl's work was the earliest, being contained in his thesis presented in April 1938, but it was not otherwise published. The papers of Frenkel and Band both appeared early in 1939 in The Journal of Chemical Physics. Although the ideas are quite straightforward and would have been understood many years previously, the stimulus in all three cases seems to have come from Mayer's work on the activity and virial expansions where the coefficients are determined by the famous cluster integrals. (In Mayer's theory, however, the "clusters" cannot be identified directly with real physical clusters as in the droplet picture. This seems to be the cost of making a fully rigorous and complete expansion.)

The essential correctness of the droplet explanation of the origin of condensation in a system with short range forces may be seen from the discussion of the existence of spontaneous magnetization in the plane Ising model ferromagnet given already in 1936 by Peierls.[11] The ideas of the surface tension and the entropy of clusters (in this case of 'overturned spins') play a crucial part in his argument although they are not expressed explicitly in this language. More recently R. B. Griffiths has adapted Peierls analysis to give a fully rigorous proof that condensation takes place in the Ising model.[12] The nature of the singularity, however, is still unrevealed. Furthermore the proof uses in an apparently essential way the symmetry of the Ising model under a change of sign of the field. Unfortunately more general models do not have such an exact symmetry although experimentally the shape of coexistence curves indicates a fairly precise symmetry between simple liquids and their vapours.

To clothe the droplet or cluster theory in mathematical form and to explore its consequences — and its difficulties — let us construct the classical configurational partition function for a cluster of l molecules in a domain of volume V. We have

$$q_l(\beta, V) = \frac{1}{l!} \int d\mathbf{r}_1 \cdots \int d\mathbf{r}_l \exp(-\beta U_l) \tag{11}$$

where $U_l = E_l$ is the potential energy of interaction and where the integrations are restricted to configurations in which the l molecules form a cluster. For definiteness we may suppose the pair interaction potential $\varphi(r)$ has an attractive tail of range b so that $\varphi(r) \equiv 0$ for $r \geq b$ and then define a "clustering distance" c so that if $|\mathbf{r}_i - \mathbf{r}_j| < c$ the ith and jth molecules belong to the same cluster. It is, of course, natural to take $c = b$. We will simplify further by assuming the potential has an infinite hard core and an attractive square well, that is

$$\begin{aligned}\varphi(r) &= \infty & &\text{for } r < a, \\ &= -\varphi_0, & &a \leq r < b \\ &= 0, & &r \geq b.\end{aligned} \tag{12}$$

In this case the energy E_l of any configuration is an integral multiple of $-\varphi_0$. Furthermore if b/a is not too large the binding energy in the dense phase in three dimensions will be $F_0 = \frac{1}{2}(12\varphi_0) = 6\varphi_0$ since hard spheres will pack with twelve nearest neighbours. The surface energy W may then be defined precisely through (8). Taking $w = \varphi_0$ so that $W = \varphi_0 s$ then defines the surface "area" s and shows it is an integer as in the lattice case. Inserting these relations in (11) we may write

$$q_l(\beta, V) = V \sum_s g(l,s) e^{l\beta E_0 - \beta w s} \tag{13}$$

where the combinatorial factor $g(l,s)$ is the number (or more correctly the volume in $d(l-1)$ dimensional configuration space) of configurations of l indistinguishable molecules with a fixed centre of mass which form a cluster of surface area s. (For a lattice gas, of course, the configurations are discrete; conversely in a more general model one would have an integration over a continuous range of s.) The factor V comes from the integration of one coordinate over the volume with the neglect of boundary effects.

Following De Boer's treatment[13] consider now the coefficient of z^N in

$$\mathscr{S} = \exp\left[\sum_{l=1}^{\infty} q_l z^l\right]$$

$$= 1 + \frac{1}{1!}\sum_{l=1}^{\infty} q_l z^l + \frac{1}{2!}\left[\sum_{l=1}^{\infty} q_l z^l\right]^2 + \cdots \tag{14}$$

Evidently this coefficient consists of the sums of products of cluster partition functions formed by decomposing N identical molecules into clusters in all possible ways. If we now assume that:

(A) *the effects of excluded volume between clusters may be neglected*, then this coefficient is simply the total configurational partition function $Q_N(\beta, V)$. Thus

$$\mathscr{S} = \sum_{N=0}^{\infty} z^N Q_N(\beta, V) \tag{15}$$

which is recognised as the grand partition function $\Xi(z, \beta; V)$ for the system. For a large system the pressure is given in the standard way, by

$$p/kT = (1/V) \ln \Xi(z, \beta; V)$$

$$= \sum_{l=1}^{\infty} (q_l/V) z^l \tag{16}$$

provided the series converges (see below), and hence the density is

$$\rho = \langle N \rangle / V = \sum_{l=1}^{\infty} l(q_l/V) z^l . \tag{17}$$

One may readily check (for example by introducing separate activity coefficients z_l for clusters of size l) that the partial number densities of l-clusters, which are proportional to the probability of finding an l-cluster, are just

$$\rho_l = (q_l/V)z^l. \tag{18}$$

By neglecting the interaction, that is essentially the excluded volume, between clusters the calculation of the equation of state has thus been reduced to the calculation of the single-cluster 'internal partition function', q_l/V. One feels that at low densities the excluded volume effects should not be important (although nearer the critical point they might be). Nevertheless the assumption (A) is obviously an important and potentially far reaching one and we will return to it later. To proceed further, however, the cluster partition function must be analysed in greater detail.

4. Cluster Partition Function and Condensation

Even for a lattice gas a full analysis of the combinatorial factor $g(l,s)$ entering in (13) seems very difficult. This is an important theoretical task central, as we will see, to the study of condensation. To see where the essential difficulties lie let us put

$$G_l(\beta) = \sum_s g(l,s) e^{-\beta w s} \tag{19}$$

$$= e^{-l\beta E_0} q_l(\beta, V)/V \tag{20}$$

and observe firstly that the bulk entropy per particle S_0 in a large cluster may be defined by

$$S_0(\beta) = k \lim_{l \to \infty} (1/l) \ln G_l(\beta). \tag{21}$$

The existence of the limit in (21) can be proved rigorously by a generalised subadditive argument based on decomposing a cluster of l particles into two clusters of l' and $l - l' + 1$ particles with one particle in common. This justifies the assumption of a bulk contribution to the entropy.

Secondly notice that the surface of a cluster cannot exceed some constant multiple of the number of particles (as is achieved by a cluster in the form of a string of beads, for example) nor can it be less than some minimum surface attained by some approximately spherical cluster. In d dimensions we thus have

$$a_1 l^{1-(1/d)} \leq s \leq a_2 l, \tag{22}$$

where a_1 and a_2 are appropriate constants. Since the terms in (19) are all positive the standard maximum term argument shows that

$$\max_s\{g(l,s)e^{-\beta ws}\} \leq (G_l\beta) \leq a_2 l \max_s\{g(l,s)e^{-\beta ws}\}. \tag{23}$$

Suppose that the maximum is attained for $s = \bar{s}$ so that $\bar{s} = \bar{s}(l;\beta)$ is the *most probable* or, loosely, *mean surface area*. Then as l becomes large

$$\ln G_l(\beta) = \ln g[l,\bar{s}(l;\beta)] - \beta w \bar{s}(l;\beta) + O(\ln l). \tag{24}$$

For the present we now avoid the full weight of the combinatorial problem by arguing that at low temperatures the most important configurations will be those relatively compact, roughly globular arrangements which have surface areas not vastly greater than the minimum possible and hence, for large l, increasing more slowly than l. We thus assume

(B i) *the mean surface area $\bar{s}(l,\beta)$ satisfies*

$$\bar{s}(l;\beta)/l \to 0 \quad \text{as } l \to \infty, \tag{25}$$

which is almost tantamount to the meaning of any well defined "surface". By virtue of the lower limit in (22) we may also assume that

(B ii) $$\bar{s}(l;\beta)/\ln l \to \infty \quad \text{as } l \to \infty. \tag{26}$$

These two assumptions are all that is essential for the main conclusions of the theory but in view of the bounds (22) and what has been said it is natural to expect more specifically that

$$\bar{s}(l,\beta) \approx a_0 l^\sigma \quad (l \to \infty) \tag{27}$$

with $a_0 = a_0(\beta)$ and $\sigma = \sigma(\beta)$ satisfying

$$0 < \sigma < 1. \tag{28}$$

In particular at low temperatures one might expect the exponent σ to be equal to (or close to) the value $\frac{2}{3}$ for $d = 3$ while for $d = 2$ it should be $\frac{1}{2}$. These specific assumptions were in fact made by the early workers. It seems quite possible however, that an "effective mean surface area" which took account of the interference between clusters at finite densities due to the excluded volume might lead to a smaller exponent. Conversely the vastly larger number of configurations which can occur with a larger area might tend to increase the value of σ.[14] Consequently even if we accept (27) and (28) as a convenient and concrete expression of the basic assumptions (25) and (26), it seems better to leave the value of the exponent σ as an open question (but see further below).

From (21), (24) and (25) we see that $k \ln g[l, \bar{s}(l)]$ varies as lS_0 for large l so that the difference defines a residual or "surface entropy". In fact this entropy will be associated directly with the many possible configurations of a section of the surface of a cluster (large in itself but small compared with the total surface). It is thus natural to assume finally that

(B iii) *the residual entropy satisfies*

$$k \ln g[l, \bar{s}(l)] - lS_0 \approx \omega \bar{s}(l) \quad (l \to \infty). \tag{29}$$

The entropy per unit of surface ω is supposed finite but it could in principle be zero. (It might evidently, also, depend on temperature.)

We may summarize these considerations by writing

$$\ln G_l(\beta) \approx l(S_0/k) - \beta(w - \omega T)\bar{s}(l) - \tau \ln l + \ln q_0, \tag{30}$$

in which we have also specifically recognized the existence of higher order terms proportional to $\ln l$ and of order unity. A logarithmic term was not included by the earlier workers and is not essential to our main considerations. However many studies of related combinatorial problems (in particular self-avoiding random walks) have shown that the asymptotic expansion should be expected to have this form with τ a positive number of magnitude unity or greater but depending principally on the dimensionality.[15,16]

Substituting with (30) and (20) into the expression (16) for the pressure, and adopting the form (27) purely for simplicity, yields finally

$$p/kT = \pi(\beta, z) = q_0 \sum_{l=1}^{\infty} l^{-\tau} x^{l^{\sigma}} y^l \tag{31}$$

where

$$y = z \exp[\beta E_0 + (S_0/k)], \tag{32}$$

and

$$x = \exp[-a_0 \beta(w - \omega T)], \tag{33}$$

so that y is proportional to the activity and x essentially measures the temperature (approaching zero as $T \to 0$). This completes the mathematical derivation of the theory on the basis of assumptions (A) and (B). Let us turn now to the consequences.

Consider the probability of finding a cluster of size l. By (18) and (30) this is proportional to

$$\rho_l = q_0 x^{l^{\sigma}} y^l / l^{\tau}. \tag{34}$$

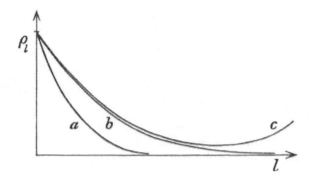

Figure 4. Density of clusters of size l at an activity (a) below the condensation point, $y < 1$, (b) at the condensation point, $y = 1$, and (c) just above condensation at $y > 1$.

At low temperatures x is small and at low activities (and hence low densities) y is also small. In these circumstances ρ_l decays rapidly to zero as l increases (see Figure 4). As y approaches unity, however, the decay becomes slower. When $y = 1$, ρ_l still decays to zero but only as $\exp[-\text{const.}\, l^\sigma]$ (assuming, of course, that $x < 1$, that is, $T < T_c = w/\omega$). On the other hand when y is slightly larger than unity the probability at first decreases because of the factor x^{l^σ} but finally increases when the exponent $(\ln y)l$ dominates the exponent $-|\ln x|l^\sigma$ or, more generally, $\bar{s}(l)$ (see Figure 4). The large (divergent) probability of very large clusters indicates, as observed before, that condensation has taken place. Consequently we identify $y = y_\sigma = 1$ as the condensation point so that

$$z_\sigma = \exp[-\beta(E_0 - TS_0)], \tag{35}$$

while the chemical potential at condensation is

$$\mu_\sigma(T) = -E_0 - TS_0 + \tfrac{1}{2}kTd \ln(h^2/2\pi mkT). \tag{36}$$

If y is only slightly greater than y_σ, that is if $\mu - \mu_\sigma(T)$ is very small, the minimum value of ρ_l can be extremely small since

$$l_{\min} \simeq [a_0(w - \omega T)/(\mu - \mu_\sigma(T))]^{1/(1-\sigma)} \tag{37}$$

will be very large. Consider a system in equilibrium at a chemical potential just less than $\mu_\sigma(T)$ in which μ is suddenly increased slightly (or, equivalently, in which $\mu_\sigma(T)$ is decreased slightly by cooling). Evidently even if the condensation point is passed the cluster distribution ρ_l for $l < l_{\min}$ should scarcely have to change as the system "relaxes" to its new equilibrium. Furthermore if the primary mechanism by which a cluster can grow is through a binary collision with a relatively small cluster or single molecule (as will surely be the case at low densities) we see that there is a "free energy barrier" at $l = l_{\min}$; a cluster with a large value of l less than $l_{\min}$ will tend to "evaporate" or break up into smaller clusters rather than grow further. Only occasional chance fluctuations, which will be extremely infrequent when $\mu - \mu_\sigma(T)$ (the "degree of supercooling") is small, will carry the system over the barrier. Even a single cluster with $l > l_{\min}$ will, however, tend to grow, increasingly rapidly as l increases, and thus consitutes a "nucleus" of the condensed phase. (Impurities, such as dust particles, may fulfill a similar role.) We can thus understand why a rather long lived metastable state may be observed when a pure gas is slightly supercooled. The extreme "limit of metastability" might be defined, somewhat arbitrarily, by the condition $l_{\min} \simeq 1$, which yields

$$|\mu - \mu_\sigma(T)|_{\max} \simeq a_0(w - \omega T). \tag{38}$$

Evidently metastability, and in fact condensation, cannot occur if the temperature exceeds

$$T_c = w/\omega \tag{39}$$

which may thus be identified as the critical temperature. (Recall that ω might depend to some extent on temperature).

Apart from certain refinements and generalizations we have so far essentially reproduced the droplet theory of condensation (and metastability!) as set out, for example, by Frenkel in his book.[8] Let us now enquire into the analytic properties of the theory with Mayer's conjecture in mind.

5. Analytic Character and the Critical Point

For fixed x, that is fixed temperature, the radius of convergence of the series (31) is given by

$$y_0 = \lim_{l \to \infty} |l^{-\tau} x^{l^\sigma}|^{-1/l} = \lim_{l \to \infty} |x|^{-1/l^{1-\sigma}} = 1, \tag{40}$$

for *all* x. The last equality follows from the assumption $\sigma < 1$. More generally the same result follows for any $\bar{s}(l, \beta)$ from the assumption (B i). Since the terms in (31) are positive the point $y_0 = 1$ must be a singularity of the function. Furthermore we see that it coincides with the previously identified condensation point $y_\sigma = 1$. For this model, therefore *Mayer's conjecture is verified*: the condensation activity z_σ is a singularity of the analytic function $p(z)/kT$.

What is the behaviour of the density, the compressibility etc. *at* the condensation point? To answer this question note that these variables can be expressed in terms of the derivatives

$$\pi^{(n)}(\beta, z) = \left(z\frac{\partial}{\partial z}\right)\pi = \left(y\frac{\partial}{\partial y}\right)\pi = q_0 \sum_{l=1}^{\infty} l^{n-\tau} x^{l^\sigma} y^l. \tag{41}$$

Thus the density and compressibility are given by

$$\rho(\beta, z) = \pi^{(1)}(\beta, z), \quad kT\rho^2 K_T = \pi^{(2)}(\beta, z). \tag{42}$$

(The energy, specific heat, etc. may similarly be expressed in terms of derivatives with respect to x.) At the condensation point we find

$$\pi_\sigma^{(n)}(\beta) = \pi^{(n)}(\beta, z_\sigma) = q_0 \sum_{l=1}^{\infty} l^{n-\tau} x^{l^\sigma}. \tag{43}$$

Using the assumption $\sigma > 0$ or more generally (B ii), it is not difficult to show that this series converges for *all* n provided only that

$$x < x_c = 1, \quad \text{that is, } T < T_c = w/\omega. \tag{44}$$

(For $x > 1$ the series (43) is evidently always divergent.) Thus, in particular, the compressibility remains finite on the condensation curve up to a temperature $T = T_c$ which may hence be identified as the *critical point* in agreement with our previous conclusion.

The fact that *all* the derivatives with respect to z (for real z) remain finite at $z = z_\sigma$ even though we have established that this is a singulatity of $\pi(z)$ might seem surprising at first sight. It merely means, however, that the singularity at $z = z_\sigma$ is an *essential singularity*. (One may recall the function $\exp(-1/x^2)$ for which all derivatives along the real axis vanish at $x = 0$). Evidently such a singularity can hardly be detected by direct thermodynamic measurements in the homogeneous phase since none of the thermodynamic functions will exhibit any infinities or similar "anomalies"![17]

Since $\pi(z)$ has an essential singularity at $z = z_\sigma$ it is impossible to analytically continue the function *through* z_σ to find some real "metastable continuation" of the isotherm. This may be seen clearly if we estimate the derivatives at z_σ by approximating the sum in (43) by an integral, which is valid for x near unity. Thus for $x < 1$ and $n > \tau - 1$ we have

$$\pi_\sigma^{(n)}(\beta) \approx q_0 \int_0^\infty l^{n-\tau} \exp[-(\ln x^{-1})l^\sigma] dl, \tag{45}$$

and making the substitution $t = \theta l^\sigma$, where

$$\theta = \ln x^{-1} = \frac{a_0 \omega}{kT}(T_c - T) \qquad (46)$$

yields

$$\pi_\sigma^{(n)}(\beta) \approx (q_0/\sigma\theta^{(n-\tau+1)/\sigma})\Gamma\left(\frac{n-\tau+1}{\sigma}\right). \qquad (47)$$

For large n the gamma function varies as $(n!)^{1/\sigma}$. Consequently if the derivatives are used to construct a Taylor series expansion about z_σ in powers of $(\ln z - \ln z_\sigma)$ the coefficient of the nth term will vary as $(n!)^{(1-\sigma)/\sigma}$ for large n. Since, by assumption, σ is less than unity and since $(n!)^\varepsilon$ diverges faster than ξ^n for any ξ when ε is positive it follows that the Taylor series will never converge; that is, it has a *zero* radius of convergence. (For $z < z_\sigma$ and low enough T it might, however, have an asymptotic character.)

Although one cannot find a real analytical continuation of $\pi(z)$ through $z = z_\sigma$ one may hope to construct a continuation of $\pi(z)$ by passing *around* the singularity at z_σ. This task is considered in Appendix B where it is shown that $\pi(z)$ as defined by the expansion (31) extends into a function meromorphic in the entire complex z plane except for a cut along the real z axis from $z = z_\sigma$ to ∞. Across this cut the imaginary part of $\pi(z)$ has a discontinuity which varies as

$$\pi(z + i\varepsilon) - \pi(z - i\varepsilon) \approx C(z - z_\sigma)^{\tau^*} e^{-D/(z-z_\sigma)^{\sigma^*}}, \quad (z \text{ real}) \qquad (48)$$

when $z = \mathscr{R}\{z\} \to z_\sigma+$, where C and D are constants and

$$\sigma^* = \sigma/(1-\sigma), \quad \tau^* = (\tau - 1 + \tfrac{1}{2}\sigma)/(1-\sigma). \qquad (49)$$

Evidently the discontinuity *and* all its derivatives vanish as $z \to z_\sigma$ along the cut, which accounts for the extreme weakness of the singularity.

Our analysis thus shows that the cluster model of condensation which, as we have argued, should be valid at low densities (and hence low temperatures) implies a singularity at the condensation point and hence the nonexistence of a well defined real isotherm beyond condensation. Although one can be less confident of the correctness of the assumptions at higher temperatures it is of interest to pursue the consequences of the model in the vicinity of its critical point. We may hope to throw some light on the general theory of critical phenomena and on Mayer's well known hypothesis of some sort of extended critical region often referred to as the "Derby hat" phenomena.

Firstly note that from (41) and (42) the critical point density will be

$$\rho_c = q_0 \sum_{l=1}^{\infty} l^{1-\tau} \tag{50}$$

and for this to be finite we must have $\tau > 2$. From (42), (46) and (47) we see that provided $\tau < 3$ the compressibility at condensation diverges to infinity (as expected) like $(T_c - T)^{\gamma'}$ with

$$\gamma' = (3 - \tau)/\sigma. \tag{51}$$

(The notation γ' for this exponent follows the standard scheme.[18]) Evidently the nature of the critical point singularities in other variables will equally depend only on the two parameters σ and τ. Thus the shape of the gaseous side of the "coexistence curve" is found to be

$$\rho_c - \rho_\sigma(T) \sim (T_c - T)^\beta \tag{52}$$

where

$$\beta = (\tau - 2)/\sigma. \tag{53}$$

The specific heat at constant critical density, which derives from $p_\sigma(T)$, diverges with an exponent

$$\alpha' = 2 - (\tau - 1)/\sigma \tag{54}$$

where a value $\alpha' = 0$ must be interpreted as meaning a logarithmic law rather than a power law.

The shape of the critical isotherm is given by

$$p_c - p \sim (\rho_c - \rho)^\delta \tag{55}$$

where, by putting $x = 1$ in (31), one easily finds

$$\delta = 1/(\tau - 2). \tag{56}$$

By a more complicated analysis one may also investigate the behaviour of the isotherms as the critical point is approached from *above* at constant density ρ_c. One discovers that the specific heat and compressibility exponents α and γ above T_c, satisfy the symmetry relations

$$\alpha = \alpha', \quad \gamma = \gamma'. \tag{57}$$

The general appearance of the isotherms implied by the cluster model when $2 < \tau < 3$ is sketched in Figure 5. Apart from the absence of the "liquid" sections of the isotherms the overall behaviour is surprisingly like that of real fluids.

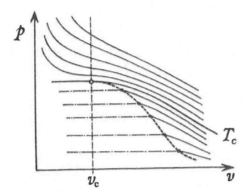

Figure 5. Isotherms following from the cluster theory of condensation showing the gaseous side of the coexistence curve and the critical point.

Since all the critical point exponents depend on only two parameters any three of them must be related. Thus we find, for example,[19]

$$\alpha' + 2\beta + \gamma' = 2 \tag{58}$$

and

$$\gamma' = \beta(\delta - 1). \tag{59}$$

These and other similar relations are satisfied exactly by the two-dimensional lattice gas with nearest neighbour interactions (Ising model) and, as far as the values are known accurately, by the three-dimensional model also.[20] They seem to be valid for real fluids although at present the experimental data are not sufficiently precise for a stringent test. Interestingly they hold also for a van der Waals-like gas which may be regarded as an infinite-dimensional system.

Of course using known values for γ, δ and β etc. we may evaluate the parameters σ and τ and compare them with our expectations. For the plane Ising model we find from

$$\sigma = 1/\delta\beta = 1/(\gamma' + \beta) \quad \text{and} \quad \tau = 2 + (1/\delta) = 2 + \beta/(\gamma' + \beta) \tag{60}$$

the values

$$\sigma = \tfrac{8}{15} = 0.53\dot{3} \quad \text{and} \quad \tau = 2\tfrac{1}{15}. \tag{61}$$

This value of σ is surprisingly close to the value $\sigma = \tfrac{1}{2}$ expected at low temperatures on naive geometrical grounds and tends to increase one's confidence in the underlying cluster picture. Presumably the difference

$\Delta\sigma = 0.033$ is due to the neglected effects of excluded volume and to the statistical geometry of noncompact clusters. For the three-dimensional lattice gas there is some uncertainty due to lack of precise knowledge of the critical exponents. Present information is consistent with the ranges

$$\sigma = \tfrac{16}{25} = 0.640, \quad \tau = 2\tfrac{1}{5} \tag{62}$$

to

$$\sigma = \tfrac{8}{13} = 0.6153\ldots \quad \tau = 2\tfrac{5}{26} = 2.192\ldots \tag{63}$$

These values of σ fall *below* the geometrical value $\sigma = \tfrac{2}{3}$ but only by $\Delta\sigma \simeq 0.027$ to 0.051. This is, however, a sure indication of the importance of the excluded volume effects since single cluster geometrical factors would be expected only to increase σ. The "classical" or van der Waals-like limit corresponds to

$$\sigma = \tfrac{2}{3} \quad \text{and} \quad \tau = 2\tfrac{1}{3}. \tag{64}$$

This seems difficult to understand in a simple fashion in view of the expectation $\sigma \simeq 1$ for $d \to \infty$, and may be an indication of a more thoroughgoing breakdown of the cluster picture in the critical region.

Finally note that in all these cases we have a unique critical point rather than a critical region. Reference back to equation (43) and the accompanying argument shows that this follows quite generally from the two assumptions (B iii) that the surface entropy is proportional to the surface energy ($\omega = 0$ would give $T_c = \infty$), and (B ii) that the mean surface of a cluster of l atoms increases *more rapidly* than $\ln l$. Conversely it is easy to see that if the surface entropy increased *faster* than the surface energy, the effective surface tension would be negative at all temperatures; small clusters would have no stability and a sharp condensation could not occur. Secondly suppose the surface energy and entropy increased only as fast as $\ln l$. Then the convergence of the series (43), that is of the nth derivatives with respect to z *at* condensation, would *depend on the order of the derivative* considered. This in turn would mean that the temperatures at which a volume discontinuity appeared, at which the compressibility became infinite, etc. would all be different. In other words the critical phenomena would take place over a range of different temperatures rather as suggested by Mayer. That this does *not* happen in practice is thus a reflection of the fact that the geometrical surface of a cluster of l particles (in two or more dimensions) increases faster than ln l (and probably as l^σ with $\sigma < 1$). A fuller description of this type of anomalous critical region will be given in the following sections in connection with an exactly soluble model displaying condensation to which we now turn.

6. An Exactly Soluble Model

As we have shown the cluster theory of condensation rests mainly on two assumptions: (A) that the interactions (in the form of excluded volume) between clusters can be neglected and (B) that the mean surface of a cluster has the expected properties and characterizes sufficiently well all clusters. In my opinion the second assumption is probably the more difficult one to justify. To judge the validity of the first assumption we will describe a one-dimensional model in which the excluded volume may be rigorously taken into account.

It is sometimes asserted that a one-dimensional model cannot display a phase transition (excluding, that is, the procedure of taking some special limit after the thermodynamic limit as in the Kac-Uhlenbeck-Hemmer model[3]). This has only been justified, however, for systems with pair interactions of *strictly finite* range b and with similar three-body and many-body interactions up to some *finite* order; that is, all the forces vanish identically for separations greater than b.[21] Indeed it seems likely that a one-dimensional fluid of particles interacting with a pair potential decaying only as $1/r^\varepsilon$ *will* exhibit condensation if $1 < \varepsilon < 2$. This is suggested by the cluster argument, as we will outline, and also, as has been remarked by Kac,[22] because the corresponding "sphericalized" lattice gas model still has a transition whereas the normal tendency of "sphericalization" seems to be to destroy transitions. At present, however, no such pair interaction model has been rigorously solved. We will consider instead a model in which the forces are of strictly finite range but many-body interactions of indefinitely great order are present. By strictly finite range we mean, as above, that if any group of $N_1 + N_2$ particles with total potential energy $U_{N_1+N_2}$ is separated into a group of N_1 particles and a group of N_2 particles with a minimum separation R between particles in different groups exceeding b, then there is no mutual interaction, between the groups, that is

$$\Phi_{N_1,N_2}(R) = U_{N_1+N_2} - (U_{N_1} + N_{N_2}) \equiv 0, \quad R > b. \tag{65}$$

Explicitly we take a pair interaction potential

$$U^{(2)}(r) = \varphi_2(r) = +\infty, \quad r \leq a,$$
$$\geq -w_A, \quad a < r \leq b, \qquad (66)$$
$$= 0, \quad r \geq b,$$

so that the particles have a hard core of diameter a. We suppose for simplicity that the range satisfies

$$b \leq 2a \qquad (67)$$

so that the pair interactions arise only between nearest neighbours. In the interval $a < r \leq b$ the potential may be arbitrary provided it is bounded below. To define the many-body interactions we introduce, as in Section 3, a clustering distance c which might be taken equal to b although this is not necessary. Labelling the particles in sequence we say that particles j to $j+k$ belong to the same cluster if $|r_{j+1} - r_j| \leq c$, $|r_{j+2} - r_{j+1}| \leq c, \ldots |r_{j+k} - r_{j+k-1}| \leq c$. The s-body potential is then taken as ($s \geq 3$)

$$U^{(s)}(r_1, \ldots r_s) = \varphi_s, \quad \text{if } |r_i - r_j| \leq c$$
$$\text{for } s-1 \text{ pairs } i = 1, \ldots s \neq j = 1, \ldots s. \qquad (68)$$
$$\equiv 0, \quad \text{otherwise}.$$

Thus there is a constant s-body interaction energy coming into play between any succession of s-particles which belong to the same cluster but no interaction between particles in different clusters. For simplicity we will assume the φ_s are all negative or zero (i.e. the many-body forces are attractive).

Consider the total energy of an (isolated) cluster of l particles. This will be

$$E_l = \sum_{i=1}^{l-1} \varphi_2(r_{i+1} - r_i) + (l-2)\varphi_3 + (l-3)\varphi_4 + \cdots + \varphi_l.$$

$$= \sum_{i=1}^{l-1} \varphi_2(r_{i+1} - r_i) + \sum_{s=3}^{l}(l+1-s)\varphi_s. \qquad (69)$$

Evidently the energy per particle in an infinite cluster satisfies

$$e_\infty \geq -w_A + \sum_{s=3}^{\infty} \varphi_s. \qquad (70)$$

For thermodynamic stability we must require[5] $e_\infty > -\infty$ or

$$\Phi = \sum_{s=3}^{\infty}(-\varphi_s) < \infty \tag{71}$$

so that $|\varphi_s|$ must decrease faster than $1/s$. It should perhaps be stressed that the condition (71) together with the strictly finite range property (65), is sufficient to guarantee rigorously the existence of a limiting free energy with the usual thermodynamic properties even when φ_s does not vanish for any s.[5]

Now we may rewrite (69) in the form

$$E_l = -l\Phi + \sum_{i=1}^{l-1}\varphi_2(r_{i+1} - r_i) + W_l \tag{72}$$

$$\geq -l(\Phi + w_A) + W_l + w_A, \tag{73}$$

where the first term is recognised as a *bulk energy* while the *surface energy* is (neglecting a constant contribution)

$$W_l = W(l) = \sum_{t=1}^{l}\sum_{s=t+1}^{\infty}(-\varphi_s) = \sum_{s=3}^{\infty}\min\{s-1,l\}(-\varphi_s). \tag{74}$$

It is easy to show that

$$W_l/l \to 0 \quad \text{as } l \to \infty \tag{75}$$

as expected [compare with (25)].

Notice at this point that if, in place of the many-body potentials, we had considered long-range pair interactions, we would have found precisely the same results (69) to (75) for the energy of an *isolated* cluster of l particles of uniform spacing d but with φ_s replaced by $\varphi_2[(s-1)d]$. Thus our many-body forces accurately imitate the effects of long-range pair interactions within a single cluster. They do not, however, reproduce the long-range attractive forces between different clusters that arise in the pair case; neglect of these would, of course, be expected to weaken any tendency towards condensation. On the other hand the repulsions, or excluded volume effects, between different clusters are given precisely by our potentials.

If we accepted the arguments of the droplet theory of condensation we would conclude from (72) and (75) that the model would display condensation provided that

$$W_l/\ln l \to \infty \quad \text{as } l \to \infty. \tag{76}$$

There should then be no critical point since there is evidently no surface contribution to the entropy of *one*-dimensional clusters (except perhaps for a *constant* contribution from the ends of the cluster). Consequently the entropy per unit surface, ω would vanish and by (44) we would have $T_c = \infty$. If W_l varied *as* $\ln l$ for large l we would expect the Derby-hat type of phenomena while if, on the other hand,

$$W_l / \ln l \to 0 \quad \text{as } l \to \infty \tag{77}$$

there would be no phase transition. As we will show these conclusions are confirmed in detail by the exact solution of the model!

Before describing the solution note that from (73) we obtain

$$\varphi_s = W_s - 2W_{s-1} + W_{s-2} \tag{78}$$

so that the conditions (75) and (76) become

$$s\varphi_s \to 0, \quad \text{as } s \to \infty \tag{79}$$

and

$$s^2 \varphi_s \to \infty, \quad \text{as } s \to \infty, \tag{80}$$

respectively. For potentials decaying faster than $1/s^2$ no transition is thus expected.

7. Analysis of the Model

We will sketch, without entertng into full details, the solution of the model with the potentials (66) to (69). As with most one-dimensional models it is advantageous to start with the grand partition function $\Xi(\beta, z; L)$ for a length L and to compute its Laplace transform

$$\Psi(\beta, z, s) = \int_0^\infty e^{-sL} \Xi(\beta, z, L) dL, \tag{81}$$

which can be regarded as a generating function for all possible sets of clusters of all possible sizes and spacings. If we write

$$J(\beta, s) = \int_0^c e^{-sr} e^{-\beta \varphi(r)} dr \tag{82}$$

and

$$K(\beta, s) = \int_c^\infty e^{-sr} e^{-\beta \varphi(r)} dr \tag{83}$$

and define the generating function

$$H(\beta, z, s) = \sum_{l=1}^{\infty} z^l e^{-l\beta\Phi} [J(\beta, s)]^{l-1} e^{-\beta W(l)}, \qquad (84)$$

which enumerates all possible single clusters with their "internal" Boltzmann factors, we can construct $\Psi(\beta, z, s)$ from the series

$$\Psi(\beta, z, s) = s^{-1} + s^{-1}Hs^{-1} + s^{-1}HKHs^{-1} + s^{-1}HKHKHs^{-1} + \cdots \qquad (85)$$

Here the first term accounts for all lengths of line with no clusters, the second term for all possible single clusters at all positions along the line, the third term for all pairs of clusters and so on. The solution of the problem is thus given formally by

$$\Psi(\beta, z, s) = s^{-1} + s^{-2} \sum_{m=1}^{\infty} [H(\beta, z, s)]^m [K(\beta, s)]^{m-1}. \qquad (86)$$

The thermodynamic behaviour is obtained by noting, from the definition (81), that the abscissa of convergence, $s_0 = s_0(\beta, z)$, of the transform $\Psi(s)$ determines the grand canonical potential since

$$\beta p = \pi(\beta, z) = \lim_{L \to \infty} (1/L) \ln \Xi(\beta, z, L) = s_0(\beta, z). \qquad (87)$$

The possibility of a phase transition may be seen immediately since, from (84) and (86), the breakdown of the convergence of $\Psi(s)$ for small $\mathscr{R}\{s\}$ is determined either by the

Interior condition

$$ze^{-\beta\Phi} J(\beta, s) = u(\beta, z, s) = 1, \qquad (88)$$

or by the

Exterior condition

$$H(\beta, z, s)K(\beta, s) = 1. \qquad (89)$$

If, as s decreases from ∞, one of these conditions is always encountered before the other there is no phase transition; a change over from one condition to the other will correspond to some sort of phase change.

The second condition can be conveniently rewritten in terms of the "master function"

$$\Upsilon(\beta, u) = \sum_{l=1}^{\infty} u^l e^{-\beta W(l)} \qquad (90)$$

defined for $|u| < 1$, which bears an obvious resemblance to the final form (31) of the grand potential in the droplet theory. In particular notice that $\Upsilon(\beta, u)$ is always singular at $u = 1$. The exterior condition then becomes

$$\Upsilon(\beta, u) = J(\beta, s)/K(\beta, s) = Q(\beta, s) \tag{91}$$

where the function $Q(\beta, s)$ is easily shown to increase monotonically (and strictly) in s from the value $Q(\beta, 0) = 0$.

Since s_0 is only determined implicitly by the equations it is simpler to choose $s = \beta p$ as an independent variable and to ask for the activity (or chemical potential or Gibbs free energy) as a function of β and s, that is, T and p. One readily finds from (88)

$$\ln z(\beta, s) = \ln u(\beta, s) - \ln J(\beta, s) + \beta \varphi \tag{92}$$

and the equation of state is then

$$v = v(\beta, s) = (\partial/\partial s) \ln u(\beta, s) - (\partial/\partial s) \ln J(\beta, s). \tag{93}$$

Now for small $s = \beta p$ the function $u(\beta, s)$ is determined through (91), the series (90) converging absolutely. As s increases u increases alytically towards unity. If the series (90) diverges at $u = 1$, that is if

$$\Upsilon(\beta, 1-) = \sum_{l=1}^{\infty} e^{-\beta W(l)} = \infty, \tag{94}$$

one easily sees that $u(\beta, s)$ remains less than unity for all s (i.e. all p) and there is *no transition*. If, on the other hand, $\Upsilon(\beta, 1-)$ is *finite* then a *transition occurs* at a pressure $\beta p_\sigma = s_\sigma$ determined by

$$Q(\beta, s_\sigma) = \Upsilon(\beta, 1-). \tag{95}$$

For $s \geq s_\sigma$ the function u has the constant value unity so that, by (92) the thermodynamic behaviour is governed entirely by the "internal function" $J(\beta, s)$.

We have thus found the conditions for a phase transition in our model. By studying the convergence of the series (94) it is not difficult to show there is

(a) *no transition at any temperature*

$$W_l / \ln l \to 0 \quad (l \to \infty) \quad \text{or} \quad -s^2 \varphi_s \to 0 \quad (s \to \infty); \tag{96}$$

(b) *a transition at all temperatures*

$$W_l / \ln l \to \infty \quad (l \to \infty) \quad \text{or} \quad -s^2 \varphi_s \to \infty \quad (s \to \infty); \tag{97}$$

(c) *a transition below a critical temperature*

$$\text{if } W_l / \ln l \quad \text{or} \quad -s^2 \varphi_s \text{ approach nonzero limits}. \tag{98}$$

These results agree with the guesses based on the droplet theory. They may be summarized conveniently by the formula for the critical temperature, namely,

$$kT_c = \lim_{l \to \infty} W_l / \ln l \quad \text{or} \quad \lim_{s \to \infty} -s^2 \varphi_s, \tag{99}$$

since cases (a) and (b) correspond merely to $T_c = \infty$ or $T_c = 0$, respectively.

The nature of the transition when it occurs depends, of course, on further details of W_l. From (93) the volume discontinuity at the transition is seen to be

$$\Delta v = Q'(\beta, s_\sigma) / \Upsilon_1(\beta, 1) \tag{100}$$

where $Q'(s) = \partial Q / \partial s > 0$ and the derivatives of the master function are defined by

$$\Upsilon_k(\beta, u) = \sum_{l=1}^{\infty} l^k u^l e^{-\beta W(l)}. \tag{101}$$

In case (b), which corresponds to the behaviour of surface energy expected in two or more dimensions, there is *always* a volume discontinuity at the transition although it diminishes as the pressure increases. On the other hand the compressibility and all higher derivatives along the isotherm remain finite as the pressure increases towards p_σ (since all the series (101) converge at $u = 1$). Nevertheless it is clear that the condensation point is a singular point of the isotherm, as Mayer conjectured although, as we would have anticipated, it is an essential singularity. This leading conclusion of the cluster theory is thus not invalidated by the full inclusion of the excluded volume effects — at least not in this simple but rigorous model!

The absence of a critical temperature in the case (b) stems as we mentioned it would, directly from the absence in the expression for the master function of any entropy factors with the same behaviour as W_l. We could at the cost of no longer having a model with a definite Hamiltonian, arbitrarily "assign" extra phase space to each cluster of l particles in such a way as to lead to an additional factor $\exp[(\omega/kw)W_l]$ in the lth term of $\Upsilon(\beta, u)$. We would then find a unique critical point at a temperature $T_c = w/\omega$, just as in the cluster theory except that the model now *also* describes the liquid phase. Apart from the liquid side of the coexistence curve (which is concave rather than convex due, essentially, to the neglect of "bubbles" in the liquid) the overall pattern of isotherms resembles that found in practice. In particular it is possible to pass continuously and smoothly from gas to liquid over the "top" of the critical point. This transition, however, is not fully analytic since one finds that the critical isotherm for $v < v_c$ is a line of essential singularity. Nothing of this is "visible", however, since all temperature or pressure derivatives are continuous through this isotherm! This means that in the model (no longer, we stress, a true Hamiltonian model) there is an *absolute* distinction between a gas and a liquid. Such a distinction does not, of course, occur with a van der Waals-like equation of state. The existence of some absolute difference between liquid and gas has often been conjectured for real systems but has never been established convincingly. If the nature of the transition between the states were of such an "infinite order" this is perhaps hardly surprising. I do not feel, however, that our result adds much plausibility to the speculation for realistic models although it serves as a warning of what *could* happen!

Finally let us investigate briefly the borderline case (c) where the model with a proper Hamiltonian does have a critical temperature. The main features follow from the observation that the *number* of derivatives (101) of the master function which remain finite at the transition point $u = 1$ *depends on temperature*. As soon as T drops below T_c, as defined in (99), all sufficiently high derivatives diverge which means that high derivatives along the isotherm are discontinuous across a *singular curve.* in the (p, v) plane. However there is *no* volume discontinuity until the pressure drops to $T'_c = \frac{1}{2}T_c$! On the other hand the compressibility remains continuous across the singular curve only down to a temperature of $\frac{2}{3}T_c = \frac{4}{3}T'_c$ below which point it becomes infinite as the transition is approached from the low density side. It remains infinite at the transition down to a temperature of $\frac{1}{3}T_c = \frac{2}{3}T'_c$ but becomes finite again as in a normal condensation process, at lower temperatures! One might indeed say that the "order" of the transition varies continuously with temperature.

The general shape of the isotherms when

$$W_l = w \ln l + \tfrac{1}{2}(1 - \lambda) \ln \ln l, \quad (\lambda > 0), \tag{102}$$

is illustrated in Figure 6. In this case the critical pressure is infinite while the gas side of the coexistence curve varies as $(T'_c - T)^\lambda$ when T approaches T'_c from below. It is quite possible, however, to have a "flat-topped" coexistence curve as suggested originally by Rice.[23]

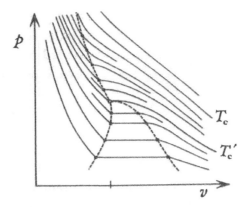

Figure 6. Sketch of the isotherms of the exactly soluble one-dimensional model in the borderline case (c) when the surface energy is given by equation (102).

This bewildering variety of peculiar possibilities is a further reflection of the artificially weak nature of the surface energy in case (c). Such a logarithmically increasing surface energy can arise, I believe, only in one dimension. Geometry alone will lead to surface energies satisfying (97) that is condition (B ii) of Section 4, in two or more dimensions.

The one-dimensional model can be generalized in various directions (in particular, the liquid and gaseous states may be treated more symmetrically) but the description of these developments must await another occasion.

8. Conclusions

My primary purpose in this lecture has been to show that the physical ideas of the droplet theory of condensation still deserve further exploration. The theory itself contains interesting and, I believe basically correct, implications regarding the nature of the condensation point and the critical point which do not seem to have been noticed previously. The predictions of an essential singularity at the condensation point, of the uniqueness of the critical point and of the inter-relations between the critical point singularities throw light on a number of long standing problems and conjectures.

I hope by reformulating and extending the cluster theory I have exposed the most important problems concerning its foundations. The exactly soluble model which I described, although it is evidently artificial in a number of respects, lends support to the validity of the conclusions and underlying assumptions of the droplet theory. There seems to be a real possibility of establishing these foundations on a more rigorous basis. The subtle and complex possibilities already revealed by our analysis suggest that this may not be an easy task; it is, however, a worthwhile and important one.

Acknowledgements

The ideas described in this article have been developed over the course of a number of years. I have been greatly encouraged and stimulated in this period by Professor Mark Kac, and more recently, by Professor G. E. Uhlenbeck. It is a pleasure to thank them. I am also indebted to Professor N. G. van Kampen and Professor J. S. Langer for informative discussions in connection with Appendix B. The partial support of the National Science Foundation is gratefully acknowledged.

Appendix A

Proof of Theorem on Condensation Singularities

Suppose we are given, at fixed temperature, the pressure as a function of the specific volume v, that is,

$$p = P(v) \tag{A.1}$$

and we wish to determine the grand potential

$$\beta p = \pi(\ln z) \tag{A.2}$$

and the density

$$\rho = 1/v = 1/v(\ln z) = \partial \pi / \partial \ln z \tag{A.3}$$

as functions of the activity z. We may revert (A.3) in the form

$$\ln z = L(v) \tag{A.4}$$

and then solve formally for π by integrating to obtain

$$\pi(\ln z) = \int^{\ln z} [1/L^{-1}(\ln z')]) d\ln z' . \tag{A.5}$$

Conversely from (A.3)

$$1/v = \beta(\partial P(v)/\partial v)(\partial v/\partial \ln z) \tag{A.6}$$

so that solving for the last derivative and integrating yields

$$\ln z - \ln z_a = \beta \int_{v_a}^{v} v' \left(\frac{\partial P}{\partial v}\right) dv' = \beta \left[pv - p_a v_a - \int_{v_a}^{v} P(v') dv'\right], \quad \text{(A.7)}$$

which determines the function $L(v)$ up to an additive constant.

This result may be checked directly by thermodynamics since, in standard notation,

$$G_N = pV_N + F_N \quad \text{(A.8)}$$

while the chemical potential is

$$\mu = G_N/N = pv + F_N/N$$
$$= pv - \int^{v} p(v') dv' \quad \text{(A.9)}$$

since

$$p = -(\partial F_N/\partial V_N) = -(\partial [F_N/N]/\partial v). \quad \text{(A.10)}$$

Recalling that $\ln z = \beta\mu + \text{constant}$ shows the equivalence of (A.9) to (A.7).

Now suppose that $P(v)$ is analytic at $v = v_\sigma \neq 0$ so that the power series

$$p = p_\sigma + \sum_{n=1}^{\infty} a_n (v - v_\sigma)^n \quad \text{(A.11)}$$

is convergent for small enough $(v - v_\sigma)$. Substitution in (A.7) with $z_a = z_\sigma$ etc. yields the convergent series

$$\ln z = L(v) = \ln z_\sigma + \beta \sum_{n=1}^{\infty} n^{-1}[n a_n v_\sigma + (n-1)a_{n-1}](v - v_\sigma)^n \quad \text{(A.12)}$$

where $a_0 \equiv 0$. Thus $L(v)$ is analytic near $v = v_\sigma$.

Provided the coefficient of $(v - v_\sigma)$ in (A.12), namely $\beta a_1 v_\sigma$, does not vanish, the function $L(v)$ may be reverted to yield v as a function of $L = \ln z$, which is *analytic* in the neighbourhood of $L_\sigma = \ln z_\sigma$. The condition $a_1 \neq 0$ is equivalent to the finiteness of $\partial v/\partial p$ and hence of the compressibility, at v_σ. Since $v_\sigma \neq 0$ the reciprocal is also analytic and consequently the integration in (A.5) yields $\pi(z)$ as an analytic function of z near z_0.

If $a_1 = a_2 = \cdots = a_{k-1} = 0$ and $a_k \neq 0$ so that the compressibility <u>is</u> infinite at the condensation point one sees similarly that $L^{-1}(\ln z)$ is analytic in the variable $(\ln z - \ln z_\sigma)^{1/k}$ so that $\pi(z)$ has a simple branch point at $z = z_\sigma$.

Appendix B

Analytical Continuation of Droplet Model Grand Potential

From (31) to (33) and (46) the grand potential, $\pi = p/kT$, for the droplet model is given by

$$\pi(z) = q_0 \sum_{l=1}^{\infty} l^{-\tau} e^{-\theta l^\sigma} y^l, \quad y = z/z_\sigma. \tag{B.1}$$

Following van Kampen[24] we attempt to find a function $f(t)$ such that

$$\pi(z) = \int_0^\infty \frac{y}{e^t - y} f(t) dt. \tag{B.2}$$

Expanding formally in powers of y under the integral sign one finds that $f(t)$ should satisfy

$$\int_0^\infty e^{-lt} f(t) dt = q_0 l^{-\tau} e^{-\theta l^\sigma}. \tag{B.3}$$

Inverting this Laplace transform yields

$$f(t) = \frac{1}{2\pi i} \int_{c-i\infty}^{c+i\infty} q_0 p^{-\tau} e^{-\theta p^\sigma} e^{pt} dp \tag{B.4}$$

for any $c > 0$. For $\theta > 0$ and $0 < \sigma < 1$ the integrand in (B.4) at $p = c + is$ has the bound

$$q_0 (c^2 + s^2)^{-\tau/2} e^{-\theta |s|^\sigma \cos(\pi\sigma/2)} e^{ct}.$$

Consequently $f(t)$ is defined by (B.4) for all real t and in turn has a bound of the form

$$f(t) < A e^{ct}, \quad A = A(c, \sigma, \tau, \theta). \tag{B.5}$$

Since c may be chosen less than unity the integral in (B.2) exists for all y except for $y = \mathscr{R}\{y\} \geq 1$, and, in fact, defines a function of y meromorphic in the cut plane.

One may now re-expand (B.2) in powers of y and check that the remainder after n terms, namely

$$R_n(y) = y^{n+1} \int \frac{e^{-nt}}{e^t - y} f(t) dt, \tag{B.6}$$

vanishes as $n \to \infty$ provided $|y| < 1$. Thus the integral (B.2) does represent $\pi(z)$ as defined by the series expansion (B.1). Furthermore it evidently continues $\pi(z)$ to the whole z (or y) plane except for a cut along the real axis from $z = z_\sigma$ to $+\infty$. On the cut the real part of $\pi(z)$ is obtained by taking the principal value integral in (B.2). On the other hand the discontinuity of the imaginary part of $\pi(z)$ across the cut at $z = \xi > z_\sigma$ is given by

$$\pi(\xi + i\varepsilon) - \pi(\xi - i\varepsilon) = 2\pi i f[\ln(\xi/z_\sigma)], \quad \varepsilon \to 0. \tag{B.7}$$

To evaluate this discontinuity as $\xi \to z_\sigma+$ or $y \to 1+$, we use the method of steepest descents to perform the Laplace inversion (B.4) for small t. Making the substitution $p = u/t$ brings the integrand to the form

$$u^{-\tau} \exp[u - Xu^\sigma], \tag{B.8}$$

where $X = \theta/t^\sigma$ is now the "large parameter". To sufficient approximation the saddle point is at

$$u = u_0 = (\sigma X)^{1/(1-\sigma)} = (\sigma\theta)^{1/(1-\sigma)} t^{-\sigma/(1-\sigma)}. \tag{B.9}$$

Finally changing variables again by putting $u = u_0(1 + iv)$ leaves, after removal of a constant factor, an integrand ρf the form

$$(1+iv)^{-\tau} \exp\{(\sigma^{-1} - 1)u_0[(1+iv)^\sigma - 1 - \sigma iv]\} \approx \exp[-1/2(\sigma^{-1} - 1)u_0 v^2], \tag{B.10}$$

which peaks sharply when $t \to 0$ so that $u_0 \to \infty$. In all, we find as $t \to 0$,

$$2\pi f(t) \approx [2\pi/\sigma(1-\sigma)]^{1/2} (t/\sigma\theta)^{\tau^*} e^{-g/t^{\sigma^*}} \tag{B.11}$$

where

$$\sigma^* = \sigma/(1-\sigma), \quad \tau^* = (\tau - 1 + 1/2\sigma)/(1-\sigma), \tag{B.12}$$

and

$$g = (1-\sigma)\sigma^{\sigma^*} \theta^{1/(1-\sigma)}. \tag{B.13}$$

Footnotes and References

1. L. Onsager, *Phys. Rev.* **65**, 117 (1944).
2. C. N. Yang and T. D. Lee, *Phys. Rev.* **87**, 404 (1952).
3. M. Kac, G. E. Uhlenbeck and P. C. Hemmer, *J. Math. Phys.* **4**, 216 (1963), See also G. A. Baker, Jr. *Phys. Rev.* **122**, 1477 (1961) and N. G. van Kampen *Phys. Rev.* **135**, A362 (1964).
4. This result has been proven recently with great generality by J. L. Lebowitz and O. Penrose, *J. Math. Phys.* **7**, 98 (1966) and E. Lieb, *J. Math. Phys.* **7**, 1016 (1966).
5. See, for example, M. E. Fisher, *Archs Ration. Mech. Analysis* **17**, 377 (1944).
6. See the account in Mayer's book: J. E. Mayer and M. G. Mayer, *Statistical Mechanics*, Chapter 14. John Wiley, New York (1940).
7. S. Katsura, *Adv. Phys.* **12**(48), 391 (1963). In this connection reference should also be made to the work of: K. Ikeda, *Prog. Theor. Phys., Osaka* **11**, 336 (1954); **19**, 653 (1958); **16**, 341 (1956); **26**, 173 (1961).

8. J. Frenkel, *J. Chem. Phys.* **7**, 200 (1939): letter to the Editor dated November 1938; *ibid* **7**, 538; *Kinetic Theory of Liquids*, Chap. 7. Oxford University Press (1946).
9. W. Band, *J. Chem. Phys.* **7**, 324 (1939): paper received 6th March 1939; *ibid* **7**, 927 (1939).
10. A. Bijl, *Discontinuities in the Energy and Specific Heat*, Doctoral Dissertation, Leiden, presented 29th April 1938.
11. R. Peierls, *Proc. Camb. Phil. Soc.* **32**, 477 (1936).
12. R. B. Griffiths, *Phys. Rev.* **136**, A437 (1964).
13. J. de Boer, *Theorie de la Condensation* in *Changements de Phases*, Comptes Rendus 2^e Reunion de Chimie Physique, Paris (1952).
14. In this connection it is interesting that B. J. Hiley and M. F. Sykes, *J. Chem. Phys.* **34**, 1531 (1961) have shown by numerical studies that for polygons of length s and area l on the plane triangular and square lattices the mean area appears to vary as $\langle l \rangle_s \sim s^\theta$ with $\theta = 1.50 \pm 0.04$. This corresponds roughly to $\sigma = 2/3$.
15. This term was introduced in J. W. Essam and M. E. Fisher, *J. Chem. Phys.* **38**, 802 (1963), who demonstrated its significance in the critical region (see below).
16. See, for example, M. E. Fisher and M. F. Sykes, *Phys. Rev.* **114**, 45 (1959) and later papers.
17. The conclusion that the condensation point should be an essential singularity of the activity series was advanced at the I.U.P.A.P. Conference on Phase Transitions at Brown University in June 1962 and was reported briefly by Katsura (on page 416 of the reference of Footnote 7) and in the 1964 University of Colorado, Boulder, Lectures in Theoretical Physics. More recently A. F. Andreev, *Soviet Phys. JETP* **18**, 1415 (1964) has reached the same conclusion. It should also be mentioned that some of our analysis and conclusions have been closely foreshadowed in an interesting but apparently little appreciated paper by J. E. Mayer and S. F. Streeter, *J. Chem. Phys.* **7**, 1019 (1939). More recently a rigorous proof has been published for d-dimensional Ising models (at sufficiently low temperatures) by S. N. Isakov, *Commun. Math. Phys.* **95**, 427–443 (1984).
18. See the reference in Footnote 15 and M. E. Fisher, *J. Math. Phys.* **5** 944 (1964).
19. The equality (58) was first advanced in Reference 15. Without entering into details here it should be mentioned that (59) was proposed independently by B. Widom, *J. Chem. Phys.* **41**, 1633 (1964). Furthermore G. S. Rushbrooke, *J. Chem. Phys.* **39**, 842 (1963), has proved that the inequality obtained from (58) by replacing = by ≥ is a thermodynamic necessity and R. B. Griffiths, *Phys. Rev. Lett.* **14**, 623 (1965) has shown the same is true of the corresponding inequality obtained by substituting for γ' with (59).
20. D. S. Gaunt, M. E. Fisher, M. F. Sykes and J. W. Essam, *Phys. Rev. Lett.* **13**, 713 (1964). For a more recent assessment see M. E. Fisher, *J. Appl. Phys.* **38**, 981 (1967) and *Rep. Prog. Phys.* **30** (1967).
21. L. van Hove, *Physica* **16**, 137 (1950).
22. At the I.U.P.A.P. Conference on Phase Transitions at Brown University, June 1962.
23. O. K. Rice, *J. Chem. Phys.* **15**, 314 (1947); *J. Phys. Colloid Chem.* **54**, 1293 (1950). See also B. H. Zimm, *J. Phys. Colloid Chem.* **54**, 1306 (1950).
24. N. G. van Kampen, private communication.

3

The States of Matter — A Theoretical Perspective*

Michael E. Fisher

Baker Laboratory, Cornell University, Ithaca, New York

LADIES and Gentlemen: I want to start by dedicating this lecture, this talk, which is, in fact, going to be a rather informal discussion, to Lars Onsager, whom I consider the greatest theoretical chemist of this century and who certainly has made the greatest contributions to the area that I am going to be talking about: namely, the states of matter and our theoretical understanding of them. Onsager got the Nobel Prize in chemistry, as most of you will know, in 1968 and died three years ago, on 5 October 1976, still at the height of his powers. He was a major inspiration to my own scientific career.

Dr. Michael E. Fisher (1979)

I would also like to take this opportunity of making a more personal remembrance of somebody who would undoubtedly be here today if he had not been cut off prematurely. This is Zevi Salsburg, a leading American theoretical chemist, a student of J. G. Kirkwood, who was a professor here in Houston at Rice University. Unfortunately, Zevi died at the age of only 41 in 1970. It is a particular personal sorrow to me that he is not with us now. I know he would be taking an active part in the discussions at this meeting, were he still alive.

---

*An address presented before "The Robert A. Welch Foundation Conferences on Chemical Research XXIII. Modern Structural Methods," which was held in Houston, Texas, November 12–14, 1979. Reprinted with permission from the *Proceedings*, Ed. W. O. Milligan (R. A. Welch Foundation, Houston, 1980) pp. 74–145, and discussion contributions, pp. 146–163.

Figure 1. Professor Lars Onsager, theoretical chemist and physicist, whose profound researches have contributed fundamentally to our current understanding of the various states of matter and their inter-relation.

The Theorist's Approach

Now, it is clear that all the speakers at this symposium have had to approach the problem of how they would address a distinguished audience but one with a rather broad range of backgrounds. As the only theorist on the program, I decided to adopt a strategy in which I would try to give some feeling for the, perhaps, peculiar ways theorists currently look at some of the old and some of the newer problems concerning the states of matter and their inter-relations. Basically I am planning not to *explain* anything! The Welch Foundation, in their wisdom, have arranged for only one main talk each afternoon, to be followed by a period for discussion and cross-examination and for more detailed explanations of those points that members of the audience may find interesting or tantalizing. Thus I will not claim to explain or justify everything I say. If some things prove hard to understand, please forgive me: but my aim is to present to you some views, some perspectives, to mention some of the words, the ideas, the jargon if one likes, that theorists currently use in thinking about the equilibrium states of matter; the sort of questions that they have been asking in the last five to ten years that are different from the questions they asked before. Some of these are old questions but asked in a fresh way; some of them are new questions.

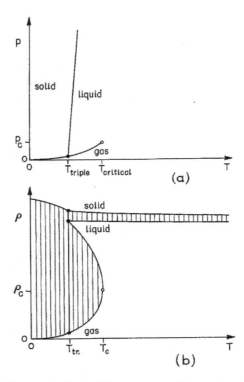

Figure 2. (a) The phase diagram, i.e. the (p, T) or pressure vs temperature diagram for a simple substance; (b) the corresponding density, ρ, vs temperature plot in which the vertical hatching represents tie-lines connecting coexisting phases.

The traditional way to ask about the states of matter is to look at Figure 2(a), which shows, for a simple substance, the pressure versus temperature diagram in which we see the traditional phases known to the Greeks: the solid, the liquid, and the gas. The same information is represented in Figure 2(b), but with the density, ρ, plotted as the ordinate while the temperature is still carried by the horizontal axis. At low temperatures, beneath the triple point temperature, T_{triple}, the gas coexists with the solid phase. As one passes the triple point, one has a gas which coexists with liquid. At higher pressures, on the melting line, the liquid can coexist with the solid. The vertical "tie lines" in the (ρ, T) plot remind us that one is in a two-phase region consisting, say, of liquid and solid. The traditional view of matter was exhausted by these three distinct, separate phases.

However, 110 years ago, Andrews discovered that really there was only one, single, fluid phase, namely, the fluid "gas-or-liquid" phase. One can indeed go smoothly and continuously from ordinary liquid to the usual gas or vapor. There is really no fundamental difference between them, although there is one last temperature below which one may still say if a substance is in a gaseous or vapor phase or, conversely, if it is a liquid. That point is the *critical point* at $T = T_c$. Because of its very special and unusual features, I will regard the critical point as a further, new state of matter. In the phase diagram the critical point looks like just one unique point. People tend to think, "Well, we can forget about a mere point," But, as a matter of fact, what happens there affects strongly the whole surrounding region. So now when we look at the phase diagram we see that there is just one fluid phase, gas and liquid being really the same thing. There is also a solid phase, and, finally, there is a critical state of matter or a "near-critical" state of matter. Then, the questions one starts to ask are rather different.

Figure 3 illustrates this quite forcefully. I want to introduce this picture to you and to present some of the associated ideas that have become very important in thinking theoretically about the states of matter. What is of importance here to the theorist and why? The theorist asks first: "In what spatial dimensionality are you interested?" The dimensionality, d, is plotted on the horizontal axis. The theorist maintains that the value of d is a crucial question. Of course one's first response is that the world is three-dimensional so that one always has $d = 3$. But recall Dr. Hagstrum's address this morning about solid surfaces. He was really talking about two-dimensional phases so that $d = 2$ also matters to us. One-dimensional, chain-like systems can also be looked at in the real world. But the theorist is even free to think of four dimensions, five dimensions, or six dimensions: indeed I have a figure to show later in which the value $d = 6$ plays a crucial role.

Furthermore the theorist has recently found it is useful to think of the dimensionality, d, not just in terms of its integral values, some of which we can study in the real world, but also as a parameter which can be varied continuously! At the bottom of Figure 3 is displayed the equation

$$d = 4 - \epsilon. \tag{1}$$

This says, "Think carefully about the dimensionality $d = 4$:" indeed, four dimensions turns out to be a very special case. Then consider a small deviation in dimensionality, ϵ, below four dimensions. Thus plotted on the figure is the small parameter ϵ that theorists have used to make perturbation expansions. Indeed later on I will show this same picture again but

The plane of Dimensionality and Symmetry

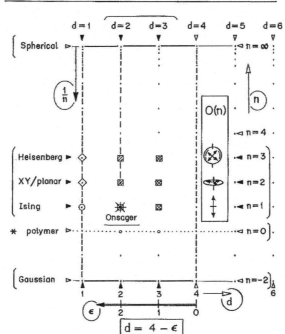

Figure 3. The plane of theory showing dimensionality, d, and symmetry number, n, and the continuous dimensionality parameter $\epsilon = 4 - d$. [After *Rev. Mod. Phys.* **46**, 597 (1974).]

with information that has been gained by the perturbation theory starting in four dimensions. Above four dimensions many things are simple; but in the real world of three and two dimensions they are less simple. However by using continuous dimensionality as an idea, we can make some progress.

Now, what about the vertical axis or ordinate in Figure 3? This is the axis of *symmetry*. When a theorist currently looks at the states of matter, the second question he asks after determining the dimensionality, is "What is the symmetry of the basic or 'fundamental quantity'?" This basic, fundamental physical quantity for a state of matter I will call by the now traditional name, the "order parameter." What is the symmetry of the order parameter? The symbol, n, in Figure 3 simply indicates the degree of symmetry in the simplest cases: thus $n = 1$ corresponds to a scalor order parameter like the density in a fluid. The symmetry then, if one likes, is simply a reflection symmetry of $+1$ into -1. Next $n = 2$ corresponds to the symmetry of a circle: the order parameter is then like a two-component vector free to point in any direction in a plane. Likewise $n = 3$ describes the symmetry of a sphere for which case the order parameter has three components. And, as theorists, we can go all the way up to an infinite number of components $n = \infty$, which turns out to be a limit where one can still say something intelligent and, indeed, precise! In case there may be some polymer chemists In the audience, let me also point out that the case of zero components, $n = 0$ — such is the beauty of the life of a theorist — describes what happens to a long polymer molecule in dilute solution. So nothing is wasted, not even the behavior of a vector of a zero number of components! Furthermore, down at the bottom of Figure 3 at $n = -2$ components, it turns out that one has another theoretically interesting (and "exactly soluble") limit which none of the experimentalists are going to be looking at, even though the case $n = 0$ is open to experimental attack.

In summary, Figure 3 makes the point that important questions we ask center around the dimensionality and center around the symmetry.

Fluctuations, Correlations, Cumulants

Now, let us go further forward and enquire in more detail how a theorist thinks about things. Dr. Baker, in his introductory remarks to the conference, made the very crucial point, which everybody knows but often conveniently forgets, namely that what we call the "structure" of any material at finite temperature — and nobody has seen a material at the absolute zero of temperature — what we regard as the "equilibrium structure", is an *average*. Sometimes we can legitimately forget about that fact because the fluctuations about the average are very small. But at other times we must not; and theorists have increasingly become aware of the fact that if they are going to think intelligently and ask new and penetrating questions, they must remember that equilibrium is a dynamic process in which the simple average structure is only one part of the story. Thus I want to emphasize that the theorist now has a prime, major interest in the fluctuations and the associated correlations. What is it that the theorist wants, more specifically to study? Consider a fluid: really, the crucially important feature is the density of the fluid. Thus I will define local densities or local "operators". "Operator" is a jargon term: the only reason that one tends to talk about local "operators" — and I may slip into that often — is that one likes, occasionally, to remember that quantum mechanics has some uses (although surprisingly few direct uses in the study of many states of matter)! In quantum mechanics, the energy, the density, and other such things, as you well know, are represented by operators. So please forgive me and other theorists when we sometimes refer to important locally-defined physical quantities as "operators".

Thus consider some local densities or operators $A(\mathbf{r})$, $B(\mathbf{r})$, defined at a point $\mathbf{r}$ (in d-dimensional space). As mentioned, in a fluid we have first of all the local particle density

$$\Psi(\mathbf{r}) = \rho(\mathbf{r}), \qquad (2)$$

which is clearly a scalar. Furthermore, for the fluid or gas-liquid, I identify $\rho(\mathbf{r})$ as the *order parameter* and hence use the symbol $\Psi(\mathbf{r})$ to distinguish it especially. Thus in Figure 3 the fluid corresponds to

$$n = 1 \quad \text{or} \quad \text{"Ising-like" behavior}. \qquad (3)$$

Notice the point $(d = 2, n = 1)$ in Figure 3 has been labelled "Onsager". That is because Onsager told us essentially everything we need to know, or a large fraction of what we need to know, about systems characterized by scalars in two dimensions. And so Onsager's name rightly appears there: that was one of his great contributions, technically, his solution of the two-dimensional Ising model in 1944 and following years.

Another quantity of clear interest in a fluid is the local energy density, which might be written

$$E(\mathbf{r}) = \int d\mathbf{r}_{12} \rho\left(\mathbf{r} - \frac{1}{2}\mathbf{r}_{12}\right) \phi(\mathbf{r}_{12}) \rho\left(\mathbf{r} + \frac{1}{2}\mathbf{r}_{12}\right). \qquad (4)$$

This formula simply expresses the fact that a first molecule at $\mathbf{r}_1 = \mathbf{r} - \frac{1}{2}\mathbf{r}_{12}$ will interact with another at $\mathbf{r}_2 = \mathbf{r} + \frac{1}{2}\mathbf{r}_{12}$, at separation $\mathbf{r}_{12}$, via a potential energy function $\phi(\mathbf{r}_{12})$; however, I present this detailed equation only for the sake of being concrete. I will make no special use of it.

Now, if one has a local density, say $A(\mathbf{r})$, one can take its average value, say,

$$G_A = \langle A(\mathbf{r}) \rangle = \langle A \rangle, \qquad (5)$$

which I may call a first order *cumulant*. However, of its own, an average density or first order cumulant in a fluid is not very interesting, since (except near a wall or surface) it will be independent of the position $\mathbf{r}$. Of more significance are the second order or two-point cumulants

$$G_{AB}(\mathbf{r}_1, \mathbf{r}_2) = \langle A(\mathbf{r}_1)B(\mathbf{r}_2)\rangle_{\text{cum}}$$
$$= \langle A(\mathbf{r}_1)B(\mathbf{r}_2)\rangle - \langle A\rangle\langle B\rangle, \qquad (6)$$

where the subscript cum means "cumulant average" (defined in this case by the second line). This clearly depends upon how a density fluctuation at one point correlates with one at another point. Thus one is beginning to get at structural questions: the value of the cumulant depends on the separation vector $\mathbf{r}_{12} = \mathbf{r}_2 - \mathbf{r}_1$.

Now, in any system the most important two-point cumulant, or pair correlation function, is that for the order parameter, namely,

$$G(\mathbf{r}_{12}) = G_{\Psi\Psi}(\mathbf{r}_{12}), \qquad (7)$$

which in the case of a fluid is simply

$$G(\mathbf{r}_{12}) = G_{\rho\rho}(\mathbf{r}_{12}) = \langle \delta\rho(\mathbf{r}_1)\delta\rho(\mathbf{r}_2)\rangle$$
$$= \rho^2[g_2(\mathbf{r}_{12}) - 1], \qquad (8)$$

where $\delta\rho(\mathbf{r})$ is the local fluctuation in density, while $g_2(\mathbf{r}_{12})$ is the radial distribution function. As is well known, the radial distribution function can be observed via scattering experiments using light, or X-rays, or neutrons. In fact, the scattering intensity, $I(\mathbf{q})$, at wave vector $\mathbf{q}$ [with $q = (4\pi/\lambda)\sin\frac{1}{2}\theta$ where θ is the scattering angle] measures the Fourier transform of $G(\mathbf{r})$ according to

$$I(\mathbf{q}) \propto \int e^{i\mathbf{q}\cdot\mathbf{r}} G(\mathbf{r})d\mathbf{r}. \qquad (9)$$

In the case of a magnetic material one expresses the order parameter $\Psi(\mathbf{r})$ in terms of the local magnetization density, $M(\mathbf{r})$, but by using neutrons to do the scattering the corresponding relations hold. Conversely, when one does an X-ray or neutron scattering experiment, all one is really doing is measuring these averages. One is not truly observing the structure, just the average structure, as far as it is revealed by the pair correlation functions.

Now, the theorist may go on and has the freedom of looking at, let us say, three different densities at three different places as entailed in the three-point cumulants

$$G_{ABC}(\mathbf{r}_1, \mathbf{r}_2, \mathbf{r}_3) = \langle A(\mathbf{r}_1)B(\mathbf{r}_2)C(\mathbf{r}_3)\rangle_{\text{cum}}. \qquad (10)$$

To devise methods of measuring such three-point cumulants represents a profound physical problem. Regrettably, I cannot lay claim to any good ideas, but I throw out the question as a challenge to the experimentalists: what methods do we have of really observing the correlations between, let us say, the density at three distinct points? (What one desires, of course, is the net correlation that is left over after one has subtracted off the effects that can be assigned merely to actions of the separate pair correlations.) Later on I will point out that a special principle, the so called principle of *conformal covariance*, makes some surprising predictions about $G_{ABC}(\mathbf{r}_1, \mathbf{r}_2, \mathbf{r}_3)$ that have been tested on theoretical models but have never been checked experimentally.

Finally, as far as the theorist is concerned, there is never too much of a good thing, so I can go on to four-point cumulants, G_{ABCD}, five-point correlations, and more, and more! In any event, I must now discuss, first and foremost, what we know about the correlations and fluctuations and how we think about them.

Disordered States of Matter

Consider first the simplest situation, the case of *disorder*. I am going to classify the states of matter into the *disordered states*, the *ordered states*, and the *critical states*. In the disordered states of matter, the crucial feature is that the correlation functions decay fast. If one separates the two points $\mathbf{r}_1$ and $\mathbf{r}_2$ from one another, the correlation, $G_{AB}(\mathbf{r}_{12})$ decays to zero as $r_{12} \to \infty$. There are no long range correlations. We know this is true in the gas; we know it is true in the liquid. That is part of reason why there is no difference between a gas and a liquid. What is important, and has been realized to be increasingly important in recent years, is that there is a definite scale for this decay; in the simplest cases, one has an exponential decay of correlation $G(r) \sim \exp(-r/\xi)$. But, even if the decay is more complex, there is a scale, there is a *correlation length*, nowadays always called ξ. Now ξ depends on the state of the system; it depends on the temperature, the pressure, the magnetic field, the chemical potentials of the various species, and so on. Thus, even in a disordered state, the first question that the theorist asks Is: "What is the value of $\xi(T, p, \vec{H}, \mu_i, \ldots)$, the correlation length?" Is it just an Angstrom or two? Is it five or ten Angstroms? Is it a hundred Angstroms? Is it climbing up to reach a thousand Angstroms? Indeed, once ξ is measured in hundreds of Angstroms, one is already moving into what is effectively a new state of matter.

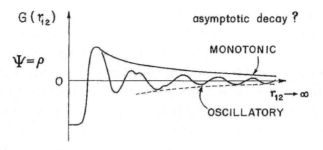

Figure 4. Schematic plot of the decay of the correlation function, $G(r_{12})$, in disordered states of matter. For a fluid one has $G(r_{12}) = \rho^2(g_2(r_{12}) - 1)$, where $g_2(r)$ is the usual radial distribution function.

Returning to the gas and the liquid, in the phase diagram in Figure 2 we see that one can go smoothly from one to the other by passing above and around the critical point. Both liquid and gas are disordered: the correlations always fall off to zero. There is really no profound distinction: the theory of liquids must just be the theory of dense gases. But actually we can say a little bit more because, although the correlation functions must decay to zero, as indicated in Figure 4, we can look at large distances and ask if the decay is *monotonic*, say always downwards from above the axis, or, alternatively does $G(r_{12})$ decay in an *oscillatory* fashion? If one considers a dilute gas, it is easy to convience oneself that the decay is always monotonic. In Figure 5, which is really the same diagram as Figure 2, the expected region of monotonic decay thus includes the whole gaseous region; conversely in a dense liquid the decay is expected to be oscillatory as indicted. So there is, if you like, a certain rough distinction between gas and liquid one can make. Some years ago Professor Widom and I thought about this question and produced this picture (Figure 5) suggesting roughly where the dividing line should lie. As far as I know this has never been checked out experimentally although I still feel it is a theoretical prediction worth testing. However I should, perhaps, mention some of the evidence on which it is based. Unfortunately, neither Widom nor I can solve the statistical mechanics involved in a realistic, three-dimensional fluid! Never the less we could do the calculations exactly for one-dimensional fluids. Figure 6 shows some of the results for one-dimensional models of a gas with infinitely hard repulsive cores and short range "square well" attractions. Both the continuum model and the lattice gas model can be solved. The latter model might well be appropriate to describe the one-dimensional motion of molecules along

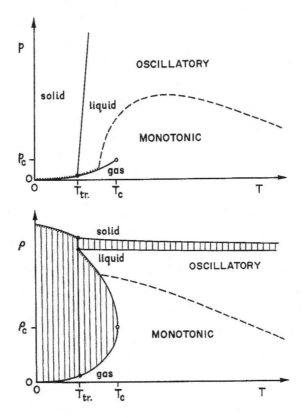

Figure 5. Phase diagram and coexistence curves for a simple substance (compare with Figure 2), showing the regions in the fluid or gas-liquid phase in which the asymptotic decay of the correlation function is believed to be monotonic or oscillatory, respectively. [After M. E. Fisher and B. Widom, *J. Chem. Phys.* **50**, 3756 (1969).]

channels on some of the (110) surfaces which we heard about in Dr. Hagstrum's lecture. We find unambiguously a division between monotonic and oscillatory asymptotic correlation decoy. By thinking through the analogy to the three-dimensional situation we reach the conclusion embodied in Figure 5.

But in a deep sense there remains really only one thing to say about disordered phases from the modern viewpoint (assuming one is not interested in the gory details — and I maintain theorists should not be too interested in the gory details: if you really want to know a particular answer, ask a good experimentalist to tell you!) The modern statement is simply that liquids and gases are basically the same and not that fascinating. So let us come to new things, particularly the question of *order*.

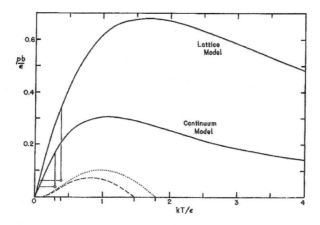

Figure 6. Pressure vs temperature diagram for one-dimensional continuum and lattice gases with hard core repulsions and square well attractive potentials, showing the loci (solid curves) separating asymptotic oscillatory decay of correlation, at high pressures, from asymptotically monotonic decay at low pressures. The circles represent the corresponding van der Waals (or mean field theory) critical points. The dotted and dashed curves represent the loci of the maxima in $\rho K_T = (\partial \rho / \partial p)_T$ (for the lattice and continuum models, respectively), which, below the effective critical temperatures, can be regarded as approximating the vapor pressure curves in the corresponding three-dimensional system. [After M. E. Fisher and B. Widom, loc. cit.]

Ordered States of Matter

Consider once again the same basic phase diagram but as drawn in Figure 7 where I have added something a little different inside the solid phase. One may regard the diagram now as describing a ferromagnetic material, for instance like nickel. When one first cools molten nickel it freezes, and then, on further cooling it becomes ferromagnetic. The little arrows heads in Figure 7 indicate the ferromagnetic region of nickel where all the electronic spins, or, more accurately, a large fraction of the spins point in the same direction, say "upwards". Note that there is a sharp boundary that divides the ferromagnetic from the paramagnetic states of the solid phase. Now a ferromagnet is surely an ordered state of matter! Furthermore we know that we can characterize that order in a rather clean way if we look at the appropriate correlation function. The first thing to do is identify the order parameter $\vec{\Psi}(\mathbf{r})$, simply as the local spin $\vec{S}(\mathbf{r})$, or if you prefer, as the local magnetization, $\vec{M}(\mathbf{r})$, as indicated in the upper part of Figure 7. Next notice that a spin has three components; $S^x(\mathbf{r})$, $S^y(\mathbf{r})$, and $S^z(\mathbf{r})$. Thus we associate this order parameter with the symmetry number $n = 3$, which corresponds to the third row up on the theorist's plane of dimensionality and

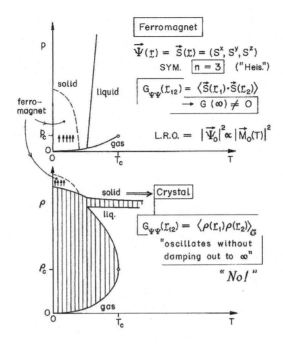

Figure 7. Phase and coexistence diagrams for a magnetic material (like nickel or iron) showing the appearance of a ferromagnetic phase bounded by a sharp phase boundary (dashed curve) within the solid region.

symmetry exhibited in Figure 3. More specifically, in ferromagnets like nickel the spins may point in any direction: i.e. the energetics leave the order parameter vector essentially unconstrained in orientation; such $n = 3$ systems are often said to be Heisenberg-like (as indicated in Figure 3).

Now consider the basic correlation function between the spins namely $G_{\Psi\Psi}(\mathbf{r}_{12}) = \langle \vec{S}(r_1) \cdot \vec{S}(r_2) \rangle$. In the paramagnetic region of solid nickel or, for that matter, in molten or fluid nickel, the spin–spin correlation function decays to zero as $r_{12} \to \infty$. But in the ferromagnetic region when $r_{12} \to \infty$, $G_{\Psi\Psi}(r_{12})$ approaches some finite, nonzero value, say $G(\infty)$, which we call the *long-range order*. Thus we say that what characterizes the ferromagnetic state is the presence of long-range magnetic or spin order. When one follows this through, one finds that the long range order, $G(\infty)$ is simply proportional to the square of the spontaneous magnetization, $M_0(T)$, which can be measured directly by reducing the applied magnetic field to zero (on a suitably shaped specimen). However one can also observe the long-range ferromagnetic order directly, via neutron scattering. Thus in characterizing ferromagnets one has a rather clear and straightforward situation.

But what about the solid itself? We all know that a solid, or at least a nice crystalline solid, is characterized by a regular arrangement of molecules, atoms, or ions. Furthermore the literature of chemical physics and physics tends to suggest that all one need do theoretically is to look at the density–density correlation function $G_{\rho\rho}(\mathbf{r}_{12})$. Now certainly the density is a crucial physical variable: there are no disagreements about that. Thus consider the density–density correlation function. We saw in a fluid that it decays away to zero: However now we say, "Yes, but in a crystal it will oscillate: it will not damp out to zero at large distances." We believe that it will oscillate because of the regular arrangement of particle after particle. So one gets the impression that one need only look at the pair correlation function to check that it has undamped oscillations out to infinity and that will tell us we have a crystal. But that statement in Figure 7 has been marked "No!". Let me re-emphasize that although such implications appear in the literature, I believe that they do *not* represent the truth; they are not the correct theoretical answer! Actually, the true situation is widely "known" but it is a piece of knowledge that many do not admit to. One way to probe the issue is to recognize that we normally think of characterizing a good crystal by Bragg spots or peaks of sharpness limited only by instrumental resolution. If one asks if one has a single crystal the answer is "Surely so, because if I had a powder of many microcrystals, I would not get sharp Bragg spots; I would get a different pattern." But if you had studied a single crystal, then you must, in fact, have held it in some definite orientation.

If one uses equilibrium statistical mechanics and thinks about the correlation functions, one starts by putting the crystal in a container; but unless I tell you I am holding it somewhere definite it is going to float around and rotate freely. If one lets a crystal loose in a spaceship and then diffracts X-rays from it (taking a long exposure) one will not see Bragg peaks. This is why it is inadequate to consider only $G_{\rho\rho}(\mathbf{r})$ or $g_2(\mathbf{r})$. One might indeed ask — I have never asked it too seriously, but I present it as an example of an interesting, unanswered theoretical question — what it is that one actually needs in terms of the correlation functions of a solid to decide if one really has a crystal and to determine what class of crystal it is? I think that one needs to look at the five-point correlation function! Perhaps that is a bit extravagant, and there is someone who can convince me one need only look at the four-point correlation function to be sure one has a crystal and to tell its type. But recall that we know of crystals which are quite regular

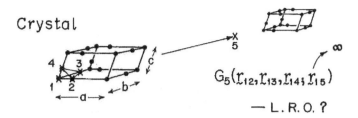

Figure 8. A view of some atoms and lattice cells in a triclinic crystal of low symmetry, illustrating how four points 1, 2, 3, and 4 are needed to position and orient the crystal, while a fifth point, 5, is needed to explore the crystalline propagation of order as $r_{15} \to \infty$.

but which, nevertheless, do not have very high symmetry. Thus consider, as illustrated in Figure 8, a crystal with a triclinic unit cell of low symmetry with, say, four (or more) atoms per cell. I then argue that one needs one, two, three, four points, as marked in the figure, to "hold" the basic tetrahedron in place and to hold the whole original cell pointing the "right way". Finally one needs a fifth point to explore the system at long distances away from the original cell to see if the implied order is maintained out to infinity. I conclude that one probably does need a five-point correlation function to be certain theoretically in the most general case.

If this is so, why are we normally satisfied with merely a two-point correlation function? The reason, as I have indicated, is that one procures for oneself a single crystal, and then holds it in a controlled orientation. But nobody whoever performs neutron diffraction studies or X-ray studies on fluids ever gets themselves a single droplet and holds it carefully oriented in one position: one laughs at the thought! Clearly, in dealing with crystals we take advantage of an important, additional property of an ordered state; namely, that there is a greatly enhanced sensitivity to the boundary conditions imposed on the system. Indeed it is true that if one sets up the boundary conditions correctly, then it does suffice to look at the pair correlation function, $g_2(\mathbf{r})$, and one will see Bragg spots or peaks just as everybody asserted.

But there is still a caveat: we now know that if one has a two-dimensional crystal,* one *cannot* see sharp Bragg peaks even if one imposes boundary conditions appropriate to a perfectly ordered and oriented array. There are deep mathematical theorems on this point. The result may

---

*"Two-dimensional" does not mean infinitesimally thin: thus a freely suspended crystalline film of, say, six atomic layers thickness is perfectly "two-dimensional".

be understood, if one likes, as due to the fact that the Debye–Waller factor describing the thermal vibrations is too big when $d = 2$ and thus washes out the Bragg peaks. This means that two-dimensional crystals are rather different objects than ordinary, three-dimensional crystals and the simple idea of long-range density–density order is not satisfactory to characterize them. One is just beginning to explore these features experimentally.

What these considerations really mean is that one has to think about such things as rigidity modulus and shear modulus, which then seem to be perfectly good criteria for specifying ordered states of matter in all dimensions but particularly for $d \neq 2$, where the simple long-range order, the simple Bragg peak, does not work (and does not appear unless one has some externally imposed bulk ordering field).

It is appropriate to mention here superfluid helium, where, in my view, Onsager (with Oliver Penrose) made one of his greatest contributions: he identified the nature of the order parameter, $\Psi(\mathbf{r})$, and showed it was essentially a *wave-function*, $\Psi(\mathbf{r})$. I want to stress this because the wave-function, as we all know but often forget, is a complex number. So it has a real part and an imaginary part, $\psi = \psi' + i\psi'$, which means one can think of a wave-function as a two-component vector, $\vec{\psi} = (\psi', \psi')$. So when the modern theorist thinks of superfluid helium, he does not think of helium atoms knocking around in a container: What he thinks of is either a two-dimensional system, if it is a helium film, or a three-dimensional system, if it has significant bulk, and he thinks of the two-component, or cylindrical, O(2), symmetry associated with the fact that we cannot tell the phase of a wave-function: changing the phase throughout the system leaves the physics unaltered. This corresponds to the circular symmetry level labelled $n = 2$ (or "XY" or "planar") in Figure 3, "the plane of theory".

What makes helium become a superfluid? Well, as Onsager and Penrose pointed out, an important role is played by the long-range order, sometimes called by the jargon phrase, *off-diagonal long-range order*; however it is just long-range order, but in the wave-function, not in the density. So the order parameter of helium is a wave-function and that is really the one crucial place where quantum mechanics enters. Again it turns out that in bulk, three-dimensional helium the existence of off-diagonal long-range order explains why there can be a superfluid phase. However, we now know a lot about superfluid films also: in particular we know that one cannot have off-diagonal long-range order in a film, but one can still have a nonzero

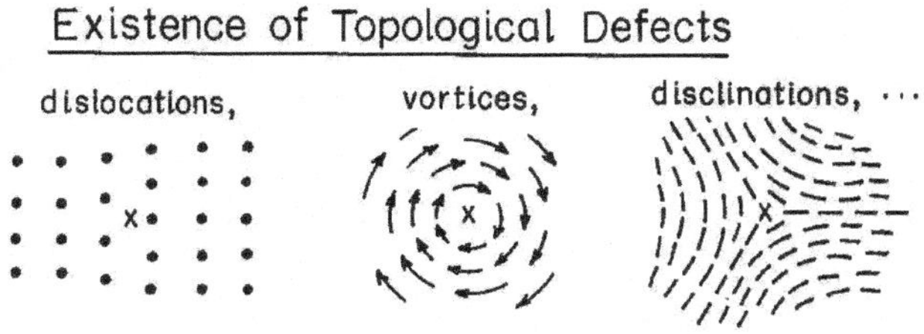

Existence of Topological Defects

dislocations,   vortices,   disclinations, ···

Figure 9. Schematic illustration of the nature of some topological defects in different types of ordered states. (In a nematic liquid crystal, in which disclinations can occur, the order parameter may be regarded as a headless arrow: the corresponding symmetry is not precisely one of those included in the "plane of theory" of Figure 3.)

superfluid density, $\rho_S(T)$. Further we understand that to say that one has a superfluid density is essentially the same thing as to say that a crystal has a shear modulus or displays rigidity. So we regard these various systems as particular but analogous cases, represented by neighboring point in the "plane of theory", Figure 3.

Finally, theorists are also increasingly coming to recognize that there is another important way to think about ordered states. This reflects the fact that in an ordered material one can have *stable topological defects* as illustrated in Figure 9. In a crystal one can have a *dislocation* in the pattern of atomic positions; in superfluid helium may have a *vortex* in the "flow fields" generated by the order parameter vectors, a stable vortex; in a liquid crystal, as indicted in Figure 9, for example, one can have *disclinations*. Note that it is the existence of the ordered state that allows one to have these defects and to define them. Conversely, to talk about a disclination in a liquid crystal which has undergone a phase transition from the ordered nematic phase to the disordered, isotropic fluid phase is nonsensical! So the existence of topological defects is again a criterion for order which the theorists are getting to grips with and which is realized to play a particularly important role in low dimensionalities, especially for $d = 2$.

So now I have talked — and perhaps at too great a length — about *disorder* and about *order*. Let us turn to *criticality* and the approach to criticality.

The Approach to Criticality

Please refer once more to the basic phase diagram as displayed in Figure 10, and note the gas–liquid critical point. As one approaches that critical point,

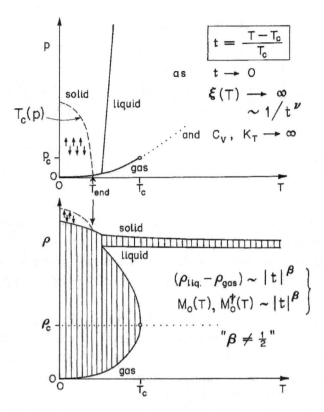

Figure 10. Phase diagram and coexistence curves illustrating the approach to criticality at the gas–liquid critical point and the presence of an antiferromagnetic solid phase and corresponding critical line, $T_c(p)$. The reduced temperature deviation, t, and the critical exponents, ν and β, are defined in the displayed formulae.

a lot of rather striking things happen. Figure 11 shows some very beautiful pioneering measurements taken by the Russian experimentalist, Voronel, of the constant volume specific heat $C_V(T)$, actually of argon, along the critical isochore, the line of constant density $\rho = \rho_c$. As the critical point is approached, the specific heat diverges [see Figure 11(b)]; we believe it increases and, in fact, becomes infinite just *at* the critical point, $T = T_c$. The compressibility, K_T, we know also becomes infinite. Likewise, the correlation length, $\xi(T)$, which we saw was characteristic of the disordered regions of liquid or fluid, also approaches infinity as one approaches criticality.

To characterize the nature of these divergences, we measure the deviation from criticality by the reduced temperature interval, $t = (T - T_c)/T_c$, as stated in Figure 10. The strength of the divergence is then measured

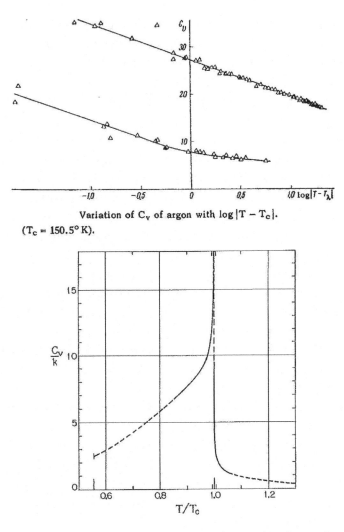

Figure 11. The specific heat at constant volume of argon on the critical isochore $\rho = \rho_c$: (a) the original data of Voronel and coworkers [after M. I. Bagastskii et al., Zh. Eksper. Teor. Fiz. **43**, 728 (1962), Soviet Phys. — JETP **16**, 517 (1963)] plotted on a logarithmic temperature scale with $T_\lambda \equiv T_c$; (b) the data plotted linearly versus T/T_c [following M. E. Fisher, Phys. Rev. **136A**, 1599 (1964)].

by determining the appropriate *critical exponent*. For the correlation length, this is termed ν and we write $\xi \sim 1/t^\nu$ as $t \to 0$ ($\rho = \rho_c$). Other critical exponents, which we need not define here, specify the behavior of the specific heat and of the compressibility.

In drawing Figure 10, it has been supposed that the solid is a magnetic material such as MnF_2, which exhibits an *antiferromagnetically ordered phase* in which alternate microscopic magnetic moments point "up-down", as illustrated in the figure. In that case, as for the ferromagnet, there is a sharp phase boundary, marked $T_c(p)$, which is actually a line of critical points. As this line is approached, the corresponding magnetic correlation length diverges, and, by analogy, we still call the exponent ν, Similarly, there is a divergent specific heat on the locus $T_c(p)$.

Now, there are various questions to ask. One of them, of course, is, "What are the values of the various critical exponents?" I want to focus particularly here on the shape of the coexistence curve for gas and liquid as one approaches the fluid critical point. Thus, consider the difference $(\rho_{\text{liquid}} - \rho_{\text{gas}})$ as a function of temperature as $t \to 0-$. As is evident from Figure 10, the difference vanishes at the critical point, and the behavior may be characterized by a critical exponent β.

If one asks the analogous question for a ferromagnet, one must recall that the order parameter, Ψ, is the local spin or magnetization, so that one asks for the variation of the spontaneous magnetization $M_0(T)$ and describes that similarly by the power law $|t|^\beta$ as $t \to 0-$. However, in the case of an *anti*ferromagnet, one must identify the order parameter, Ψ, with the "staggered" or "sublattice" spontaneous magnetization, which may be denoted $M_0^\dagger(T)$ [see Figure 10].

Now, what value should we expect for the coexistence or order parameter exponent, β? Well, we may appeal to the van der Waals theory, or to many other more elaborate but basically equivalent theories, all of which assert that the coexistence curve should be parabolical as $T \to T_c$: this simply means $\beta = \frac{1}{2}$. Similar theories for ferromagnets and antiferromagnets make the same, quite unequivocal prediction, namely, that the answer on approach to criticality should be $\beta = \frac{1}{2}$.

However, the modern theorist says, "I am sorry, but that is just not correct for the world in which we live!" What the modern theorist thinks is displayed by the $\beta(d,n)$ contours in the plane of theory as shown in Figure 12. From this diagram we conclude that, if van der Waals have lived in a four-dimensional or, preferably, a five-dimensional world, we would still revere him, not just for his deep insights into the nature of fluids, gases, and liquids, but also, more specifically, for his equation. But we know now that, as one comes down below four dimensions, the van der Waals and similar equations are wrong in a crucial and irreparable way. In particular, thanks to Onsager — who, you recall, has solved for us the

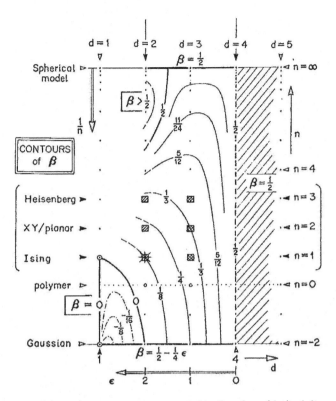

Figure 12. Contours of the order parameter exponent β in the plane (d,n) of dimensionality d and symmetry number n, as suggested by exact results (for $d = n = 2$) and $\epsilon = 4 - d$ expansion calculations [after M. E. Fisher, *Rev. Mod. Phys.* **46**, 597–616 (1974)]. More recent work requires modification of the diagram for $d = 2$ and $n \geq 2$.

two-dimensional scalar system, $d = 2$, $n = 1$, we know that in that case β is not $\frac{1}{2}$: instead, Onsager tells us that $\beta = \frac{1}{8}$ (exactly!).

Now, we would like to know the answers for β for three-dimensional systems. The case $n = 1$ in Figures 3 and 12 corresponds to the gas; the $n = 2$ situation should describe the superfluid; the $n = 3$ case is for isotropic ferromagnets. The contours drawn in Figure 12 are obtained mainly by the recent theoretical methods based on the $\epsilon = 4 - d$ dimensionality expansion starting near four dimensions, where ϵ is small, and expanding downwards, pretending that the dimensionality is continuous! We discover that, for $d = 3$ (or $\epsilon = 1$) and $n = 1$ to 3, the contours are almost vertical, and the value of β is close to $\frac{1}{3}$ — a little less than $\frac{1}{3}$ for $n = 1$ and a little more than $\frac{1}{3}$ for $n = 3$. So that is what the theorist thinks now. The old ideas that the coexistence curve is a simple parabola are surely wrong, he says; but

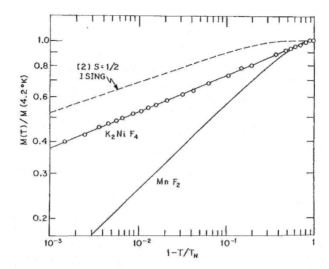

Figure 13. Log–log plot of the order parameter (spontaneous magnetization or sublattice magnetization) versus t (with $T_N \equiv T_c$ for MnF_2, for K_2NiF_4, and for the two-dimensional Ising model as found by Onsager [after R. J. Birgeneau, H. J. Guggenheim, and G. Shirane, Phys. Rev. Lett. **22**, 720 (1969), Phys. Rev. B**1**, 2211 (1970)].

what do the experiments say? Well, I would have loved to have the time to show you fifteen or more beautiful experiments addressing this question, but, in fact, I am going to tell you about only two experiments; their results are shown in Figure 13.

Consider first the curve marked MnF_2, which represents a log–log plot of the order parameter — here the sublattice magnetization, $M_0^\dagger(T)$, as measured by neutron scattering — versus the reduced temperature deviation $t = 1 - (T/T_c)$. (Recall, as I mentioned, that MnF_2 is a typical bulk three-dimensional antiferromagnet.) The slope of the linear part for $t \leq 2 \times 10^{-2}$ determines the exponent β. It is easy to see, by looking at the scales, that the slope is less than $\frac{1}{2}$ and that, in fact, as you may check, it is very close to $\frac{1}{3}$. Thus, the epsilon expansion or dimensionnlity expansion prediction comes out with the right value: rather than a parabolic coexistence curve, one has something closer to a cubic law.

But what about Onsager's prediction for two dimensions? The actual calculated value of $M_0(T)$ [or $M_0^\dagger(T)$] for the two-dimensional spin-$\frac{1}{2}$ Ising model is also plotted in Figure 13. The slope for small t is, of course, $\frac{1}{8}$ rather than $\frac{1}{3}$. Can this be tested in the real world? Indeed it can, as indicated by the data curve labeled K_2NiF_4. For this material the plot in Figure 13 comes

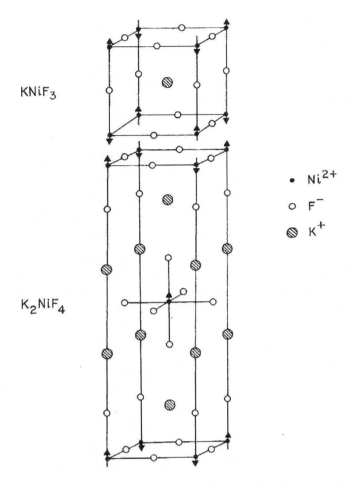

Figure 14. The crystal structure of the ordinary three-dimensional antiferromagnetic KNiF$_3$, and of the related "two-dimensional" or layered antiferromagnet K$_2$NiF$_4$: note that solid dots and arrows denote the nickel ions and their spins in an ordered configuration [after Birgeneau *et al.*, *loc. cit.*].

out quite parallel to the Onsager result. However, the measurements are not made on a freely suspended magnetic film; rather, our inorganic chemistry colleagues have done some pretty chemistry to help us. In order to appreciate this, consult Figure 14, which shows the crystal structure of KNiF$_3$ and of the related compound, K$_2$NiF$_4$, which contains an extra potassium and fluorine ion in each enlarged unit cell. The former compound is a perfectly ordinary three-dimensional antiferromagnet. Note the nickel ions, marked by black circles, which are the magnetically active centers: evidently they form a simple cubic lattice, and each ionic spin interacts strongly with its closest neighbors in all directions and, hence, is aware that it sits in a three-dimensional magnetic system. Measurements of the sub-lattice magnetization thus come out close to those for MnF_2 and, again, yield $\beta \simeq \frac{1}{3}$.

Conversely, it is clear that the structure of K_2NiF_4 consists of separated magnetic layers of nickel ions that can still interact closely in two dimensions, forming square lattice planes, but which are comparatively widely spaced vertically (along the crystalline c-axis). Furthermore, there is a horizontal offset between neighboring layers which results in still further magnetic decoupling between layers. Thus, creative chemists have very kindly built for us what turns out to be an excellent approximation to a two-dimensional magnet, even though, crystallographically, it is still fully three-dimensional. (This latter fact ensures a high spatial concentration of magnetic ions, which greatly eases the problems of experimental observation relative to those entailed in measuring the magnetization of a thin film.) In any event, the measurements for K_2NiF_4 recorded in Figure 13 speak for themselves. A direct analysis indicates $\beta \simeq 0.13$ to 0.14, but we must bear in mind that, even in the best cases, it is hard to measure a critical exponent with a precision much better than ± 0.02. Thus, as far as I am concerned, the result is $\beta \simeq \frac{1}{8} (= 0.125)$, and it constitutes a beautiful verification of Onsager's prediction, which antedated the experiments by more than 20 years.

At this point, I hope I have convinced you that this almost-crazy-sounding theoretical approach of focusing practically exclusively on dimensionality and on symmetry does, after all, correspond quite closely to striking aspects of reality.

Criticality

Now let us consider in a little more detail the state of criticality itself. I want to make some special theoretical points which will constitute some of my most technical remarks. Thus, if you are not mathematically inclined, please bear with me, and try to discern the central issues I am aiming to expose.

How should we think about critical behavior? We have already seen some of its peculiarities: there are certain power laws that are very special; the critical states are very sensitive to dimensionality (as, but to a lesser degree, were the ordered states themselves); various quantities become infinite at criticality. How should all this be analyzed? Well, the theorist says, "First look at the correlations." What is so special about the correlations at criticality? The main point is that the correlation length is infinite:

$$\xi_c = \infty. \tag{11}$$

Hence the correlations no longer die away rapidly; they still decay to zero at large distances, but they fall off very slowly. Thus, criticality implies slow decay, which, in fact, means power law decays.

Consider, then, the basic two-point order–order correlation function, i.e. the density–density correlations for fluids or the spin–spin correlations in a ferromagnet. At criticality we anticipate

$$G_{\Psi\Psi}(r_{12}) \sim 1/r_{12}^{d-2+\eta}, \quad \text{as} \quad r_{12} \to \infty, \tag{12}$$

where, by simple theories, one is told that the new decay exponent η vanishes: however, more generally, we must ask what is the value of this exponent; what ultimately determines it; can we understand it? As a matter of fact, one finds that η is numerically quite small, but, for $d = 2$ or 3 and $n < \infty$, it does *not* vanish.

The next crucial point is to ask how many distinct or independent *critical operators* or *critical densities*, $Q_k(\mathbf{r})$, one must consider. We recall that at a normal gas–liquid critical point the compressibility $K_T \propto (\partial \rho/\partial p)_T$ diverges: it is not hard to see that this is directly related to the slow decay (12) of the order–order correlation function, since the compressibility really measures just the fluctuations in the density. Recall, however, that the specific heat at constant volume also displays a singularity. The specific heat measures the fluctuations in the energy density, $E(\mathbf{r})$. Thence we conclude that, if we take the first critical operator as $Q_1 \equiv \Psi$, we must take a second one as $Q_2 = E$. It turns out, however, that we can then stop: at an ordinary critical point there are *just two* basic critical operators to worry about — the order parameter and the energy density. This holds true for ferromagnetic critical points, antirerromagnetic critical points, the superfluid critical (or lambda) point, and so on.

Shortly we will discuss *multicriticality*, which is a newer concept less well explored both theoretically and experimentally. But, although the physical and chemical situations involved will often look very different, it turns out that the most crucial new feature is simply the *number of critical operators* that must be considered. (In the current jargon these are often termed the "relevant" or "relevant and marginal" critical operators to distinguish them from all other "irrelevant" operators or densities that do not play a leading role at or near the multicritical point.) Some of the sorts of multicriticality I will mention, and the corresponding numbers of critical operators, are set out in Table 1. Thus in *tricriticality* there are *four* important operators or densities: one is the order parameter (which will be a sublattice magnetization in some cases we will look at); a second is the energy; but then there are two more on which we will touch only indirectly.

In bicriticality there are as many as 7 or 8 critical operators. I say "7 or 8" because there is one theorist I greatly respect who says that there are only 7, but I cannot understand all his arguments, and in my own manipulations I can reduce the number needed down only to 8. So there is an open theoretical question — at least open in my mind, so far as bicritical points go.

Notice also in Table 1 what I call "protocriticality", where it transpires that there is only one critical operator. It would be nice to be able to study that situation experimentally, and maybe one day we will be able to. Currently, however, I regret that I cannot tell you the experiment to do. A theorist can do what is required, but he has to go out into the complex magnetic field plane or deal with imaginary chemical potentials to do it — then he can study protocriticality as I will explain briefly later.

A fascinating situation one comes across not infrequently has been labeled "near-criticality". As one approaches a transition point, it gives many signs of criticality, very large order fluctuations, long correlation lengths, large specific heat anomalies, etc.: but then, for some special reasons, the system does not quite make it and, at the last moment, jumps by a discrete but small first-order transition to the new phase. Liquid crystals frequently exhibit such behavior.

Going down the list, I would, given time, want to talk extensively about "continuous criticality", which, as we now understand it, is of particular relevance to the theory of many real two-dimensional systems. Continuous criticality is characterized by the existence of one, or sometimes more, "marginal critical operators". To explain how one identifies a marginal operator or critical density brings us to the third central point in discussing criticality, namely, the fact that associated with each critical operator, Q_k, is a single basic number or exponent, ω_k, which is the main thing one

Table 1. Types of multicriticality and corresponding critical operators and dimensions.

(i)	Ordinary criticality	$Q_1 = \Psi, Q_2 = E, \omega_1 \equiv \omega_\Psi, \omega_2 \equiv \omega_E$
(ii)	Tricriticality	$Q_k, \omega_k, k = 1, 2, 3,$ and 4
(iii)	Bicriticality	$Q_k, \omega_k, k = 1, 2, \ldots, 7$ or 8 (?)
(iv)	Protocriticality	Q_1, ω_1 (only !)
(v)	Near-criticality	
(vi)	Continuous criticality	There is a Q_m with $\omega_m = d$ (especially in $d = 2$)
(vii)	Lifshitz criticality	Spatially modulated phases
	Associated principles	Dilatational or scaling covariance conformal covariance

needs to know about the operator. The ω_k are sometimes referred to as the "dimensions" or "anomalous dimensions" of the critical operators, for reasons that will emerge below. Knowing the ω_k enables one to determine all the critical point divergences and the other critical exponents such as β and ν, at which we looked. At an ordinary critical point, then, we require only $\omega_1 \equiv \omega_\Psi$ and $\omega_2 \equiv \omega_E$. The way one derives the other critical exponents involves the concept of "dilatational" or "scaling" covariance, which we will address next. Further details and restrictions follow from the principle of "conformal covariance", which I will also explain briefly.

Returning, momentarily, to "continuous criticality", we can now detect it by the criterion listed in Table 1, namely, that there be a "marginal critical operator", Q_m, defined explicitly by the condition that ω_m equal d, the spatial dimensionality. Then one must be alert to some special and peculiar things that may happen, some of which I will describe below in the context of superfluid helium films.

Finally, I will touch on the last line in Table 1, that is labeled "Lifshitz criticality". I hesitate somewhat, in addressing an audience largely composed of experimentalists, to mention this because theorists have been thinking hard for some years to find or to devise a real system in which one might actually see Lifshitz criticality. As far as I know, we do not yet have a good example, but, because the topic ties in which some of the questions that have already come up following Dr. Hagstrum's presentation — specifically the question of spatially modulated phases and what controls the wave length of the spatial modulation, I want to mention it. Finally, in that connection, I will also make an important point about the existence and nature of various phases of matter by exhibiting one very simple, almost trivial theoretical model in which, nevertheless, one can prove the occurrence of an infinite number of phases, corresponding to an infinite number of distinct modulated states of matter, with an infinite number of different wave lengths, most being rather sensitively related to the external parameters.

Interlude

Table 1 represents, in essence, a summary of the rest of what I would like to say and is thus a good point at which to break my address and pause. If there are questions or objections to the things I have said so far, I would be pleased to hear them. Perhaps members of the audience would appreciate a chance to shift in their seats, or, if anyone wants to leave, I will turn around and face the blackboard!

Dr. Paul A. Fleury (Discussion Leader), **Bell Laboratories:** I would like to ask you to comment on the question of the onset of simple mean-field or van der Waals type critical behavior above four dimensions, which thus appears as a special dimensionality, because it is my understanding that certain systems, while they are not four-dimensional, still exhibit mean-field-like behavior, so that their critical or marginal dimensionality is different from $d = 4$.

Dr. Michael E. Fisher: Dr. Fleury suffers from the advantage of knowing too much! It is true that, had van der Waals concentrated on *tri*critical points rather than on ordinary critical points, he would have been more or less correct for three-dimensional systems. However, he discussed ordinary critical points. For ordinary critical points, four dimensions is the dimensionality at which the simple theory becomes correct, i.e. the simple mean-field or van der Waals theory.

One of the fascinating things about these multicritical points is that, in the case of the tricritical points, three dimensions is the borderline dimensionality. And so, in fact, rather new and exciting things happen there which I regret I will not be able to say much about, but I will present one experiment which was stimulated by the prediction that the simple theory should work rather well in three dimensions. On the other hand, as regards protocriticality, I will display a graph which shows that one then need *six* dimensions as the dimensionality above which the simple ideas are all right but below which they are not.

So, in truth, the symmetry and a certain few other details do affect the actual borderline dimension. In this overview, I am consciously, but with a good conscience, oversimplifying somewhat. But I appreciate the question, because it is, again, one of the fascinating things for the theorist and for the experimentalist concerned, to find out where the simple theories work, where they do not, and when one is just on the borderlines, as in some of these cases.

Scaling Covariance

Let us go on, then, to review some of the more technical theoretical ideas about criticality I may, incidentally, succeed in conveying what is actually true, namely that one can have some great difficulties in doing calculations at critical points!

A few years ago there was a tremendous breakthrough initiated at Cornell University by Kenneth G. Wilson, a high-energy physicist and the son of the well-known chemist, E. Bright Wilson. The new idea goes by the code name "The Renormalization Group". All I will say is that it has almost nothing to do with group theory (it is really just a handy name), but some of the concepts we will now discuss really find a natural theoretical home there. On the other hand, as often with good experiments and good theory, one gains an intuitive feel for certain things and can make certain guesses which turn out to be right, even if one does not necessarily have the full proof. And so, historically, in fact, these concepts came first as guesses. Thus, I do not really need to apologize for telling you about them without the background of deeper theoretical structure.

Now the scaling or dilatational covariance concept is a purely group-theoretical idea at one level, but we can approach it physically by recalling that the correlation length, ξ, is what sets the physical length scale for the fluctuations and, hence, for the correlations *near* criticality. *At* criticality, however, ξ has gone to infinity, as stressed above, so that we have lost our natural scale of length. One reflection of this fact is that one finds only slow, power law decays at criticality. Since the scale is indefinite, let us ask what might happen if we look at things on a different scale, or, equivalently, if one changes the scale of distance by a factor ℓ, according to

$$\mathbf{r} \Rightarrow \mathbf{r}' = \ell \mathbf{r}, \qquad (13)$$

where r is, say, the distance between two points at which we are comparing the fluctuations in two critical densities, say $Q_a(0)$ and $Q_b(\mathbf{r})$. But, since

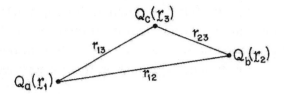

Figure 15. Configuration of critical operators or densities involved in a three-point cumulant which will satisfy a scaling relation.

there was no physical length scale originally, things should look the same on the new scale or, at least, should transform or rescale in a simple way. Mathematically, one can express this idea in terms of the correlations or cumulants of the critical operators via the postulate of covariance or invariance, namely, by supposing

$$G_{Q_aQ_b\cdots}(\mathbf{r}_1, \mathbf{r}_2, \ldots) = G_{Q'_aQ'_b\cdots}(\mathbf{r}'_1, \mathbf{r}'_2, \ldots), \tag{14}$$

where the primes on $\mathbf{r}_1$, $\mathbf{r}_2$, etc., should be noted on the right-hand side,* while the operators or densities themselves transform according to

$$Q_k(\mathbf{r}) \Rightarrow Q'_k(\mathbf{r}') = \ell^{-\omega_k} Q_k(\mathbf{r}). \tag{15}$$

Now, the implications of such a dilatational or scale invariance are quite surprising. For concreteness, consider three critical densities, $Q_a(\mathbf{r}_1)$, $Q_b(\mathbf{r}_2)$, and $Q_c(\mathbf{r}_3)$, at points $\mathbf{r}_1$, $\mathbf{r}_2$, and $\mathbf{r}_3$ as illustrated in Figure 15; Q_a and Q_b might both be the order parameter, say the density, for gas–liquid criticality, while Q_c might be the energy. In an appropriate experiment, we may then hope to observe the corresponding three-point cumulant $G_{abc}(\mathbf{r}_1, \mathbf{r}_2, \mathbf{r}_3; T)$, where now we allow for the temperature to deviate slightly from criticality. Then from (13) to (15) we may conclude that G_{abc} must assume, asymptotically as $T \to T_c$ and $r_{ij} \to \infty$, the form

$$G_{abc} \approx \frac{Y_R(r_1, r_2, r_3; T)}{R^{d-2+\eta_{abc}}}, \tag{16}$$

where R is a free scale parameter while the amplitude function has the scaling form

$$Y_R = Y\left(\frac{r_{12}}{R}, \frac{r_{13}}{R}; \frac{\xi(T)}{R}\right). \tag{17}$$

---

*There are some grounds for generalizing this postulate, to relax the so-called "hyperscaling" constraint, by including a factor $\ell^{-\omega^*}$ on the right-hand side of (14), where ω^* is an extra exponent which has been called the "anomalous dimension of the vacuum": see e.g. M. E. Fisher, pp. 16–37 in *Proc. Nobel Symp.* **24**, Eds. B. Lundqvist and S. Lundqvist (Academic Press, 1974), where other aspects of the topic are also treated in more detail.

The Inclusion of the term $d-2$ in the exponent in (16) is conventional, but the central conclusion forced by scaling covariance is the *addition law**

$$d - 2 + \eta_{abc} = \omega_a + \omega_b + \omega_c, \tag{18}$$

which, indeed, applies, with obvious modifications, to *all* the critical cumulants. Note, furthermore, that (17) implies that, if the triangle defined by the three points $\mathbf{r}_1$, $\mathbf{r}_2$, and $\mathbf{r}_3$ in Figure 15, retains its shape as its size changes, then the amplitude of the critical point decay of the cumulant remains fixed: this clearly embodies the idea of invariance under scale change. Lastly, when T deviates from T_c, the scale of variation is again set by the correlation length $\xi(T)$, but its exponent of divergence with temperature, as introduced in Figure 10, is now given explicitly by

$$\nu = 1/(d - \omega_E); \tag{19}$$

that is, it is fixed by the exponent or dimension ω_E, characterizing the critical energy density, the energy in turn being the variable thermodynamically conjugate to the temperature. [Incidentally, by substituting (18) in (16) at the critical point and recalling the cumulant definition (10), one sees that each critical operator, Q_k, can be regarded as contributing a factor $1/r^{\omega_k}$ to the asymptotic decay and, in this sense, having dimensions of (length)$^{-\omega_k}$.]

Now, further *exponent relations* or *scaling laws* also follow. If I describe the divergence of specific heat, as observed for fluids by Voronel, by an exponent α, we find

$$C_V \sim t^{-\alpha} \quad \text{with} \quad \alpha = (2 - \eta_{EE})\nu, \tag{20}$$

where only ω_E enters into the determination, as is reasonable since the specific heat is, as mentioned, just a measure of the energy fluctuations. Likewise, the divergence of the compressibility, K_T, of IIfluid along the critical isochore can be described by

$$K_T \sim t^{-\gamma} \quad \text{with} \quad \gamma = (2 - \eta)\nu, \tag{21}$$

where $\eta \equiv \eta_{\Psi\Psi}$ is just the exponent introduced in (12) which is related to ω_Ψ via the analog of the addition law (18). The van der Waals equation tells us that $\gamma = 1$, but we know now that this is wrong. In fact, by measuring α and the coexistence curve exponent β (defined in Figure 10), I can use the further scaling law

$$\alpha + 2\beta + \gamma = 2, \tag{22}$$

---

*In case hyperscaling (see the previous footnote) is not assumed, the additional term $-\omega^*$ should appear on the right-hand side of (18).

to predict the value of γ. One finds $\gamma = 1.2$ to 1.3 (depending somewhat on the symmetry number n for the different classes of system), and this prediction can be checked readily against direct observations with excellent results. Thus, three quite distinct experiments are successfully linked together through (22). Likewise, scattering experiments can measure $G_{\Psi\Psi}$ and confirm in detail the more extensive scaling predictions analogous to (16) and (17). I would enjoy displaying some of the experimental evidence for your delectation, but that would distract too much from our main theme, which is the power of a simple but general theoretical idea: invariance under scale change when no natural scale is left.

Conformal Covariance

The next theoretical idea about critical point correlations is even more "far out", as the current slang has it! Remember the games which group theorists like to play with vectors or points in Euclidean space. There are *translations* and *rotations*, with which we are all familiar and under which we take for granted the invariance of the correlation functions and cumulants of a fluid. Then there are the *dilatations*, which we have just explored. But the *full conformal group*, which preserves the angles between curves or lines meeting at all points, also encompasses *inversions* such as are illustrated in Figure 16 or expressed mathematically by

$$\mathbf{r} \Rightarrow \mathbf{r}' = \mathbf{r}/r^2. \tag{23}$$

By following an inversion by a fixed translation specified by a vector $\mathbf{b}$, and then reinverting, one obtains the group of *special conformal transformations* expressed by

$$\mathbf{r}'/(r')^2 = \mathbf{b} + \mathbf{r}/(r)^2, \tag{24}$$

or, more explicitly but more obscurely, by

$$\mathbf{r} \Rightarrow \mathbf{r}' = \frac{\mathbf{r} + r^2 \mathbf{b}}{1 + 2\mathbf{r} \cdot \mathbf{b} + r^2 b^2}, \tag{25}$$

which represents a nonuniform stretching distortion.

What should one make of all this? Well, let us be bold and, following the Russian theorist Polyakov, who drew his inspiration from modern developments in quantum field theory, postulate that the critical point cumulants should also be conformally covariant, i.e. the relations (14) and (15) should

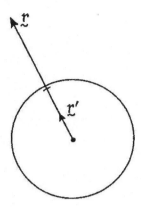

Figure 16. Illustration of an inversion operation on points **r** in Euclidean space: the circle represents a sphere of unit radius in d dimensions.

hold under the transformation (25). It may be crazy, but again a series of remarkable predictions follow: let me recount them briefly. First, there is a fairly straightforward orthogonality relation between two-point cumulants for different critical operators, say Q_a and Q_b, namely,

$$G_{ab} = \langle Q_a Q_b \rangle^c_{\text{cum}} \approx 0, \quad \text{if} \quad \omega_a \neq \omega_b. \tag{26}$$

Second, and more striking, consider a general three-point critical cumulant $G_{abc}(r_1, r_2, r_3)$ as illustrated in Figure 15: one discovers that it must have a *completely unique* form, namely,

$$G_{abc} \approx D_{abc} \frac{r_{12}^{2\omega_c} r_{13}^{2\omega_b} r_{23}^{2\omega_a}}{(r_{12} r_{13} r_{23})^{\omega_a + \omega_b + \omega_c}}. \tag{27}$$

This may be compared with (16) to (18), and we see that, up to the constant amplitude factor D_{abc}, the form of the critical point scaling function, $Y(x_{12}, x_{13}; \infty)$ in (16), is now *fully determined* by the conformal covariance requirement! Note that only the dimensions ω_a, ω_b, and ω_c of the critical operators concerned enter.*

A similar, although not quite so strong, prediction follows for the four-point cumulant illustrated in Figure 17. This takes the form

$$G_{abcd} \approx \frac{r_{12}^{\omega_c + \omega_d} r_{34}^{\omega_a + \omega_b} D_{abcd}(x_{123}, x_{124})}{r_{13}^{\omega_a + \omega_c} r_{24}^{\omega_b + \omega_d} r_{14}^{\omega_a + \omega_d} r_{23}^{\omega_b + \omega_c}}, \tag{28}$$

---

*Here we accept the hyperscaling realtion $\omega^* = 0$: see *Proc. Nobel Symp.*, *loc. cit.* for a further critique.

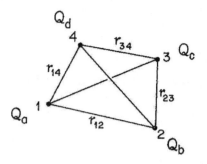

Figure 17. Configuration of critical operators or densities entering a four-point cumulant: note the meaning of the cross-ratio $x_{124} = r_{12}r_{34}/r_{14}r_{23}$.

where x_{123} is the so-called "cross-ratio" $(r_{12}r_{34}/r_{13}r_{24})$, and x_{124} is defined similarly — such an elaborate and restricted form following from a rather simple covariance postulate!

So far as I know, there are no direct experimental tests of these remarkable predictions for G_{abc} and G_{abcd}. However, thanks to the beautiful work of Onsager and many subtle and difficult subsequent theoretical developments, we can test these expressions in two dimensions (for $n=1$) against the results of very, very heavy calculations. And, when one emerges at the end of a long, detailed calculation, one discovers one has these same explicit answers, but one cannot see at any state of the work why it must always come out that way!

One of the frustrating things about a fair bit of recent theory is that there is some beautiful mathematics which one cannot really understand. Then there are some simple ideas which seem to predict the answers ahead of one's getting there, but the two are hard to combine convincingly. So the theorists currently have their frustrations: this is one of them.

Finally, apologies again to all those who found this mathematical excursion a little arduous. I hope at least to have conveyed the flavor of the theorists' thoughts. Now, however, I would like to return to more qualitative matters and illustrate concretely in different physical systems the nature of some of the multicritical states of matter listed in Table 1.

Tricriticality

Let us start by picking up the solid phase illustrated in Figure 10, which has been supposed to exhibit antiferromagnetism. More specifically, consider a

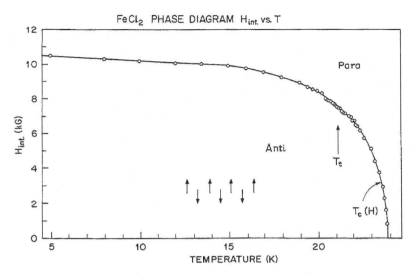

Figure 18. The (T, H) phase diagram of FeCl$_2$ (with $H \equiv H_{\text{internal}}$) exhibiting the separation between paramagnetic and antiferromagnetic phases [after R. J. Birgeneau, G. Shirane, M. Blume, and W. C. Koehler, *Phys. Rev. Lett.* **33**, 1098 (1974) and private communication].

relatively simple magnetic crystal like FeCl$_2$, and examine its (T, H) phase diagram, as illustrated in Figure 18. As one cools the crystal in zero field, $H = 0$, it passes through a critical point often called the Neel point T_N, and enters an antiferromagnetic phase with long-range up-down order in the spin system. On cooling the system in a nonzero, but not too large field, the same thing happens but at a lowered critical temperature marked by the data points labeled $T_c(H)$. The phase boundary separating the paramagnetic, or magnetically disordered, state from the antiferromagnetic state continues quite smoothly to high fields with no visible "anomaly", but, at the temperature marked T_t on the figure, something special happens: there is still a sharp transition, but, if we look at Figure 19, which shows the same experimental data, but with magnetization M now plotted versus temperature T, we see that a striking event has occurred. At low magnetizations and temperatures close to $T_c(0) \equiv T_N$, the transition is continuous and critical, as asserted: however, below T_t, the transition becomes of first order. As the field increases, there is a discontinuous jump in magnetization and, in fact, one can see in the laboratory* a mixed state with both antiferromagnetic and paramagnetic domain coexisting in the same crystal.

---

*For this purpose, one controls the external applied field, $H_{\text{app.}}$, as against the internal field, $H_{\text{int.}}$, plotted in Figure 18.

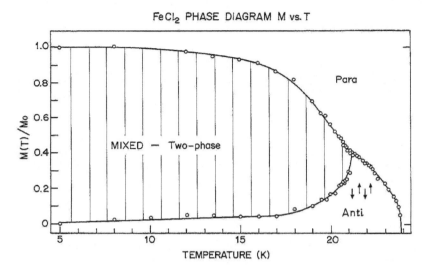

Figure 19. Coexistence plot of magnetization, M, versus T for FeCl$_2$ displaying a first order transition and associated mixed phase ending at the tricritical point (T_t, M_t): compare with Figure 18 [after R. J. Birgeneau, et al., loc. cit.].

At the *tricritical point* itself, (T_t, H_t, M_t), we identify a new state of matter at which in the plane of two field variables, (T, H), a curve or line of critical points turns abruptly into a first order transition boundary.

It is interesting to look at the tricritical vicinity more closely in another magnetic material, namely, dysprosium aluminum garnet or DAG. Figure 20 shows some beautiful experimental data taken by a young experimentalist, Nicholas Giordano, working in the laboratories of W. P. Wolf at Yale. For smaller magnetizations one sees the antiferromagnetic phase, which above $T_t \simeq 1.81°$K transforms via a critical state into the disordered paramagnetic phase: note how straight the phase boundary is as $T \to T_{t^+}$. Below T_t the mixed phase is bounded by two lines which are again remarkably straight. The jump $\Delta M = M_{\text{para}} - M_{\text{anti}}$ is shown in the inset and apparently vanishes linearly with $|T - T_t|$. As a matter of fact, this question might have been addressed by van der Waals, although, to my knowledge, it was not. However, the great Russian theorist Lev D. Landau did discuss the problem using simple phenomenological ideas and, indeed, reached the conclusion that ΔM should vanish linearly with $(T - T_t)$.

The reason for his success, as mentioned above (during the Interlude), is that the tricritical state in $d = 3$ dimensions is just *at* the borderline dimensionality where one can get away with the simple theories. [Actually,

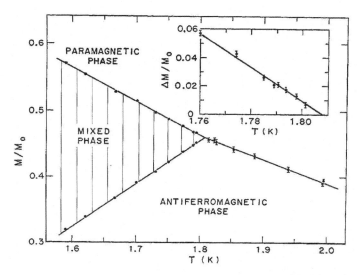

Figure 20. The tricritical region of dysprosium aluminum garnet (DAG) in the (T, M) plane: the inset shows the jump in magnetization, ΔM, on a larger scale below T_t [after N. Giordano and W. P. Wolf, Phys. Rev. Lett. **35**, 799 (1975)].

this is not quite true, first because, as seen in the figure, the critical line if extended below T_t extrapolates *into* the mixed phase region, whereas, according to the simple predictions, it should be *tangent* to the upper, paramagnetic coexistence boundary; second, modern theory predicts subtle but definite logarithmic deviations from simple linear variation with $(T - T_t)$.]

Earlier, I mentioned superfluid helium, and I want to return to it again to show you the power of analogy in theorists' eyes. Thus, regard Figure 21, which portrays essentially the same situation, although it also illustrates one difference, as it were, between chemists and physicists, namely, that the latter tend to prefer to plot the temperature horizontally, while the former usually prefer plotting it vertically! But, if you will, in your mind's eye rotate or, more properly, reflect Figure 21 so that temperature varies in a horizontal direction, you will see that it becomes quite analogous to Figure 20. To understand the situation in more detail, recall that, as one cools down helium-four below about 2.18°K, it becomes a superfluid, as marked in the figure. If one dilutes the helium-four with a little helium-three, which is what the figure represents, it still remains a superfluid. But, if too much helium-three is added, the mixture becomes a normal fluid: that is, the precise analog of varying the magnetization in $FeCl_2$ or DAG. However, at low enough temperatures, instead of going through the transition continuously

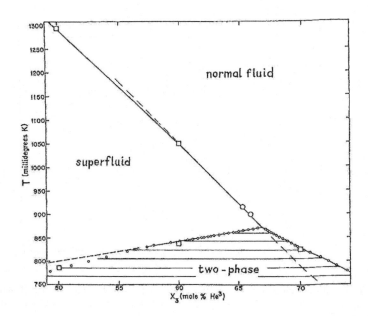

Figure 21. Normal fluid, superfluid, and mixed phases in helium three-four mixtures illustrating a tricritical point at $T_t = 0.872°K$ and $x_3 \simeq 0.67$ [after E. H. Graf, Cornell University Ph.D. Thesis 1967 and E. H. Graf, D. M. Lee, and J. D. Reppy, *Phys. Rev.* **19**, 417 (1967)].

via the critical or "lambda" line, the system does it by a first order jump. Then, in fact, at intermediate concentrations one has in the same container a helium three-rich normal fluid phase coexisting with a helium four-rich superfluid phase.

The tricritical point again marks the state where a critical line becomes a first order transition boundary. Notice that in Figure 21 all the phase boundaries also appear asymptotically straight as the tricritical point is approached — as predicted by simple theories (even though the critical line still extrapolates *into* the two-phase region). Once again, we see that the prime importance of the spatial dimensionality has emerged. Different, nonlinear power law behavior is expected in two-dimensional tricriticality; but we may conclude that dysprosium aluminum garnet and helium three-four mixtures are, as far as their tricritical behavior is concerned, essentially the same problem.*

Last, I would like to show you briefly, tricriticality as the chemist sees it when he studies appropriate multicomponent fluid mixtures. In that

---

*The $n = 2$ character of superfluid criticality, as against $n = 1$ for uniaxial magnetic criticality, will yield some, relatively small differences.

language it is, indeed, a little easier to see why one needs four basic variables or critical densities to describe tricriticality. My colleague Ben Widom and his associate John Lang at Cornell have invesigated in detail the four-component mixture: benzene, ethyl alcohol, water, and salt (actually, ammonium sulfate) under atmospheric pressure but with varying temperature, T. In the range $T = 20°C$ to $40°C$, one finds that for most compositions the mixture separates into *three* coexisting equilibrium fluid phases. The lightest in density is rich in benzene with alcohol as the principal second component, and floats on the top; the middle phase is mainly an alcohol+water-rich phase, while the heaviest phase is a salt-rich aqueous solution. As the temperature is raised, one can, by more carefully adjusting the overall composition of the mixture, still maintain three-phase equilibrium; but, as T approaches $48.9°C$, which, in fact, indentlfies the tricritical temperature for the system, the available three-phase composition range shrinks. Furthermore, the three phases become more and more alike in composition and density. What one witnesses, in fact, is three phases *simultaneously* becoming critical: this is one reason for applying the adjective "tricritical" to the final state at which all three phases do become identical.

The appearance of the three-phase system near tricriticality, as viewed in diffuse transmitted light at a tenth of a degree or so below T_t, is shown in Figure 22. It should be explained that, at temperatures lower by only a degree, all three fluid phases are quite transparent and colorless: but in the tricritical region they appear with a brownish-orange hue — somewhat reminiscent of a sunset in Los Angeles on a smoggy day! Those of you familiar with the appearance of critical opalescence in ordinary fluids or in binary fluid mixtures will recognize this characteristic appearance. The correlation length, ξ, and hence the scale of the intrinsic density fluctuations is of the order of hundreds to thousands of Angstroms in each phase: the shorter wavelengths at the blue end of the spectrum of visible light are being scattered sideways through wide angles (by essentially the same mechanism responsible for the blue color of the sky), leaving, on direct transmission, mainly the longer wavelength, red-orange end of the spectrum.

If one raises the temperature of the solution shown in Figure 22 by just a tenth of a degree, everything clears: one goes through the tricritical point and completely loses the three coexisting phases: they all disappear together. Thus, in chemistry we can now understand why we need four basic

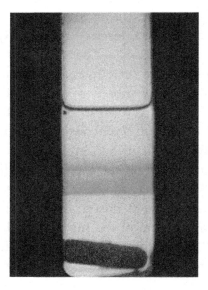

Figure 22. Three-phase critical opalescence viewed in transmitted light near the tricritical point of the benzene — ethyl alcohol — water — ammonium sulfate system [from the work of J. C. Lang, Jr., and B. Widom reported quantitatively in *Physica* **81A**, 190 (1975)]. Note that the volume above the uppermost, dark meniscus represents vapor at atmospheric pressure: the middle fluid phase is narrowest in extent and displays the strongest opalescence.

variables to specify tricriticality: one is temperature (or its conjugate, energy); the others are the three independent relative composition variables (say, the mole fractions of benzene, alcohol, and water). So that is how the fourfoldedness of tricriticality appears in the laboratory.

Bicriticality

Next in our overview of multicriticality we should talk about bicriticality. Magnetic systems provide many of the simplest situations for the theorist because of their underlying symmetries. Chemistry is harder, because there is not as much symmetry as one would like — the fact that reversing the sign of a magnetic field, $\vec{H}$, leaves the physics unaltered helps a lot.

With this in mind, consider again the antiferromagnet manganese fluoride. Figure 23 shows the basic observations regarding the (T, H) phase diagram. On cooling in low fields, one goes into a simple antiferromagnetic phase where the spins point up-down-up-down parallel to the crystalline c-axis. On increasing the magnetic field parallel to the c-axis at low temperatures, the system undergoes a first order transition, a sudden jump, which

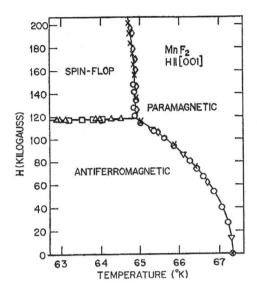

Figure 23. The $(T, H_\parallel)$ phase diagram of MnF$_2$ showing paramagnetic, antiferromagnetic, and spin-flop phases and their boundaries meeting at a bicritical point [after the ultrasonic attenuation measurements of Y. Shapira and S. Foner, *Phys. Rev. B* **1**, 3083 (1970)].

is traditionally called the spin-flop transition. In the new spin-flop phase the spins now order sideways, i.e., normal to the *c*-axis. Neighboring spins still orient in almost opposite senses, but their direction perpendicular to the field is not restricted. Thus we have here a symmetry change from, in the theorist's eyes, a simple $n = 1$ or scalar phase in low fields, to an $n = 2$ or XY-like situation (representing the rotational freedom of the spin ordering) in high-fields. Increasing the temperature at high fields carries one back into the disordered paramagnetic phase via a critical line, as seen in Figure 23.

Evidently, we have a situation where there are two critical lines of different character "parallel" ($\parallel$, $n = 1$) in low fields and "perpendicular" ($\perp$, $n = 2$) in high fields. These competing critical lines meet one another at the *bicritical point* (T_b, H_b); evidently, they also meet the first-order (spin-flop) boundary separating the two ordered phases below T_b. The disposition of the critical lines and the symmetries of the ordered phases is shown schematically in Figure 24, where the *crossover field*, g, represents $(H - H_b)$, while, in this case, we define the reduced temperature variable by

$$t = (T - T_b)/T_b \quad \text{(bicritical region)}. \tag{29}$$

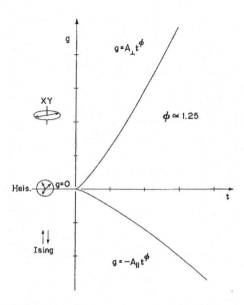

Figure 24. Schematic phase diagram showing a bicritical point in the (T,g) plane as predicted theoretically. Note that $t \propto (T - T_b)$ and, for an antiferromagnet, $g \propto (H - H_b)$ [after M. E. Fisher, A. I. P. Conf. Proc. **24**, 273 (1975)].

One might guess that at the bicritical borderline itself both symmetries combine, so that the spins are essentially free to point in any direction, $\parallel$ or $\perp$. This would correspond to the higher, $n = 3$ or Heisenberg-like symmetry, as indicated in Figure 24, and, in fact, this is what modern theory predicts. One has to know the appropriate new or extra bicritical operator, say Q, corresponding to the crossover field $g \propto (H - H_b)$ — recall, from Table 1, that there are at least 7 basic bicritical operators, of which only three are accounted for by the overall energy, E, and the order parameters, or staggered magnetizations, $\Psi_\parallel = M_\parallel^\dagger$ and $\vec{\Psi}_\perp = \vec{M}_\perp$. From the corresponding exponent or dimension ω_Q for Q, one can define, in analogy to (19), a new *crossover exponent*

$$\phi = (d - \omega_Q)/(d - \omega_E). \tag{30}$$

Now, if I may relate a success of the modern theory, it turns out that this exponent controls the influence of $g \propto (H - H_b)$ on the phase boundaries, as indicated in Figure 24. Specifically, since one finds theoretically that ϕ exceeds unity, one is forced to conclude that the two critical lines, say $H_c^\parallel(T)$ and $H_c^\perp(T)$ or $g_c^\parallel(t)$ and $g_c^\perp(t)$, should meet one another *tangentially* at the bicritical point; in other words, they should come together with zero slope with respect to the first-order (spin-flop) boundary. (This contrasts with the simple theories for which $\phi = 1$ and no such conclusion can be drawn.)

The prediction of tangency was first made in a paper submitted to Physical Review Letters (sometimes called the Journal of Telegraphic Physics by one of my colleagues), and reference was made to the data shown in Figure 23 for MnF$_2$. But the referee said that when he looked at the data he did not see much evidence of the critical lines coming together tangentially or with zero slope! Indeed, at first sight, the spin-flop-to-paramagnetic boundary seems to come down almost vertically on the plot, while the antiferromagnetic-to-paramagnetic boundary appears to leave the bicritical point with a slope of about 45° below horizontal. Nevertheless, since ϕ exceeded unity (although by only about $\frac{1}{6}$ to $\frac{1}{4}$), the theoretical prediction stood — and the referee was kind enough to allow the paper to pass and await the judgement of precise experiments which looked more closely at the bicritical region.

There was indeed a talented and painstaking experimentalist, H. Rohrer of Zurich, who was also sceptical but who was prepared to accept the challenge of checking the theory carefully. Some of his experimental data, obtained first on the antiferromagnet GdAlO$_3$, are shown in Figure 25. The upper part of this figure corresponds to the (T, H) plot of Figure 23 but at great enlargement in the bicritical region. The variable $H_\parallel^2 - H_b^2 \approx 2H_b(H_\parallel - H_b)$ is used for theoretical reasons.) Most surprisingly, one sees that the upper boundary, $H_c^\perp(T)$, comes down steeply but then swings around rapidly but continuously and does, indeed, finally come in tangentially to the spin-flop boundary (shown as a dashed line). Likewise, although less dramatically, the lower, antiferromagnetic boundary, $H_c^\parallel(T)$, also reduces its slope as $T \to T_{b^+}$ and comes in tangentially at T_b. Later experiments on MnF$_2$ itself reveal exactly the same behavior.* And, indeed, the experimentalists are very nice people: if one asks what the data say about the value of the crossover exponent ϕ, which controls the degree of tangency of the phase boundaries, the answers are embarrassingly close to the theoretical prediction of $\phi = 1.25 \pm 0.02$ (for $n = 3$ or Heisenberg bicriticality as appropriate to MnF$_2$; for GdAlO$_3$ one finds $n = 2$ is required, and then $\phi \simeq 1.18$).

It is instructive to pause and consider some of the experimental problems. I have referred glibly to the magnetic field H, which the theorist

---

*See A. R. King and H. Rohrer, *Phys. Rev.* B **19**, 5864 (1979).

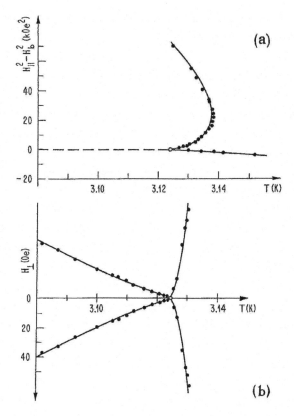

Figure 25. Data on the bicritical point in GdAlO$_3$ showing (a) the critical lines in the $(H_\parallel, T)$ plane to be compared with Figures 23 and 24, and (b) the effect of a transverse field, $H_\perp$, at fixed $H_\parallel = H_b$ [after H. Rohrer and Ch. Gerber, *Phys. Rev. Lett.* **38**, 909 (1977)].

supposes is always applied parallel to the "easy" or c-axis of the MnF$_2$ single crystal. In practice, this is not so readily achieved: furthermore, very high precision proves essential: indeed, the magnetic field vector $\vec{H}_{\text{app}}$, must be aligned to within one millidegree of arc if the bicritical point is to be properly located. Once the bicritical point is found, one can also depart from it by varying the field transversely to the easy axis. The phase behavior then observed is shown in the lower part of Figure 25. There are still two critical lines springing from the bicritical point and marking the transition to the disordered paramagnetic phase but, in addition, there are two *further* critical lines coming in, tangentially again, from *below* T_c. In this plane, therefore, the bicritical point exhibits a "tetracritical" aspect — four distinct critical lines meet together at one point.

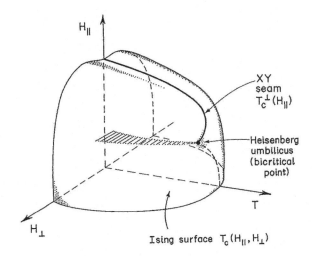

Figure 26. Bicritical phase diagram, showing Ising-like critical surface and first-order spin-flop transition "wedge" below T_b, in the $(T, H_\parallel, H_\perp)$ space [after M. E. Fisher, *A. I. P. Conf. Proc.* **24**, 273 (1975)].

This observation really represents the start of a second lesson in the anatomy of bicritical points, since we are here exploring the effects of some of the other basic bicritical operators: but such a second lesson may be inappropriate, since there has hardly been time to absorb the first lesson! Nevertheless, I cannot resist carrying the second lesson a step further by exhibiting Figure 26, which may at least be anatomically entertaining! What is depicted is the combined effect of both parallel and perpendicular components of the total magnetic field $\vec{H} = (H_\parallel, H_\perp)$. One discovers a critical surface, which is of $n = 1$ or Ising-like character except along a special $n = 2$ or XY-like seam lying in a shallow but definite furrow. The bicritical point itself sits in a sharp umbilical depression on the overall critical surface. Inside the critical surface, which bounds the ordered magnetic regions, lies a narrow pointed wedge or shelf, marking the first-order spin flop transition. The edges of this wedge represent the critical lines below T_b already seen in the lower half of Figure 25.

As the complexity of this last figure suggests, there is still a good deal more to be done both experimentally and theoretically in exploring bicritical points, even though, at this point, I intended to leave what is comparatively well known experimentally, in order to touch on other types of multicriticality.

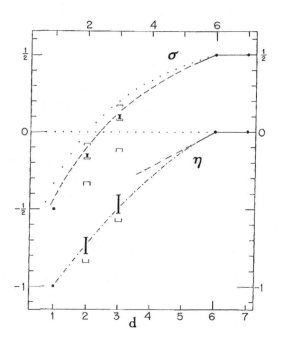

Figure 27. Variation of the "Yang-Lee edge" or protocritical exponents η and σ with dimensionality d, The black dots represent exact results for $d = 1$ and $d = 6$: the dot-dash curves and black bars represent best estimates for intermediate d [after M. E. Fisher, *Phys. Rev. Lett.* **40**, 1610 (1978)].

Protocriticality and Near-Criticality

As I threatened earlier, in the Interlude, there are critical-like situations where $d = 6$ enters as the borderline dimensionality above which simple, van der Waalsian theories are correct. One of these is what I termed *proto*criticality, where there is only one critical operator $Q_1 \equiv \Psi$. In Figure 27 are displayed the theoretical predictions for the correlation decay exponent η, defined in (12), and another related exponent σ, plotted versus the dimensionality d in the range 1 to 7. As you will notice, the exponent values break away from simple constant values below $d = 6$.

I will give a large prize — say $50 — to anyone who can carry out an experiment to check these predictions in two or three dimensions! I will not be too shocked if, within ten years, I have to pay off the $50, but, as I mentioned before, I myself have not been able to think of a direct experiment in which the appropriate exponent η or σ could be observed directly.

Theoretically, protocriticality arises in the context of a very interesting suggestion going back to C. N. Yang and T. D. Lee, who said that, in order to

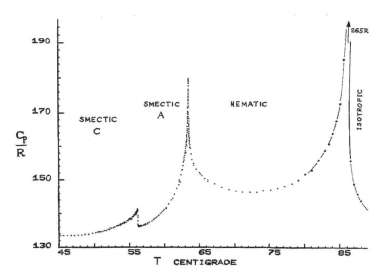

Figure 28. The specific heat of the liquid crystalline material 8S5 showing isotropic, nematic, smectic A, and smectic C regions [after C. A. Schantz and D. Johnson (1977) and *Phys. Rev. A* **17**, 1504 (1978)].

understand the origin of phase transitions at a deep level, theorists should always be prepared to venture out into the complex magnetic field or chemical potential planes — even if experimentalists cannot follow then. Out in the complex plane, one discovers, for a finite system, the so called Yang-Lee zeros of the partition function. In a very large system these zeros close up to form sharp lines, which in fact represent first-order (although complex) phase transitions. The endpoint of such a line of zeros — the "Yang-Lee edge" — represents what is really a critical point,* although mathematically it appears just as a branch point in the complex plane. The exponent σ in Figure 27 describes the behavior of the asymptotic density of zeros as the branch point-cum-critical point is approached by varying complex H (or complex T).

By contrast, let us return to the real world and some fascinating chemical systems, namely, liquid crystals. In Figure 28 are displayed some excellent measurements made by Johnson and Schantz on a material called, for short, 8S5, which really stands for $C_8H_{17}OC_6H_4COSC_6H_4C_5H_{11}$. As you will recall, liquid crystal molecules are roughly rod-shaped (although usually with flexible "tails" of varying length) and tend to line up parallel to one another. At high temperatures, in this case above 86°C, they form a

---

*See e.g. M. E. Fisher, *Phys. Rev. Lett.* **40**, 1610 (1978).

completely isotropic liquid; but, as they are cooled down, the molecules start to order by lining up locally, as indicated by the rising specific heat on the extreme right side of Figure 28. Then the isotropic liquid undergoes an abrupt transition to the so called *nematic phase*, which is a uniaxial liquid with a definite axis parallel to which most molecules line up, even though their positions in space remain disordered. On further cooling, the molecules start to form layered regions, and, via a sharp transition again, in this case at $T \simeq 64°C$, the fluid develops long range *smectic order*, in which the layers extend coherently throughout the fluid. Initially, the nematic axis, or average molecular orientation, is perpendicular to the layers: this is the *smectic A phase*; but on cooling yet further another abrupt change of symmetry occurs, at about 55°C for 8S5, and the molecular axes develop a definite tilt with respect to the normals to the planes of layering, thus forming a *smectic C phase*.

The three distinct specific heat peaks in Figure 28 corresponding to these three transitions all look, at first sight, very much like the critical specific heat peaks seen at ordinary fluid critical points (as illustrated in Figure 11) and at magnetic and antiferromagnetic transitions. In all these previous cases, the specific heat displays a sharp integrable transition, so that the entropy is continuous through the critical point — there is no latent heat, which is, of course, a signature of a first order transition. But one of the sad things when one looks very carefully at these various liquid crystal transitions is that one discovers — and I think more of us are getting convinced, although there are still points of argument — one discovers that they are only "nearly critical": they do not quite make it! The correlation lengths undoubtedly get very long, and there are large fluctuations in the various order parameters, but at the last moment the systems seem to shy away from criticality and indulge instead in first-order, discontinuous transitions.

Such near-critical transitions have not proved easy for theorists to handle. Frequently, they seem to depend on rather subtle interactions with other degrees of freedom of the system, particularly elastic modes, whose effects and magnitudes have proved hard to assess with confidence. In other cases, certain types of symmetry more complex than the simple rotational $O(n)$ symmetries with $n \leq 3$ discussed above, seem especially prone to the phenomenon of instability leading to a first order transition. Such an example is embodied in the longstanding problem posed by the antiferromagnet MnO, which, because the lattice formed by the nearest neighbor

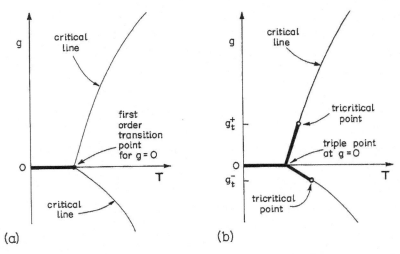

Figure 29. Possible phase diagrams for a system exhibiting a near-critical transition associated with a complex symmetry under the action of a symmetry lowering field: case (b) with close-by tricritical points seems the most likely to be realized in practice [after E. Domany, D. Mukamel, and M. E. Fisher, *Phys. Rev.* **15**, 5432 (1977)].

magnetic ions has triangles, cannot support the simplest sort of checkerboard antiferromagnetic order. The specific heat of MnO again looks quite like that of MnF$_2$ (or like Figure 11), but it ultimately exhibits a latent heat and a small but definite first order transition. Using the modern renormalization group techniques, however, one can attack these problems to some extent. Consider the imposition of some new field g, which lowers the symmetry of the system so that, for $g \neq 0$, a critical state of new symmetry might be realized. In the case of MnO, a uniaxial stress will do the trick. Two conceptual possibilities are illustrated in Figure 29: in the first, the near-critical transition immediately becomes critical. However, this seems unlikely to be the general situation; rather, as checked by explicit calculations for one model situation, the second possibility (b) of Figure 29 is probably the more realistic expectation. Here the near-critical first-order transition actually represents a triple point in the extended (T, g) plane, but there are two *nearby tricritical points* at which the transition does become continuous and critical of the anticipated character.

For MnO, this is indeed found to happen: squeezing the crystal along one axis spoils the higher symmetry, and a tricritical point is reached; thereafter, ordinary critical behavior is observed. Thus, by destroying or lowering the symmetry — something that always tends to upset out field theorist colleagues, who like to think that increasing the underlying symmetry is generally a good thing to do — by destroying the symmetry, then, we remove the near-critical first-order transition and actually simplify both the theory and, indeed, the physical system itself.

Continuous Criticality

Return for a moment to the plane of theory as exhibited in Figures 3 and 12 and consider the special point ($d = 2$, $n = 2$), lying just above the famous Onsager point ($d = 2$, $n = 1$). There one first encounters, so we believe, the phenomenon of continuous criticality, which has been all the rage amongst leading condensed matter theorists in the last few years. Let us, in particular, inquire after the critical point decay exponent η, defined in (12). Recall that the simple theories (of Ornstein-Zernike, or Landau, or Debye) say $\eta = 0$: but Onsager has told us that the value is $\eta = \frac{1}{4}$ for $d = 2$, $n = 1$. Some years ago, I could with reasonable confidence then draw the contours of constant $\eta(d, n)$ in the plane of theory:* but in the region of the special point ($d = 2$, $n = 2$) there was considerable uncertainty. Now we are rather convinced that there is a pileup of contours for $\eta \geq \frac{1}{4}$, all meeting at ($d = 2$, $n = 2$): matters are thus very different than at other normal critical points which mark transitions to states with long range order. Rather, we believe that, below the initial critical point $d = 2$, $n = 2$, where the correlation length, $\xi(T)$, first diverges as T is reduced, one enters into what are really quite new states of matter. They display no long range order of the usual sort (in accordance with the theorems discussed earlier about crystalline and similar ordering in two dimensions). Instead, they remain critical in the sense that crucial correlation functions still decay slowly, as at the initial critical point, although (and this is a central point) with an exponent $\eta(T)$ which *varies continuously with temperature*. This variation is nonuniversal in a manner quite inconsistent with what is observed in other critical states. Despite their criticality, such states are expected to exhibit various types of rigidity or shear modulus and to sustain topological defects like some of those illustrated in Figure 9.

Regrettably, much of what I am relating here is still in the realm of wondrous theoretical story-telling and creative folklore! There are comparatively few experiments on truly two-dimensional systems which meet the

---
*See e.g. Figure 4 in M. E. Fisher, *Rev. Mod. Phys.* **46**, 597 (1974).

required conditions and test the theory. However, I might mention, in particular, some most striking studies of freely suspended liquid crystal films only a few monolayers thick performed recently by Moncton and Pindak at Bell Laboratories. Such experiments will help keep the theorist at least semihonest! However, I would like to tell you about some other experiments where the theory appears rather successful. These concern very thin superfluid helium films which, even though they adhere to a three-dimensional substrate, are probably rather good representations of an ideal ($d = 2, n = 2$)-system.

Despite their two-dimensionality and consequent absence of long range off-diagonal order (mentioned originally in discussing the order in superfluid helium), such films have been found, in a variety of experiments, to display an apparently nonzero superfluid density $\rho_s(T)$, or more properly an areal superfluid density, analogous to a shear modulus in a crystal. At low temperatures, the superfluid fraction is quite high, but, as in three-dimensional bulk helium, it drops as the temperature is raised: see the schematic curves in Figure 30. In a thicker film, the overall areal superfluid density is, of course, higher, so that one gets a series of curves for different film thickness, indicated in the figure.

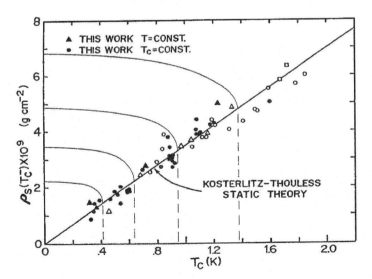

Figure 30. The areal superfluid density, $\rho_s(T)$, showing the points of discontinuity, $\rho_s(T_c-)$, estimated from a variety of experiments on helium films of different thicknesses versus the critical temperature T_c. Note that the variation of $\rho_s(T)$ shown for $T < T_c$ is purely schematic [after D. J. Bishop and J. D. Reppy, Phys. Rev. Lett. **40**, 1727 (1978)].

But what was noted sometime back, although, since the theorists had not approved it, the experimentalists dutifully kept quiet about it, was that this areal superfluid density seemed to drop *discontinuously* to zero at the transition point — as suggested by the vertical dashed lines in Figure 30. Now, while the theorists were industriously plowing their way through the rest of the plane of theory diagram, such a discontinuous jump at a critical point seemed rather implausible, and, furthermore, the experiments always displayed some "rounding" or "tailing". But now it is thought that this is just what should happen! That is, there should be a discontinuous jump in $\rho_s(T)$ when one reaches the critical point in a two-dimensional XY-like or $n = 2$ system. Furthermore, this discontinuous jump should have a magnitude that is essentially equal to the reciprocal of the universal critical exponent $\eta(T_c) = \frac{1}{4}$. That may seem a crazy way to state matters, but the analysis of Nelson and Kosterlitz[*] (based on the theory of Kosterlitz and Thouless[†] for the static equilibrium state of continuous criticality and its ultimate transition to disorder) comes out that way and makes the striking prediction

$$\rho_s(T_c-)/T_c = K, \tag{31}$$

where the constant is universal, *independent* of the film thickness, and given explicitly by

$$K = 8\pi m_4^2 k_B/h^2 \simeq 3.49 \times 10^{-9} \text{ g/cm}^2\text{K}, \tag{32}$$

where m_4 is the mass of a helium-four atom, while k_B is Boltzmann's constant, and h is Planck's constant. The data points shown in Figure 30, the most recent being those of Reppy and Bishop at Cornell who measured $\rho_s(T)$ directly by oscillating a helium film on a roll of mylar, confirm the predicted linearity between $\rho_s(T_c-)$, the critical jump in superfluid density, and T_c, from the thinnest films, where T_c is depressed to only 0.3°K, up to thick films, where T_c has attained 80% of its full bulk value. As indicated, there is, in practice, some rounding due to finite frequency and other effects which have to be corrected for: nonetheless, the agreement, if not absolutely convincing, is most encouraging and certainly suggests that the modern ideas about the special sort of continuous criticality states to be found in two-dimensional physical systems are essentially correct.

---

[*] D. R. Nelson and J. M. Kosterlitz, *Phys. Rev. Lett.* **39**, 1201 (1977).
[†] J. M. Kosterlitz and D. J. Thouless, *J. Phys. C* **6**, 118 (1973).

I hope this brief recounting gives some feel for the excitement generated by these recent developments: indeed, it leaves some of those theorists who are still concerned to puzzle about peculiarities and lack of detailed understanding in ordinary, chemical, three-dimensional, tricritical fluid mixtures already feeling a little oldfashioned!

Lifshitz Criticality and Modulated Phases

The last entry in the list of types of multicriticality in Table 1 has now been reached — although theorists and experimentalists have discovered other types,[*] which I will not review. To discuss Lifshitz criticality it is appropriate to return to the basic scattering function, $I(q)$ as defined in (9), especially since this is frequently accessible to direct experimental observation. Recall that the wave number, q, serves to measure θ, the angle of scattering. It may also be helpful to have in mind neutron scattering from a magnetic material, although the principles apply equally to other systems. Then, as illustrated in Figure 31(a), as one approaches the critical point of a typical system, say a ferromagnet, from above T_c, the scattering intensity starts to peak at low angles, displaying a maximum at $q = 0$, i.e., at $\theta = 0$. The peaking represents the buildup of the short range order and the increasing correlation length. As $T \to T_c$, the maximum, $I(\mathbf{0})$, rises to infinity, and, at the critical point itself, one finds the power law behavior

$$I_c(\mathbf{q}) \sim 1/|\mathbf{q}|^{2-\eta}, \quad \text{as } q \to 0, \tag{33}$$

where η is again the order correlation decay exponent introduced originally in (12). Below T_c, as indicated in the figure, the scattering intensity $I(\mathbf{q})$ *extrapolated* to zero angle ($q, \theta \to 0$) becomes finite again, representing residual short range fluctuations in the order, but the presence of long-range order, or phase separation, results in a Bragg peak — mathematically a delta-function — located just at $\mathbf{q} = 0 = \mathbf{0}$.

That is the standard course of events, but consider now the situation depicted schematically in Figure 31(b). In this case, the scattering intensity above T_c starts to grow *two* maxima, symmetrically positioned at nonzero

---

[*] A notable example is the so called "polycritical point" found in liquid helium-three under pressure at a temperature of a few millikelvin or so; the order parameter is then much more complex than any discussed above, but, for reasons to do with the fermion character of the system, the transitions in liquid helium-three, like those in ordinary metallic superconductors, can be successfully treated by the van der Waalsian, phenomenological, or the mean field type of theory called, in this domain, BCS theory after the pioneering work of Bardeen, Cooper, and Schrieffer.

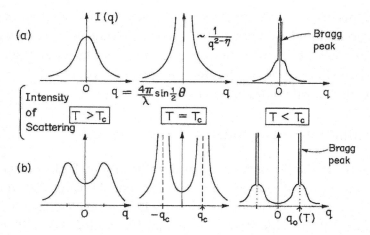

Figure 31. Schematic variation of the scattering intensity, $I(q)$, on passing through criticality (a) in an ordinary system like a ferromagnet or fluid, and (b) in a system which undergoes a transition to a spatially modulated phase characterized by a wavevector $q_0(T)$.

q: this means that wave-like correlations are developing in the material. At the critical point itself, these two peaks diverge at wavevectors $+\mathbf{q}_c$ and $-\mathbf{q}_c$: each critical peak is similar to the single peak at an ordinary critical point, so one has

$$I_c(\mathbf{q}) \sim 1/|\mathbf{q} - \mathbf{q}_c|^{2-\eta}, \quad \text{as } \mathbf{q} \to \mathbf{q}_c, \tag{34}$$

and likewise as $\mathbf{q} \to -\mathbf{q}_c$. The value of the exponent η will normally correspond to a higher symmetry number, n, than in the corresponding case represented in Figure 31(a), but otherwise the critical behavior is not really anomalous. Finally, on going below T_c in this situation, one is left with *two* Bragg peaks located at wavevectors $\pm \mathbf{q}_0$, where the wave number $\mathbf{q}_0$ need not necessarily take any simple value. Indeed, $q_0 = q_0(T)$ will frequently vary slowly and continuously (as far as can be verified) with the temperature, or the pressure, etc.

The two Bragg peaks at nonzero wavevector mean that, in the ordered state of the system, one now has a regular spatial modulation of the order parameter, say the local magnetization, $M(r)$, of wavelength $\lambda_0(T) = 2\pi/q_0(T)$. The wave fronts will, of course, be normal to the wavevector q_0, which, for simplicity, we may take as directed along the spatial z-axis. Since q_0 varies with T, one has an *incommensurate modulated state* which has not locked-in to any other periodicities of the system such as, for example, the crystal lattice spacing. What goes on microscopically is illustrated in

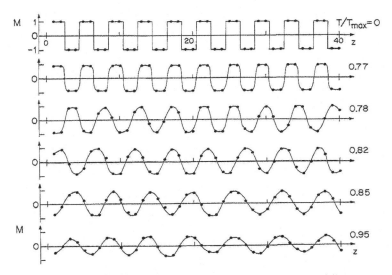

Figure 32. Equilibrium configurations of the average layer magnetization $M(z)$ at increasing temperatures (with $T_{\max} \simeq T_c$) calculated for the ANNNI model by Monte Carlo methods: each dot represents a layer magnetization, the lines being merely guides to the eye [after W. Selke and M. E Fisher, *Phys. Rev. B* **20**, 257 (1979)].

Figure 32, which shows the mean layer magnetization, $M(z)$, for various temperatures below a $T_{\max} \cong T_c$, as calculated for a model system to be explained shortly.

Now suppose I have some field or parameter, κ, which can be varied: physically, this might be the pressure or stress along some axis. Then it may be possible to go from ordinary critical behavior with uniform long-range order to the modulated or "sinusoidally" critical state, all within the same physical system. A phase diagram illustrating this is shown in Figure 33. The critical temperature, $T_c(\kappa)$, now depends on κ, but the critical state remains ferromagnetic for $\kappa < \kappa_L$ (where, for the system illustrated, $\kappa_L \simeq 0.27$). However, for $\kappa > \kappa_L$ the critical line changes character and now bounds a sinusoidal or modulated ordered phase. The critical wave vector, $q_c(\kappa)$, now also varies with κ and, so we may anticipate, approaches zero, corresponding to infinite wavelength, when $\kappa \to \kappa_{L+}$. The boundary point at $T_L = T_c(\kappa_L)$ is the long-announced *Lifshitz point*; it is marked L in Figure 33. Generally then, a Lifshitz point is a special point on a critical line at which the transition from the disordered state changes from being into a uniform ordered state to being into an incommensurate, spatially modulated state.

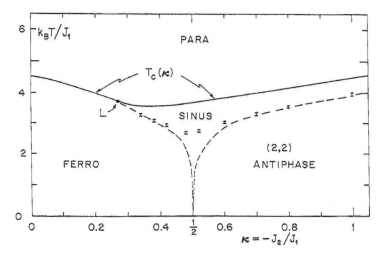

Figure 33. Phase diagram in the (T, κ) plane for the ANNNI model, exhibiting a disordered, paramagnetic phase, a ferromagnetic phase, a sinusoidal, incommensurate spatially modulated phase, a (2,2) antiphase region, and a Lifshitz point (marked L). Note the phases below the critical line $T_c(\kappa)$ are determined by Monte Carlo techniques which are inadequate at temperatures below about $2.5(J_1/k_B)$ [after W. Selke and M. E. Fisher, *Proc. 1979 Int. Conf. Magnetism, J. Mag. Mag. Maths.* **15–18**, 403 (1980)].

Since, as mentioned, I am unable at present to show you a phase diagram for a real system which exhibits a convincing, simple Lifshitz point, it is instructive to look further at the model on which the calculations leading to Figures 32 and 33 are based. This has been christened the ANNNI model, the acronym standing for the "anisotropic next-nearest neighbor Ising" model: it is a very simple spin $\frac{1}{2}$, scalar or $n = 1$ system where each spin, s_i, at a lattice site i, can point only "up" or "down" corresponding to $s_i = \pm 1$. The local environment of a typical spin is then as illustrated in Figure 34. Each spin on, say, a simple cubic lattice is coupled ferromagnetically to its in-layer nearest neighbor with strength $J_0 > 0$; along the z-axis (or crystalline c-axis), it is also coupled ferromagnetically to its two nearest neighbors with strength J_1 (which has been set equal to J_0 in the calculations displayed already); finally, along the z-axis (only), there is a competing, *anti*ferromagnetic, next-nearest neighbor coupling $J_2 < 0$. This last feature is really the only way in which the model goes beyond the nearest-neighbor Ising models considered by Onsager. The basic parameter, carrying us from uniform to spatially modulated states in then simply

$$\kappa = -J_2/J_1 > 0. \tag{35}$$

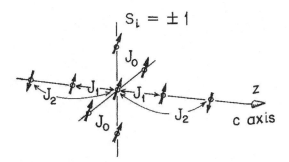

Figure 34. Illustrating the environment and interactions of a typical Ising spin in the ANNNI or anisotropic next-nearest neighbor Ising model. Note that only those spins coupled directly to the central spin are shown, although spins are located on all points of the simple cubic lattice defined by the axes drawn.

For large κ, layers of spins will want strongly to be oriented oppositely to their second-neighbor layers, as suggested by the spin orientations drawn in Figure 34. At low temperatures, this means the ground state will be a so called (2, 2) *antiphase state* in which alternate *pairs* of layers point up or down, as shown in the local magnetization plot at the top of Figure 32 — essentially a square wave of magnetization, $M(z)$. Furthermore, as indicated in the phase diagram, Figure 33, for $d \geq 3$ this (2, 2) state retains its integrity, and fixed *commensurate* wavevector, up to some finite "melting temperature". Conversely, for $\kappa < \frac{1}{2}$, it is not hard to see that the ground state is ferromagnetic and the ferromagnetic state also persists at finite temperatures.

Now large computers are very handy instruments, but they can prove to be dangerous toys! The first thing we did was to study a finite version of the model on $6 \times 6 \times 40$ and $10 \times 10 \times 40$ lattices using Monte Carlo sampling techniques. That led to the mean layer magnetization patterns shown in Figure 32 and to the dashed-line melting curve for the (2, 2) antiphase state shown in the phase diagram in Figure 33. Above that, as Figure 32 illustrates, the mean magnetization wave apparently breaks loose from the lattice and becomes remarkably sinusoidal in profile (with very small third and higher order harmonic content). Notice that we obtain accurate overall incommensurate sine waves, even though the individual layer spins can each point only fully "up" or fully "down" — this serves to emphasize that the structure that one "sees" in any scattering experiment is only an *average* structure. Because this is a computer simulation of finite time duration,

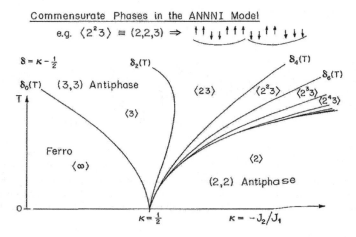

Figure 35. Schematic phase diagram for the ANNNI model at low temperatures near $\kappa = \frac{1}{2}$, showing various first-order phase boundaries, $\delta_{2k}(T)$, with $\delta = \kappa - \frac{1}{2}$, separating distinct commensurate phases, $\langle 2^k 3 \rangle$, of character indicated at the top of the figure.

and because the lattice studied is not infinite, we cannot be sure that one really enters a freely-floating incommensurate phase when one approaches the critical line, $T_c(\kappa)$, in Figure 33. However, it seems quite plausible that the model does have such a phase and does exhibit a Lifshitz point where $q_c(\kappa) \to 0$ when $\kappa \to \kappa_{L+}$, as asserted. One can then test various theoretical ideas about Lifshitz points and behavior in their vicinity. Matters seem to work out quite well, and theorists are now looking forward eagerly to more stringent tests of their ideas in real systems. (It might just be mentioned that Lifshitz points come in quite a variety, so that much remains to be explored, both on the experimental and theoretical fronts.)

But let us not trust the computer at low temperatures! More specifically, it is interesting, in trying to grasp in more detail the connection between microscopic Hamiltonians and phase transition behavior, to inquire what actually goes on near the special point $(T = 0, \kappa = \frac{1}{2})$, in the phase diagram of the model: recall that this is where the ground state abruptly changes its character. Let me bring this story to its final chapter by telling you what we can show happens. As a matter of fact, one has to do some better theory: one cannot appeal to a computer. The conclusions are illustrated schematically in Figure 35. One discovers, at nonzero temperatures, a veritable fountain of distinct spatially modulated phases: on the left for $\kappa < \frac{1}{2}$, one first has the ordinary, uniform ferromagnetic phase, marked $\langle \infty \rangle$; but via

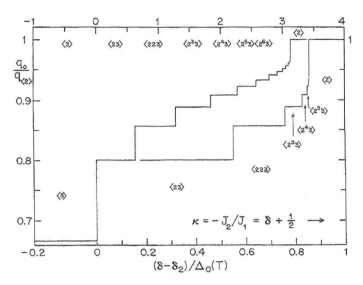

Figure 36. Variation of the wavevector $q(T,\kappa)$ in the ANNNI model for particular parameter values. Note that $\delta = \kappa - \frac{1}{2}$, $\delta_2(T)$ denotes the $\langle 3\text{-}23\rangle$ boundary (in Figure 35), and $\Delta_0(T)$ increases smoothly with T (see text). An infinite number of infinitesimal steps pile up as the limiting value $q/q_{\langle 2\rangle} = 1$ is attained.

a first order transition this goes over to a (3, 3) *antiphase state*, marked $\langle 3\rangle$, consisting of alternate triples of layers, each pointing predominantly "up", then "down", and so on; next, via another first order transition, follows a (2, 3) antiphase state, marked $\langle 23\rangle$; then a $\langle 2^2 3\rangle$ or (2, 2, 3) state, as illustrated at the top of the figure, and so on through an *infinite* sequence of distinct, commensurate phases $\langle 2^k 3\rangle$, before, finally, as $\kappa \to \infty$, the simple (2, 2) antiphase state, denoted $\langle 2\rangle$, is again reached.

If one asks how the wavevectors vary as one passes through the phase diagram, by varying κ or T, one must consult Figure 36, which portrays the situation for two different values of J_1/J_0, as embodied in the scale factor

$$\Delta_0(T) \approx (k_B T/J_1) \exp(-8J_0/k_B T). \tag{36}$$

Evidently, the wavevector $q(T,\kappa)$ varies in an abrupt staircase-like fashion with a finite jump on crossing each phase boundary. But, as the order, k, of the phase increases, the jumps get smaller, and the widths of the steps get narrower: ultimately, as one approaches the (2, 2) antiphase state, with fixed wavevector $q_{\langle 2\rangle} = \pi/2a$ where a is the lattice spacing, one reaches what might be called the "Devil's top stair"! Thus, as the (2, 2) phase melts in this region, it passes through an infinite number of infinitesimally small steps in q before the wavevector changes by a finite amount — and each step corresponds to a distinct, equilibrium, commensurate modulated phase.

In Conclusion

From the study of the very simple, almost trivial ANNNI model, we learn clearly that modulated phases of matter and their transitions can involve considerable subtlety. I hope you may now sympathize a little more with the, perhaps, overdramatic comments (in the discussion following Dr. Hagstrum's address) concerning my attitude toward predicting the occurrence of the 7×7 surface reconstruction phase of silicon. Even in this elementary ANNNI model, with merely two-valued, ± 1 variables, one discovers that a small variation in the basic interaction parameter alters abruptly the states and phases which can be realized: but I am not the Lord God Almighty! Who am I, then, to say with confidence what must happen on the [111] surface of silicon? Maybe there are braver souls around. However, I must admit that I am a cynical soul, even if I am not that courageous. So, when I see what complexity can arise in a very simple model, I do wonder about what it is reasonable to hope to assert in more complex and realistic cases. Again, I would say to you that many theorists are now becoming reluctant merely to attempt to fulfill the traditional role, which was to aim to calculate everything that an experimentalist could observe. That has often been the role in which theorists have been cast: "The successful theory is one in which you can calculate everything that can be observed". Rather, I believe a successful theory is one which enables one to understand, or to gain some understanding of what can be observed, including an understanding of what is not worth trying to calculate, of what it is better to measure directly if one desires to know it. A good theory reveals the analogies, the deep underlying analogies and similarities between what look, at first sight, like intrinsically different physical systems; a good theory suggests new experiments and new ways of looking at old experiments.

On this philosophic, but I feel appropriately humble, note, I would like to close by reminding you, through Figure 37, of an early theoretical attempt to understand the cosmos, published by Kepler in 1595. Kepler observed the orbits of the planets and tried to comprehend their relative sizes: he found that, if he took a sphere and inscribed a cube, and within that

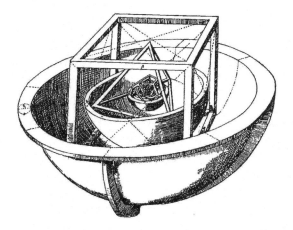

Figure 37. Kepler's construction for the radii of the planetary orbits [from his Mysterium Cosmographicum (1595)].

inscribed another sphere and then a tetrahedron, and thus through the remaining five Platonic solids: dodecahedron, octahedron, and icosahedron, then he could reproduce the orbital radii of the six planets Saturn, Jupiter, Mars, Earth, Venus, and Mercury. It is a beautiful idea. But, nowadays, I think it fair to say that, even if it has not been published before, one could not get it accepted for publication in one of the journals of planetary science. Of course, we also know there are extra planets, Uranus, Neptune, and Pluto, for which no provision seems to be made. Perhaps, however, we should not be too confident, since there certainly is some definite regularity which is seen. Nevertheless, we no longer believe that Kepler's approach is a correct way of looking at this aspect of the real world. Possibily the theoretical perspectives I have been presenting here* will suffer a similar profound change with time: but at least I believe that our theoretical imagination is going constructively forward and will continue to reveal new levels of beauty and, one hopes, truth.

Apologia. Clearly this article does not constitute a scholarly review of all the topics which were addressed in the lecture on which it was based. In particular, while the main thrust has been to convey current theoretical ideas, quite inadequate references have been given to the original theoretical literature. Apologies are particularly due to authors whose work and ideas have been mentioned without attribution even by name; certainly, no slight or implied evaluations are meant by such omissions. For the reader who would like to explore further, however, I believe the references given in the footnotes and in the figure captions will allow ready entry into most relevant aspects of the literature. It should suffice, beyond that, to draw attention to the many authoritative review articles contained in the volumes 1 to 6 (excepting 4) of the series "Phase Transitions and Critical Phenomena" edited by C. Domb and M. S. Green (Academic Press, New York 1972–1976) and to mention the review of N. D. Mermin, *Rev. Mod. Phys.* **51**, 591 (1979) concerning topological defects.

Acknowledgments

The Support of the National Science Foundation, in part through the Materials Science Center at Cornell University, for the preparation of this article and for some of the researches reviewed is gratefully acknowledged. Without the education and interactions provided by many friends and colleagues through personal contacts, via correspondence, and through the scientific literature, this article could not have been written.

Introduction and Discussion

Dr. W. O. Baker (Speaker), **Bell Laboratories:** We shall take a more formal pose, if not a more formal view, for our discussion; namely, you are invited to be seated if you feel like it. We are enthusiastically awaiting the next scenes, the next stages of the program that was so eloquently developed this morning.

Professor Michael Fisher will begin the discussion of the states of matter from a theoretical perspective. As we stated earlier, his associates — particularly Dr. Fleury and Dr. Widom, but also the discussion leaders and others in general — will join in as appropriate.

I take pleasure in introducing Dr. Michael Fisher.

[The following discussion took place after the presentation of Dr. Fisher's address.]

Dr. Baker: We thank you indeed, Dr. Fisher, for your heroic pursuit of the unobvious. We shall look forward to a chance for a discussion led by Dr. Fleury and Dr. Widom.

Dr. Paul A. Fleury (Discussion Leader), **Bell Laboratories:** Before saying anything on my own which would be somewhat presumptuous, I would like to call for any questions or comments if you are not stunned by the power of what we've just heard.

Dr. B. Widom (Discussion Leader), **Cornell University:** I would like first to congratulate Dr. Fisher on that heroic performance — as we've already heard it described — and that extremely eloquent overview of such a vast area in the theory of the structure of matter in relation to phase transitions.

I would like, if I could, to take up just one or two of the points that Dr. Fisher mentioned, and give a slightly different view of them, or go a little further in those matters in which we have had some experience in our own group.

One of them has to do with tricritical points, one of the classes of higher-order critical points that Dr. Fisher mentioned, as one sees them in mixtures of ordinary classical fluids. Dr. Fisher showed pictures of three-phase equilibrium in a multi-component system. I would like to say a little bit about the chemical motivation that leads to the construction of mixtures of that kind, and to tell you about another mixture, analogous to the one that Dr. Fisher showed you the photograph of, which shows the same kind of chemical motivation in its construction. And then I'll also put a question to Dr. Fisher about certain analogies between that way of viewing the fluid mixture and corresponding ways of viewing the tricritical phenomenon in magnets.

The example of the tricritical point in the multi-component system that Dr. Fisher called to your attention was one we worked on in my research group some years ago, as Dr. Fisher mentioned. It was a mixture of the substances benzene, ethanol, water, and a salt, which in the case that he showed you was ammonium sulfate. These four substances are related to each other in a very special way in their mutual solubility, and it's this relation among their solubilities that leads ultimately to the occurrence of that tricritical point.

You know that benzene and water are very largely insoluble in each other, but that ethanol is a very good common solvent for both: benzene is miscible with ethanol in all proportions, and likewise water is miscible with ethanol in all proportions. Suppose for the moment that you forget about the ammonium sulfate and just concentrate on those first three chemical substances: the benzene, the ethanol, and the water. Because of the immiscibility of benzene and water, and because ethanol is a good common solvent simultaneously for benzene and water, we can achieve a critical solution point; *i.e.*, at the appropriate composition of a mixture of those three substances we can go through a point at which the two previously distinct phases — the benzene-rich and water-rich phases — have merged their identity and become just one. That is a critical point in the same sense as the ordinary critical points that Dr. Fisher described at the beginning of his talk.

Now forget about that first substance, the benzene, and concentrate on the last three on the list, the ethanol, the water, and the ammonium sulfate. The ethanol and salt are very largely insoluble with each other. On the other hand, the water is a very good common solvent for both the ethanol and the salt. Therefore, still in the absence of the benzene, we could have another critical point which we would achieve by mixing appropriate amounts of ethanol, water, and ammonium sulfate: with enough of the common solvent, water, we can cause the complete mutual solubility of the ethanol and salt. So we then have a critical solution point associated with mixtures of those three substances.

The tricritical point occurred in the four-component system — and Dr. Fisher emphasized the necessity of that number four, and how basic that number four is for the occurrence of the tricritical phenomenon. In the four-component system we have enough thermodynamic degrees of freedom at our disposal (even when the pressure is fixed) to have that first critical point and the second one occur at the same moment. It is the coincidence of those two ordinary critical points that makes the tricritical point.

That, as I said, was the system we worked on some years ago. It was Dr. John Lang in our research group who did those experiments. We are now approaching the end of another series of experiments, in another four-component system, this being worked on by Mr. Peter Bocko in our group. That is the system in which the four components are n-hexane, benzene, acetonitrile, and water. We have the same mutual solubility relations among those four components as before. If for the moment you forget about the water, then hexane and acetonitrile are quite insoluble in each other, but benzene is a good common solvent for those two.

On the other hand, if you forget about the hexane, then benzene and water, once again, as in the earlier system, are largely insoluble, with acetonitrile now a good common solvent for those two. The tricritical point in this system is, then, the place where the critical solution point of the first three components and the critical solution point of the last three components coincide. So the first question that I'll be putting to Dr. Fisher when I finish the remarks that I'm now making will be whether one can understand the tricritical point in the magnetic system, or in the helium 3-4 mixture that he had mentioned, in some analogous way; that is, can one somehow look upon the tricritical point of those systems as occurring as a coincidence of two ordinary critical points, or at a place where two previously distinct kinds of critical points have become one and the same?

I would mention also that one of our motives in studying the tricritical points in these systems was to test the very important and powerful idea that Dr. Fisher emphasized in his talk: the dependence of the value of the critical-point exponents on dimensionality. We wished also to test another point, which has already come out in the discussion (this has to do with a question that Paul Fleury asked during the course of Dr. Fisher's talk): the point that the borderline dimensionality for tricritical points — the dimensionality above which critical point exponents have their classical values and below which they are dimensionality-dependent with nonclassical values — is three, according to the modern theory. By measuring the size and shape of the region of three phase co-existence in some appropriate thermodynamic space, we were able to determine the values of those critical-point exponents in these systems; not very searchingly, I'm sorry to say, but, insofar as we have been able to determine them, they are in at least rough agreement with the idea that, indeed, in three dimensions the critical point exponents have their classical values.

Now, another point I would like to make is that when studying these or any other kinds of phase equilibrium one should also be interested in the nature of the interface between the co-existing phases. This is important in fluid systems. This is important also in the magnetic systems where one would be interested in the structure of the boundary that separates co-existing magnetic domains.

In the fluid case one sees these questions about interfaces in their simplest context. Looking at a liquid-vapor interface, we can ask, what is its structure? This is now making contact with the talk that we heard this morning by Dr. Hagstrum, when we were asking questions about the detailed structure of a two-dimensional layer that was itself a transition layer between one bulk phase and another. In this case the methods by which we would study this structure, or understand it, are not nearly so well developed. So we ask rather cruder and more primitive questions.

What we mean by structure in a case such as this is merely the profile of density or composition going from one bulk phase through the interface into the other bulk phase. The question of what is the structure of that interface is equivalent to the question of what is the functional form of that profile as it goes from the density of the gas to the density of the liquid, through all intermediate densities.

If now we were to ask the corresponding questions about the neighborhood of the tricritical point where one has three-phase equilibrium and three phases coming to merge their identity, then you see we can in principle be talking about any one of three interfaces: the interface between the top phase and the middle phase, that between the middle phase and the bottom phase, and (if we remove the middle phase) the interface between the top and bottom phases. The simplest way of imagining those three interfaces to be present simultaneously is to have those three solutions not in a vertical tube, as in the photograph that Dr. Fisher showed you, but rather in a tube that has the shape of a torus. If we had that, then we could have an alpha phase, a beta phase, and a gamma phase, and see the interface between alpha-beta, beta-gamma, and this very important and special alpha-gamma interface all present at the same time. If we were to apply the same kind of theory to the understanding of the structure of this alpha-gamma interface as we did earlier to understanding the structure of the simple liquid-vapor interface — that is, the interface near an ordinary critical point — we would find a composition profile that has two steps in it, one from the composition of the bulk alpha phase to that of the bulk beta phase, and the other from the composition of bulk beta to that of bulk gamma. Amazingly, the theory says that the alpha-gamma interface knows that the alpha and gamma phases want to be in equilibrium with the beta phase, and it commemorates that fact by having a structure in which, in the intermediate portion of the interface, there is a region of composition exactly that of the bulk beta phase. This is something of which we have some practical and simple experimental realizations, but of which no really very careful quantitative study has been made. We are hoping at Cornell to undertake experiments of this kind in collaboration with Professor Webb in our Applied Physics department. We hope by an ellipsometric study of an alpha-gamma interface to find evidence for the existence of the beta film at that interface.

A last point that I would like to make in connection with the structure of interfaces and with interfacial tensions will again make contact with some of the theoretical ideas that Professor Fisher brought out in his talk.

We can always ask the question, what is the tension of such an interface? What, for example, is the surface tension of a liquid? We know that that tension vanishes as one approaches a critical point of the equilibrium of those two coexistent phases. When you go to a critical point, those phases become identical. The interface that separates them becomes infinitely diffuse and disappears. That's the way the phases become identical. The tension associated with that interface is just the free energy associated with the inhomogeneity in density or composition, and that free energy disappears as one approaches the critical point. Now we can ask with what power of the variable t that Professor Fisher used — this is just the temperature distance from the critical point — with what power, let's say mu, of that temperature displacement from the critical point, does the surface tension vanish? Now, even in an ordinary homogeneous phase, of course, there are fluctuations. These fluctuations are correlated over distances equal to the correlation length that Professor Fisher has called ξ — and he has already introduced the notation that this correlation length ξ diverges as a power nu of the inverse of the temperature-distance from the critical point. It happens also that there is a very close connection between the structure of the interface and the correlation length: the thickness of that interface, thus its diffuseness, is essentially measured by that correlation length.

But, in any case, if we look at fluctuations occurring inside an otherwise homogeneous fluid, we may ask what is the free energy associated with the spontaneous fluctuations occurring in a little element of volume, where the linear dimension of that element of volume is ξ, so that the volume is just ξ^3, or ξ^d in a general d-dimensional system. The free energy associated with such an elementary fluctuation, like the basic free energy associated with anything, is just kT, and approaches, of course, just kT_c as the temperature approaches the critical. But there is another way of viewing that basic free energy associated with the fluctuation. That fluctuation, after all, produces an inhomogeneity in the density. There is a surface free energy associated with that basic inhomogeneity in the density. We agreed that ξ^{d-1} would be the surface area of the region in which that is occurring. So this area times the surface tension is an alternative to kT_c as an expression for the basic free energy associated with the elementary density fluctuations; they are one and the same thing. But, if they are, then notice that that tells us that there must necessarily be a connection between the critical point exponent mu for the vanishing of the surface tension and the critical point exponent nu for the divergence of the correlation length. In fact, this says that mu must be the same thing as $d-1$ times nu in order that the product of surface tension and ξ^{d-1} be a quantity, kT, that neither diverges nor vanishes when approaching the critical point. There are no experiments on mu in two-dimensional systems, although we have theoretical answers there. But if you ask for the experimental values of mu and nu in three-dimensional systems, the value of nu is roughly 0.64 in the kind of system that Dr. Fisher identified as an $n=1$ system. That is an Ising system, which would correspond to ordinary critical points. The value of mu as it's found by innumerable experimental determinations of the vanishing of the surface tension as one approaches the critical point is about 1.28 in a three dimensional system; which, you see, again, means good agreement between theoretical ideas and experiments.

I would now put to Dr. Fisher the second of the questions that I was going to address to him, not specifically having to do with the value of the surface tension exponent, but, rather, to ask if in some of the magnetic systems, or in any of the other systems about which he told us, there are any analogies to these questions of the structure of an interface; that is, are there circumstances in which one can ask about the structure of the interface between co-existing phases in the neighborhood of some of the critical points that he was talking about. Let me, then, just ask over again the two questions that I'm putting to you, if I may. The first was if in the magnetic systems there are analogies to viewing the tricritical point as a point of coincidence of two ordinary critical points. The second was whether in the variety of systems you have described to us there were analogies to these questions about the structure of an interface.

Dr. Fisher: Let me take the two questions in reverse order and discuss, first of all, the relationship you have pointed out between the interfacial tension and the correlation length in a normal fluid, which illustrates dramatically the characteristic way in which we believe the dimensionality of a system enters. In fact, the existence of interfacial tensions in fluids also reflect the fact that the symmetry number, n, equals 1 in such scalar systems. The first statement is that the same should be true – or, at least we expect theoretically that it is going to be precisely true, of any of the analogous systems. Now, what are the analogous systems? Well, there are many magnetic materials in which one has a strong uniaxial anisotropy. When I mentioned nickel or gadolinium and some of the other magnetic systems, the order parameter, say, the magnetization, was free to point anywhere over a sphere; but there are other magnetic systems in which the spontaneous magnetization is constrained to lie parallel to a definite direction in the crystal. Then one can consider what happens when one goes from a domain of "up" spins to a domain of "down" spins: one has, in fact, a Bloch wall. And the Bloch wall is the complete analog of the interfacial wall or surface that one sees between liquid and vapor phases. Thus, one can certainly ask the analogous question. What I do not know — although I think there is some work in the literature — is of any really precise measurements of how the Bloch wall energy varies as one approaches the magnetic critical point, as one would need to check out the relation between correlation length and interfacial energy. I suspect that most recent researchers in the field were interested in Bloch walls mainly because of their significance for magnetic bubble memories and for other applications. In those circumstances, the critical point is rather a nuisance: one would rather stay away from it! So I believe there is still an interesting class of experiments to be done.

However, if one proceeds to discuss systems with $n = 2$, or more generally, say $n \geq 2$, components for the order parameter, the inference is that one has a significant underlying symmetry. Then there is a difficulty, because, if one goes from, let us say, a spin "up" region to a spin "down" region, the local magnetization does *not* have to vanish at some place, as would follow from the picture of a simple ($n = 1$) interface which Professor Widom drew. In the case of a uniaxial ($n = 1$) ferromagnet with, say, the z-axis as the easy axis along which the magnetization aligns, one can translate the interfacial profile for fluids to magnets by replacing the local density or composition variables by the local magnetization component $M_z(\mathbf{r}) \propto \langle S_z(\mathbf{r}) \rangle$, where $S_z(\mathbf{r})$ is the local spin variable. Then as one goes along, say, the spatial x-axis from $x = 0$ where the spins all point up, so that $M_z(x = 0) = +M_0$, to a domain at $x \geq L$ where the spins point down, so that $M_z(x = L) = -M_0$, there is clearly a point in the middle of the Bloch wall where the order parameter, M_z, vanishes. Consider now, however, for concreteness an $n = 2$ or "planar" type of magnetic system in which the local magnetization vector $\vec{M}(\mathbf{r})$ is free to point along any direction in, say, the (y, z) plane. Now, in attempting to find the magnetization configuration of lowest free energy between the "up" domain at $x \leq 0$ and the "down" domain at $x \geq L$, we must remember that the local order parameter, i.e. the local magnetization or spin, is free to twist away from the z-axis. Thus, in an isotropic ($n \geq 2$) magnet, we expect to find a helical or spiral arrangement of the spins between the domains, obeying equations like

$$M_z(x) = M_0 \cos(\pi x/L), \quad M_y(x) = M_0 \sin(\pi x/L).$$

In these circumstances the overall free energy is reduced by making the wall thickness L as great as possible: what was a Bloch wall of well-defined thickness becomes indefinitely wide and diffuse. Certainly there are experiments which show that in magnetic materials with low anisotropy (which can often be arranged by varying the temperature) the Bloch walls do become very, very wide. Thus one must first answer Professor Widom's question by saying, "No, for systems with symmetric order parameter of $n \geq 2$ components, one does *not* have well-defined interfacial profiles between complementary phases as in fluid systems."

On the other hand, one can surely ask; "But must there not be something physically significant to say about the transition region between domains where the orientation of the order parameter differs by an angle, say θ?" Now Professor Widom pointed out that in fluid or, more generally, in scalar ($n = 1$) systems the existence of an interface is associated with a definite incremental free energy (per unit area for $d = 3$, or per unit length for $d = 2$), namely, the interfacial tension $\sigma(T)$. Evidently, there must also be an incremental free energy associated with the twisting of the order parameter over a length (or thickness) L in an isotropic ($n \geq 2$) system. But to what physical observations or experiments might this incremental, "twisting" free energy correspond? Consider a crystal: we agreed that a characteristic of its order was the existence, in three dimensions, of sharp Bragg scattering peaks; but recall there are other important characterizing properties of crystals. One such property of a crystal is its shear modulus. Suppose one has a crystalline rod of length L with the crystal axes aligned parallel to the x, y, and z axes at one of the rod but that one twists the far end of the rod through some angle θ about the longitudinal or x-axis. Clearly, the crystal is strained, but the strain energy is spread through its length, and hence the incremental free energy density is proportional to $\frac{1}{2}K(\theta/L)^2$, where $K = K(T)$ is, essentially, the shear modulus. This suggests that the analog of the interfacial tension, $\sigma(T)$, is going to be something like a shear modulus measuring the resistance to a twisting deformation.

Next, consider a superfluid, where Onsager pointed out that the order parameter may be regarded as a macroscopic wave function $\Psi(\mathbf{r}) = |\Psi|e^{i\phi(\mathbf{r})}$ with phase $\phi(\mathbf{r})$. Suppose in a tube of length L the phase at $x = 0$ is fixed at $\phi(x = 0) = 0$ but that a twist through an angle θ is imposed at $x = L$. Along the tube the phase will then vary uniformly according to $\phi(x) = \theta x/L$, so that grad $\phi = \theta/L$. But the gradient of the phase of a wave function in quantum mechanics measures a velocity, in this case simply the velocity of superflow along the tube, namely, $v_s = (\hbar/m)\text{grad } \phi$ (where m is the mass of a helium atom and $\hbar$ is Planck's constant). The incremental free energy density associated with superfluid flow is normally written as

$\frac{1}{2}\rho_s v^2$, where $\rho_s(T)$ is the so-called "superfluid density" (which vanishes identically in a normal fluid). Evidently this incremental free energy density can be written $\frac{1}{2}(\rho_s \hbar^2/m^2)(\theta/L)^2$, which is clearly quite analogous to the twisting energy in the example of a crystal rod.

Thus we reach the conclusion that for isotropic systems (at least for $n = 2$) there should be a parameter analogous to the interfacial tension in scalar systems which might be termed a "helicity modulus".* It has been suggested this might be denoted generally by the beautiful Greek letter Υ, capital upsilon! For a superfluid one then has $\Upsilon(T) = (\hbar/m)^2 \rho_s(T)$. Finally, we may develop an argument directly analogous to that presented by Professor Widom for the surface tension, to discuss the critical behavior of $\Upsilon(T)$. As he said, the basic scale of free energy is set by $k_B T_c$, while the basic length scale is given by the correlation length $\xi(T) \sim t^{-\nu}$ as $T \to T_c$. Surface tension has dimensions (energy)/(length)$^{d-1}$ in d dimensions, but, as is seen directly from our arguments, the helicity modulus has dimensions (energy)/(length)$^{d-2}$, the extra power of length arising from the indefinite diffuseness of the "interfacial wall" in the isotropic ($n \geq 2$) cases. We hence conclude from the scaling argument

$$\Upsilon(T) \approx k_B T_c / \xi^{d-2} \sim t^{(d-2)\nu}.$$

For a real helium superfluid in $d = 3$ dimensions one finds $\rho_s(T) \sim t^{0.67}$ and also has $\nu \simeq 0.67$, so the agreement there is excellent! Analogous experiments are in principle possible for the appropriate class of ($n = 2$) ferromagnets, but I know of no directly relevant observations at present.

Finally, one might even turn these arguments around and focus on the existence of a helicity modulus, or shear modulus, or superfluid density, as the principal criterion of the presence of order in a phase. Then one might reinterpret the existence of an interface and an interfacial tension in a fluid as the crucial indication of the existence of order. This would apply even if, as in a hypothetical phase separation between isomeric species, the densities and entropies of two coexisting fluids were identical!

Now I must respond to Professor Widom's first question. He explained how, in the context of the traditional chemistry of fluid phase separation, one can understand tricriticality in a rather direct and intuitive way. He gave a beautiful exposition of the arguments by which one might anticipate the existence of tricriticality in four-component systems and, further,

---
*See M. E. Fisher, M. N. Barber and D. Jasnow, *Phys. Rev. A* **8**, 1111 (1973) for a more detailed discussion.

set about looking for real examples. Can we develop similar insights for other systems displaying tricriticality, in particular, for superfluids and for antiferromagnets? At present the answer seems to be "No". Basically, the reason is that in these other systems the parameters we would like to vary, the analogs of all the fluid composition variables, are not under full experimental control (or, in some cases, even under very good conceptual or theoretical control!).

In the case of the superfluid helium three-four mixtures, where we would much like to discuss the issue, what are the parameters needed? Recall the two variables we can control: namely, the temperature, T, and the mole fraction of, say, helium-three (i.e. one composition variable: note that the overall density or pressure do *not* play a significant role at this tricritical point). What is missing? Recall again that we believe that the crucial superfluid order parameter is a macroscopic wave function. If the world were built nicely for us, there would be a corresponding analog of a magnetic field which would enable us to take hold of the wave function and make its phase point this way or that way, or to stretch it and bring it into existence even in a normal fluid. Maybe someone will actually find out how to do that; and, if they do, they will surely receive a good prize for it! Physically, then, the so-called, complex "off-diagonal field", say $h_1 = h_1' + ih_1''$, is missing. This would entail an added term $h_1^* \hat{\psi}(\mathbf{r}) + h_1 \hat{\psi}^\dagger(\mathbf{r})$ in the (second-quantized) Hamiltonian for a superfluid, i.e. a term *linear* in the wave function operator $\hat{\psi}(\mathbf{r})$. In reality no one has seen such a term, although there seems to be no really deep reason in physics to explain its absence.

However, if we try to push on even further in exploring the analogy between tricriticality in many-component fluids and in antiferromagnets and superfluids, then we encounter something still more awkward! Fundamentally, from a theoretical viewpoint, in addition to the field h_1, one needs a cubic or third-order field h_3 which, roughly speaking at least, should couple to the cube of the wave function, say $\psi^3 \sim |\psi(\mathbf{r})|^2 \psi(\mathbf{r})$. That really makes life tough!

Even in the case of antiferromagnets one used to think there was no way of generating the first field, h_1, which corresponds there to the "staggered magnetic field" (pointing "up" and "down" on alternate sites). But then, in some cases at least, a way of generating such a field was found! In particular, in this beautiful crystalline antiferromagnet I told you about, dysprosium aluminum garnet or DAG, it was discovered that, owing to its subtle and complex structure and symmetry (with many magnetic ions per unit cell), one could in fact generate a staggered magnetic field on some of the sublattices by orienting the uniform applied magnetic field appropriately — something one thought could never be done! However, even in that favorable case, one does not see how to *separately* vary the cubic field h_3 (which should couple to the cube of the staggered magnetization or its equivalent). Finally, to be frank, at present our theoretical guesses as to the effects of these special fields, even if we could generate them, are not well educated as regards predicting tricriticality in real systems.

So in the study of tricriticality one has a case where I think chemistry, which at first sight looked more complicated and lacking in helpful symmetries, is actually more transparent and enables one to see and understand important overall symmetries in a deeper way. In other systems, to my mind, tricriticality is still a matter of more mystery: I do not personally have much of a feel for when there are going to be magnetic tricritical points or as to why the phase separation between helium 3 and helium 4 at low temperatures leads to a tricritical point. It would be nice to develop such a broader understanding.

Dr. Henry Eyring, University of Utah: I would like to say first that I'm very much impressed with the beauty of Professor Fisher's presentation. I will briefly outline the concept that is relevant here. The bulk liquid and vapor phases in equilibrium with each other have the same chemical potential. Accordingly, one would have intermingling of clusters of vapor in the liquid and vice versa if it were not for gravity acting on the difference in density and secondly the surface tension between two phases. Both these effects tend to disappear as the phases approach identity at the critical point. In the neighborhood of the critical point then a new degeneracy begins to matter. We find that the Helmholtz free energy, A, can be quite well represented by

$$A = -kT \ln f_{N'} - kT \ln(e^{qa'/kT} + e^{-qa'/kT}) \qquad (1)$$

At this point I want to insert a reference, Shao-mu Ma, Y. H. Yoon, S. T. Lee and Henry Eyring. "Single Partition Function for Three Phases", *Proc. Natl. Acad. Sci. U.S.A.* **74**, No. 7, pp. 2598–2601, July 1977. Consult the paper referenced for further references and more detail. I now quote a portion of this paper.

"**The Degeneracy Term Which Governs the Critical Phenomena**. Equation [1] without the

$$f_N = (f_s + f_{s'}n(X-1)e^{-aE_s/RT(X-1)})^{N/X(f_g)N(X-1/X)}$$

$$\times \left(\frac{N(V-V_s)}{V}!\right)^{-1} \cdot \left(\frac{e^{qa'/kT}+e^{-qa'/kT}}{2}\right)^{N/q} \quad [1]$$

last term is quantitatively accurate away from the critical point. However, it neglects the fact that small clusters of q molecules in the liquid can reorganize as vapor with only the expenditure of the interfacial energy and the energy of expansion and the gravitational energy arising from the change in density. Of these three energies, the surface energy will predominate near the critical point. The degeneracy factor $((e^{qa'/kT}+e^{-qa'/kT})/2)^{N/q}$ reduces to 1 at the critical point because qa' becomes zero and for $|T_c - T|$ large is allowed for in A.

We consider just how C_v changes with temperature when the volume is held constant at the critical volume, V_c. Near the critical point, every cooperative cluster of q molecules has the option of being in the liquid state with a Helmholtz free energy qa' below that of the critical structure or alternatively in the vapor state with free energy qa' higher. Accordingly, the Helmholtz free energy A may be written as

$$A = -kT \ln \left[f_{N'} \left(\frac{e^{qa'/kT}+e^{-qa'/kT}}{2}\right)^{N/q} \right] \quad [6]$$

which reduces to $A_0 = -kT \ln f_{N'}$ as it should at the critical point where $a' = 0$. If we take

$$A_1 = -kT \ln \left(\frac{1}{2}(e^{qa'/kT}+e^{-qa'/kT})^{N/q}\right), \quad [7]$$

then we have

$$A = A_0 + A_1 \quad [8]$$

and all properties are the sum of the properties arising from A_0 and A_1. If Eq. [7] is expanded into a series and the higher terms are neglected, we obtain

$$A_1 \simeq -\frac{N}{2}a'\left(\frac{a'q}{kT}\right). \quad [9]$$

This approximation requires that $qa' < kT$.

Because liquid and vapor are in equilibrium, mixing is prevented by the gravitational effect on density and by the surface tension which depends on the two principal radii of an ellipsoid. The cluster size, q, tends to remain small to keep its surface energy at a minimum, but for clusters to be stable they must not be too small. As the temperature is raised, coalescence makes clusters combine and wipes out surface energy between clusters. The growth of clusters with T is slow except in the vicinity of T_c. The net effect due to coalescence is a small decrease in a' by a factor proportional to $|\Delta T|^{-\alpha}$. Accordingly, we may write

$$A_1 = -c|\Delta T|^{2-\alpha}. \qquad [10]$$

Here the constant c may be related to the order parameter $\rho_L - \rho_g$ in a certain way. For simplicity we take it as a constant that has one value for $T < T_c$ and a different value for $T > T_c$. The slowly varying factor $q^{1/3}T$ occurring in Eq. [9] is partly responsible for the volume of α in Eq. [10].

Because the A-versus-V curve passes through a region of mechanical instability, $A_0(T, V_c)$ cannot be calculated directly from the partition function $f_N(T, V_c)$. Instead, the following method was used. First, the Helmholtz free energies for the liquid, A_L, and the gas A_g, were calculated from $f_N(T, V)$. Then we assume

$$A_0(T, V_c) = \chi A_L(T, V_L) + (1 - \chi) A_g(T, V_g) \qquad [11]$$

in which χ and $(1 - \chi)$ are the fractional amounts of the liquid and gas phases, respectively. The value of χ can be obtained from the coexistence curve

$$\chi = \frac{\rho_c - \rho_g}{\rho_L - \rho_g} = \frac{b_1|\Delta T|^\beta}{b_2|\Delta T|^\beta} = \text{constant}. \qquad [12]$$

Upon differentiating Eq. [10] twice, we obtain

$$(C_v)_1(T, V_c) = c(2 - \alpha)(1 - \alpha)T|\Delta T|^{-\alpha} \qquad [13]$$

Next, we consider the specific heat along the coexistence curve ($\rho = \rho_L$). At a given temperature the volume along the coexistence curve, V_L, differs from that along the critical isochore, V_c. To bring a molecule from the first state to the second state isothermally requires an energy that is proportional to $P(V - V_c)$. Therefore, we take

$$A_1 = -c|\Delta T|^{2-\alpha} e^{-a''|V_c - V|}$$

and

$$A(T, V) = A_0(T, V) = A_1(T, V). \qquad [14]$$

Table 2. Liquid volume, pressure, thermal expansion coefficients, and compressibility of liquid argon.*

T, K	V_L, ml Calc.	V_L, ml Obs.	P, atm. Calc.	P, atm. Obs.	$\alpha \times 10^3$ deg^{-1} Calc.	$\alpha \times 10^3$ deg^{-1} Obs.	$\beta \times 10^4$ atm.$^{-1}$ Calc.	$\beta \times 10^4$ atm.$^{-1}$ Obs.
83.85	28.36	28.20	0.652	0.680	4.66	4.37	1.98	2.03
87.29	28.82	28.66	0.979	1.000	4.72	4.49	2.17	2.27
90.0	29.19	29.0	1.310	1.319	4.85	4.58	2.36	2.45
100.0	30.74	30.38	3.33	3.20	5.73	5.09	3.40	3.42
110.0	32.73	32.05	7.08	6.58	7.25	5.9	5.27	4.8
120.0	35.44	34.15	13.26	11.98	9.78	7.8	8.96	7.8
130.0	39.46	37.31	22.5	20.0	14.7	12.0	17.74	16
140.0	46.34	42.7	35.7	31.26	28.34	22	49.32	40
143.0	49.66	–	40.56	–	38.94	–	78.3	–
145.0	52.67	–	43.82	–	52.7	–	111.92	–
T_c							–	–

*Observed (Obs.) data are from Ref. 12. More recently observed data (13) only differ slightly from these data.

The specific heat then takes the form

$$(C_v)_1(T, V) = C(2 - \alpha)(1 - \alpha)T|\Delta T|^{-\alpha -} e^{a''|V_c - V|}. \qquad [15]$$

Thermodynamic properties

Once the expression for the Helmholtz free energy is written explicitly, well-known thermodynamic relationships can be used to calculate all other thermodynamic properties.

The Translational Degree of Freedom in the Degeneracy Term. For argon we found, as was to be expected, that the best results can be obtained by assuming that only the one degree of freedom a molecule uses in entering the vacancy is translational — i.e., $y = 1$. The free volume V_f calculated is 0.254 Å^3 whereas the free length $\ell = V_f^{1/3} = 0.634$ Å which is approximately $\frac{1}{6}$ of the diameter of an argon molecule, as we expected.

Along the Coexistence Curve. Properties along the coexistence curve were calculated by using Eq. [14] with $\alpha = \frac{1}{8}$ and are given in Table 2 .... Agreement between calculated and observed values is good.

Along the Critical Isochore. Constant-volume specific heats, C_v, along the critical isochore were calculated by using Eq. [13] for $T < T_c$ as well as for $T > T_c$. The results are given in Table 3. We notice that Eq. [15] reduces to Eq. [13] when $\rho = \rho_0$. For $T < T_c$, we therefore used the same value for c (0.1) in both cases (C_v values for $c = 0.102$ are also given for $\rho = \rho_c$).

Table 3. C_v along the critical isochore ($\rho = \rho_c$).

T, K	$(C_v)_0$	$(C_v)_1$		(C_v) total		(C_v) obs.
		$c = 0.1$	$c = 0.102$	$c = 0.1$	$c = 0.102$	
132.7	3.57	15.15	15.46	18.72	19.02	17.52
140.0	3.58	17.05	17.39	20.63	20.97	20.0
145	3.58	19.08	19.46	22.66	23.04	22.18
147	3.58	20.38	20.79	23.96	24.37	25.48
149	3.58	22.65	23.1	26.23	26.68	27.90
150.84	3.58	–	–	–	–	–
		$c = 0.017$		$c = 0.017$		
150.96	3.58	5.49		9.07		10.47
151.0	3.58	5.30		8.88		9.7
151.53	3.58	4.43		8.00		8.32
152.43	3.58	4.01		7.59		7.53
153.13	3.58	3.85		7.43		7.0

*Observed (Obs.) data from Ref. 16.

However for $T > T_c$, a much smaller c is to be expected because the liquid and vapor structures have merged.

The Critical Properties. The critical properties of argon given in Table 1 were calculated by using the conditions $\partial P/\partial V = \partial^2 P/\partial V^2 = 0$. The critical pressure would be slightly lower if the dimer term were included in our gas-like partition function (17). It is known that $\rho_L - \rho_g \sim |\Delta T|^\beta$. Experimentally (18), $\beta \sim \frac{1}{3}$. With $A_1 \sim |\Delta T|^{2-\alpha} \sim (\rho_L - \rho_g)^{6-3\alpha}$, we obtain $P_1 \sim (\rho_L - \rho_g)^{5-3\alpha}$. If $\alpha = \frac{1}{8}$ and $\beta = \frac{1}{3}$ then $\delta = 5 - 3\alpha = 4\frac{5}{8}$. These values agree with the scaling law (19, 20)

$$\alpha + \beta(\delta + 1) = 2. \qquad [16]$$

Frankel (21) proposed a theory for three-dimensional cooperative phenomena using a ferromagnet and superfluid helium as examples. He used the following free energy expression:

$$-F \sim F_0 + \left(\frac{T_c + T}{T_c}\right)^{6\beta}.$$

The second term of this is similar to our Eq. [10]. F_0 is a constant in the critical region whereas our $A_0(T, V_c)$ is temperature-dependent. Hubbard and Schofield (22) applied Wilson's theory of renormalization groups (23) to a liquid-vapor critical point. However, their theory only holds for sufficiently small values of $\epsilon = (4 - d)$; their application to the case $d = 3, \epsilon = 1$ is questionable.

In this report, we have not discussed explicitly how a' is affected by the change in the cluster size that results from uniting of small clusters. This change, however, accounts quantitatively for the sign and the small value of α."

This procedure then gives you an infinite specific heat at the critical temperature. But the only way I see to make sense out of this is that clusters when they get closer and closer together will start uniting. And then the amount of surface area that you have will fall off. Accordingly, the surface energy will vary slightly slower with temperature than the first power of $T - T_c$. I'm surprised because I think these procedures for arriving at the coefficient must vary from case to case. I can understand it will be small in every case because this is good physics, but I'm not able to calculate well enough exactly how it will vary and the experiments are not precise enough to show it does not vary, at least sightly from case to case.

Dr. Fisher: I think the questions that have been raised by Professor Eyring are excellent ones. I presented, as it were, the theoretical dogma which comes from on high! As I said, in some cases, such as in the Onsager solution in the Ising model, one has actually calculated the partition function, although with very complicated mathematics, and we get the exact answers out; and nobody finds any mistakes! But still we cannot understand why the particular answers result. Now if we want to do the physical chemistry, we have to try to understand why the answers come out as they do. Furthermore, if the answers are universal, then just as Dr. Eyring rightly stressed, we should have some general understanding.

Maybe on that point I can give some feeling for the reasons for universality: the crucial issue is believed to be — and this fact, indeed, comes out of all the theories — that the correlation length $\xi(T)$ itself becomes very large in the critical region and then goes to infinity at the critical point. So that suppose we think of one case where I showed the experiments and they are very beautiful, namely the critical behavior of sulfur hexafluoride and of xenon. Those molecules are certainly very different chemically even though both are reasonably spherical in shape; but what is even more surprising is that one can actually do the same experiments and find the same universal value of the exponent β for water where we know, of course, that this is a very special material with very special structure. Now, to repeat, in all of the cases by anybody's theory we know that the correlation length becomes very large in the vicinity of the critical point. What that means is that the principal length scale on which things are happening is becoming very large compared with the atomic and molecular lengths, or even in the case of water where we expect some extended hydrogen-bonded structure, very large with respect to all the local structure. In the case of water you might well argue that since I know there is some sort of extended microscopic structure, I should have to go closer to the critical point before I see the universal behavior. But the belief, which seems well confirmed by all experiments, is that the asymptotic critical behavior is determined only by activity on the scale of the correlation length. All those things that happen on the shorter length scales are averaged out. Only certain rather few distinctive properties survive from the shorter length scales. I think that is good physics and good chemistry. The reason one obtains the same critical exponents is that once one gets up to a correlation length of a thousand Angstroms or so, it is only the fluctuations on that wavelength, and the way those density fluctuations interact with one another in a non-linear fashion that determines the critical behavior. True, something remains of the chemistry, but all that seems to be is the fact that the density is a scalar variable and that the attractive forces are of comparatively short range.

Now, in the case of magnetic phases it does apparently matter whether we have anisotropic spins or isotropic spins, because that symmetry survives through the averaging and makes a difference, although only a relatively small difference numerically. Thus I believe we do understand the basic physical reason for universality in a general way.

On other hand you also asked the question "How can one understand the characteristic divergence in the specific heat? What does it mean that the specific heat becomes infinite and how should one think about it?" I believe that the really honest answer is that nobody understands that phenomenon in any physically intuitive way! Certainly as regards the density fluctuations — the density being the order parameter — we understand those. Van der Waals understood them; Einstein understood them. Furthermore they seem to be enough, in principle, to give us everything we feel we should need to know about the critical point. But, as I pointed out, the energy density goes on its own apparently quite independent way. It fluctuates, and it fluctuates with a different and new critical exponent! Indeed, I think it is fair to say that whereas the van der Waals theory gives a first approximation, and it is a reasonably good one, for the density and for the density fluctuations and compressibility, there is no way of guessing from van der Waals' approach that the energy has got its own life. Basically that is how I would interpret your question: and after one starts to think about it further, one comes back and says "But there are only atoms there!" Somehow these atoms are clustering up against one another.

Now, there is one theoretical approach that might appeal to you in that regard which I have studied myself but finally had to give up because it does not go quite right as far as it should. So I call it a "picture" rather than a "theory". But consider the fact that as one approaches a fluid critical point, one has droplets of larger size and droplets of smaller size. If I take into account that each droplet has a free energy term associated with its closed surface I find that I should introduce two exponents one of which is, as Professor Eyring implies, rather directly related to the surface area-to-volume ratio. The other is related to the closure of the surface of a droplet in a more subtle way.* Then one does get some feeling for the universal exponents which emerge.

The difficulty one encounters is that we now know rather well that in high spatial dimensionalities the van der Waals predictions are correct. Really everybody believes that, and there is a lot of good theoretical support for it. But this mathematical-cum-physical picture of droplets does not naturally seem to say what really happens in four and more dimensions. The picture that apparently does tell me correctly what happens in four dimensions is the so-called field-theoretical picture which says "Smear everything out on the scale of the correlation length and just imagine that you have a self-interacting field of density fluctuations. Do not bother to ask what it was that gave you this density field, whether it was magnetic spins or molecules, just look at the non-linear interactions between the fluctuations and bother only about the possible symmetries." You may not like this sort of lesson but it has been more or less rammed down our throats, and it does seem that it really provides all one needs!

However one still has to go back and cut the problem at an intermediate stage and ask "But how do I fully justify this field-theoretic picture?" In a real sense there is still an important gap in the theory, if you wish. It was for that sort of reason I looked so hard at this very simple but special type of Ising model at the end of my lecture in order to try to understand

---

*The droplet picture of critical behavior is expounded in more detail in *Physics* 3, 255 (1967) and pages 49–73 in "Critical Phenomena" Proc. 1970 Varenna Summer School, Ed. M. S. Green (Academic Press, New York, 1971).

how one can get smooth sinusoidal variations with spins that can, individually, point only "up" or "down". In summary I do not claim to understand fully the specific heat singularities and I think you have put your finger on what will remain an unsatisfactory point of the general theory until we can understand or rationalize it more thoroughly.

Dr. Eyring: I think you can; it's small. But if it's the same in everything, I'm only surprised.

Dr. Fisher: There is one case where one can understand rather intuitively how the averaging works and yields universality: allow me to make a few comments about polymers in dilute solution. We know that in a polymer chain the mean square end-to-end distribution is really quite independent of the details of the polymer structure, of the detailed character of the monomeric links. It is just determined, firstly, by the principles of random flights or random walks and, secondly, by the excluded volume effects. It then does not matter for a long chain whether the polymer is polystyrene or polyethylene. If one looks at the mathematics sufficiently carefully one sees that all that really enters is the effective free length of one link of the polymer chain. Then, if you like, the law of large numbers comes to help us.

Dr. Eyring: Yes. I did that a long time ago [*Phys. Review* **39**, No. 4 (1932)].

Dr. Fisher: Now, it is really the same thing that is happening here in the theory of critical phenomena. It is no accident that this "symmetry number" n equals zero for polymer chains. Again one may say that the law of large numbers is serving to average out the microscopic details. The difficult point is to have a good feel for which details will survive, because we have agreed that some will! You feel in your bones that more details should survive; I am prepared to argue on the grounds of detailed and careful theoretical calculations that none or very few of them do! But it is often hard to be sure in ones bones. However I can relate briefly a history where we thought certain details would be important. My example, is magnetic dipolar interactions which are of very long range. We felt that by allowing for those in the theory of a ferromagnet it would make a big difference. Well with the new ε-expansion calculations we can put them in. I also thought I knew which way they would make a difference to the critical exponents since I knew the effects of simpler, purely attractive long range forces. But when we put them into the calculations (for $n \geq 2$), we found they make no visible or experimentally detectable changes. They do make some differences but they are very small: furthermore the changes are in the opposite direction to what we had anticipated! But I cannot tell you why that is so although I can show you the calculations! Naturally one feels rather frustrated, although one is at least relieved, and I am sure you also will be pleased to know, that the presence of dipolar interactions does certainly make some difference! Thus for dipolar molecules, I am prepared to think that there might be small differences in critical point behavior, at least in crystalline transitions. But, finally the amazing thing is how small the net effects are, in general, and how powerful the critical averaging process is.

Dr. Fleury: I would like to suggest that we turn the discussion to a slightly different direction at this point. However, before doing so I want to ask one other question of Professor Fisher, and that is: How long do you feel it's going to be before theory can tell us when we are actually in the asymptotic region so an experimentalist will know whether the difference he observes between what occurs in the laboratory and what occurs in theory is due to incorrect theory or merely that he simply hasn't gotten close enough to T_c?

Dr. Fisher: Well, this involves what might be called a friendly argument between theorists and experimentalists, in which I am actually on the side of the experimentalists. However it is not unrelated to the general point I have been making. In the lab we always do an experiment on a real system. When, I talk about being "at T_c" or "close to T_c" I have to say what I mean in terms of the actual physical parameters. But theorists have so far found it difficult to give many concrete physical estimates of how close is close enough and how far is far enough. There is progress being made, but it comes down to the generality of much of the basic theory: Consider the science of thermodynamics. One of the strengths of thermodynamics is that one has certain very general laws, the zeroth, first, second, and third laws, which carry one a long way; but they do not carry one the whole way on every given physical system. The sort of theory I have been presenting here today is similar. It carries you a long way but it does not carry you the whole way. You might say two things: either you might wish to rejoice in a rather general and deep theory; or you might say there is still a clear need for the sort of theory in which one has to think very carefully about the fact that here I actually have, say, nickel oxide and I want to calculate exactly what happens. Now personally I am a bit of a pessimist. I think ultimately one has to look at the theory and look at the experimental situation and one has to be reasonably skeptical on both sides. However, I do think things are improving as regards your specific question; we do have better ideas. I think that the papers you and I both know, which are over-optimistic on both sides, we just have to put down to over-optimism! I do not want to be over-pessimistic but I believe there are always going to be complex cases where one simply has to say "Well, this is the way the world is; we understand it reasonably by applying this theory but there is probably more there than meets the eye and than we can hope to understand fully."

Dr. Fleury: Well, I would like to interpret that to mean "be careful and keep going."

4

Walks, Walls, Wetting, and Melting*

Michael E. Fisher

Baker Laboratory, Cornell University, Ithaca, New York 14853

NEW results concerning the statistics of, in particular, p random walkers on a line whose paths do not cross are reported, extended, and interpreted. A general mechanism yielding phase transitions in one-dimensional or linear systems is recalled and applied to various wetting and melting phenomena in $(d = 2)$-dimensional systems, including fluid films and $p \times 1$ commensurate adsorbed phases, in which interfaces and domain walls can be modelled by noncrossing walks. The heuristic concept of an effective force between a walk and a rigid wall, and hence between interfaces and walls and between interfaces, is expounded and applied to wetting in an external field, to the behavior of the two-point correlations of a two-dimensional Ising model below T_c and in a field, and to the character of commensurate–incommensurate transitions for $d = 2$ (recapturing recent results by various workers). Applications of random walk ideas to three-dimensional problems are illustrated in connection with melting in a lipid membrane model.

Keywords: Walks (random and vicious); walls (of domains and containers); wetting (transitions in two dimensions); melting (of adsorbed surface phases).

---

*Based on the Boltzmann Medalist address presented at the 15th IUPAP International Conference on Thermodynamics and Statistical Mechanics, Edinburgh, July 1983. Reprinted from *Journal of Statistical Physics*, **34**, 667–729 (1984), © 1984 Plenum Publishing Corporation, with permission of Springer.

1. Introduction

The first lecture in a conference should stimulate rather than strain, should place a subject in perspective rather than delve into detail, should respect the history but survey recent happenings, and, ideally, should look on towards the future. In what follows I have endeavored to meet these specifications by revisiting the old but perennially alive topic of *random walks*. Some well known aspects will be recalled but fresh features will be exposed: in particular, the problem of p distinct walkers who interact will be discussed and some new asymptotic formulae, describing the "reunions of vicious walkers" and the effective forces between a walk and a wall and between walks themselves, will be presented. It transpires that these formulae can be applied to various problems on which recent researches in statistical physics and chemistry have focussed. Specifically, as I will demonstrate, one can give from a unified viewpoint, simple and easy derivations — admittedly mainly heuristic in character — for a range of exact theoretical results concerning wetting, melting, and other phenomena. The great wealth of results currently available is restricted to systems with only two infinite spatial dimensions: but such physical systems are increasingly open to penetrating experimental investigation. Nevertheless, one three-dimensional example is presented in the final section; however, the development of an analogous general approach for higher-dimensional systems remains a challenge for the future!

It should be mentioned that the primary purpose of this account of random walks and their applications is expository: however, one or two results not published (or in press elsewhere) do appear below.

2. The Return of a Drunken Walker

Consider a random walker in one dimension, say, a lone drunken Englishman who drops his house key on a long straight street and hunts for it. The displacement of the walker from the origin is measured by the coordinate x. For simplicity, we suppose that steps are taken at regular intervals, $\Delta t = c$, say, at each tick of a clock, so that the time for n steps is

$$t = nc \tag{2.1}$$

For concreteness it is useful to suppose also that each step is of length a so that, equivalently, one may regard the walker as walking on a one-dimensional lattice of spacing a; however, many results can be extended to a continuum in x and in t, and continuum language is the simplest to use, especially in discussing asymptotic behavior. It is also useful to regard "standing still" as a possible step and to associate *weights*, w_1 for a positive step, w_0 for remaining still, and $w_{-1} = w_1$ for a negative step. The total *weight function*, or *generating function* or *partition function* for a single step is then

$$e^{-\sigma} = w_1 + w_0 + w_{-1} \quad (w_{-1} = w_1) \tag{2.2}$$

Evidently the *mean square step length* is

$$b^2 = 2w_1 e^{\sigma} a^2 \tag{2.3}$$

Of course, if $w_0 = 0$, as commonly considered, one has $b = a$. In order to simplify the appearance of expressions we will sometimes set $b \equiv 1$ or, equivalently, suppose x is measured in units of b.

Now, as is well known, the total weight of all possible walks of n steps from the origin O ($x = 0$), to the point x varies, asymptotically when $n \to \infty$, as[1]

$$Q_m^0(x) \approx e^{-\sigma n} e^{-x^2/2b^2 n} / (2\pi b^2 n)^{\frac{1}{2}} \tag{2.4}$$

If one takes $w_1 = w_{-1} = 1$ and $w_0 = 0$, this corresponds simply to the *number* of distinct walks from O to x. We may regard $Q_n^0(x)$ as a partition function or a *partial partition function* for an n-step walk. The *total* weight or *total partition function* is then just

$$Z_n^0 \equiv \int Q_n^0(x) dx = e^{-\sigma n} \tag{2.5}$$

Clearly, the scale of the distribution is set by

$$\tilde{x}_n = \sqrt{\langle x^2 \rangle_n} = b n^{\frac{1}{2}} \tag{2.6}$$

Now, by (2.4), the probability of *return to the origin* on the nth step (or, in a continuum description, to within a distance b of the origin) is

$$r_n = \frac{Q_n^0(0)}{Z_n^0} \approx \frac{1}{\sqrt{(2\pi)} n^{\frac{1}{2}}} \tag{2.7}$$

If our drunken Englishman dropped his key at the origin, traditionally located at the only lamp-post on the street, we would be interested in the probability of his *eventual return* to O after he continues walking indefinitely. This, and many similar questions, can be answered by employing a simple but basic probabilistic lemma[1] which we state as follows:

Lemma. Let p_n, with generating function

$$P(z) = \sum_{n=1}^{\infty} p_n z^n \qquad (2.8)$$

be the probability that an event E occurs on the nth step (more generally, on the nth independent probabilistic trial). Then the probability, f_n, that E happens *for the first time* on step (or trial) n has the generating function

$$F(z) \equiv \sum_{n=1}^{\infty} f_n z^n = P(z)/[1+P(z)] \qquad (2.9)$$

The proof is simple and instructive: one merely notes that E either occurs for the first time on the last step, n, or happens sooner, say, for the first time on step $l < n$; in that case E happens again on the nth step with probability p_{n-l}. Summing on the possible values of l thus yields the identity

$$p_n = f_n + f_{n-1}p_1 + f_{n-2}p_2 + \cdots + f_1 p_{n-1} \qquad (2.10)$$

from which (2.9) follows directly.

Now the total probability that E eventually happens is given by

$$F_\infty \equiv f_1 + f_2 + \cdots \equiv \sum_{n=1}^{\infty} f_n$$

$$= F(1) = P(1-)/[1 + P(1-)] \qquad (2.11)$$

where the notation $z = 1-$ implies the limit $z \nearrow 1$. Let us, for concreteness, suppose, as will often be true, that p_n obeys an asymptotic power law, say,

$$p_n \approx p_0/n^\psi \quad (\psi > 0) \qquad (2.12)$$

Then $P(1-) = \infty$ if and only if $\psi \leq 1$. Consequently, the lemma yields the following.

Corollary. Let (2.12) describe the probability of event E occurring on the nth trial. Then, when $n \to \infty$,

(a) if $\psi \leq 1$, the event E occurs with probability one, i.e., is "certain" to occur (and, in fact, occurs infinitely often[1]); but

(b) if $\psi > 1$ the event E is *not certain*, failing to occur with a probability $1/[1 + P(1-)] > 0$.

It is not hard to see that in case (a) one has $f_n \sim 1/n^{2-\psi}$; conversely, $\psi > 1$ yields $f_n \sim 1/n^\psi$ (where, as sometimes below, we neglect the possibility of extra logarithmic factors in asymptotic behavior on borderlines).

Finally, returning to our lone drunken Englishman, we see from (2.7) that $\psi = \frac{1}{2}$ describes his probability of return: thus he is *certain* to return eventually and the probability that he returns for the first time on the nth step decays as $f_n \sim 1/n^{3/2}$.

3. The Reunions of p Harmless Drunks

Let us now give our drunken Englishman two more drunk but otherwise harmless companions, say an Indian and a Japanese. Suppose they start at, or near, the lamp-post at O and exchange keys cyclically (say, $E \Rightarrow I \Rightarrow J \Rightarrow E$) before wandering off in random fashion. In order that each one can recover his own key in a "fair exchange," i.e., such that all parties in an exchange are left equally happy, all three must meet together again somewhere at the same time, enjoy their *reunion*, and recycle the keys! What is the probability that such a reunion occurs on the nth step? Is an eventual reunion certain? More generally, what happens if p drunks, with coordinates $x_1, x_2, \ldots, x_p$, exchange their keys cyclically and walk off in the hopes of eventual reunion?

In formulating this question "harmless" has been used in a technical sense, specifically, to mean that our walkers may pass each other or occupy the same lattice site freely without harm, or, more explicitly, without change in the overall statistical weight. Since each walker is, thus, independent of the others the total weight of walks of n steps (by *each* walker) all starting at O and, collectively, proceeding to $\vec{x} = (x_1, x_2, \ldots, x_p)$ is simply

$$Q_n^p(\vec{x}) = \prod_{j=1}^{p} Q_n^0(x_j) \approx \frac{e^{-p\sigma n} e^{-|x|^2/2n}}{(2\pi n)^{\frac{1}{2}p}} \tag{3.1}$$

where, now, we have taken $b = 1$ for brevity. The significant feature of this formula for us resides mainly the exponent, $\frac{1}{2}p$, in the denominator.

From the basic distribution (3.1) we see directly that the probability of a reunion at $x_1 = x_2 = \cdots = x_p = \bar{x}$ on the nth step if all walkers started at

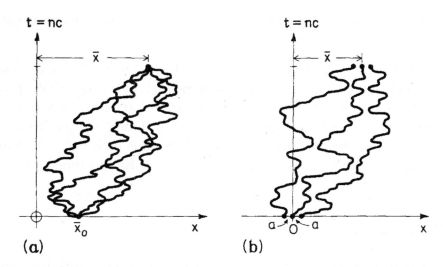

Figure 1. Illustrations of (a) a reunion of $p = 4$ harmless, or noninteracting walkers whose paths may cross; (b) a reunion of $p = 3$ interacting walkers whose paths may *not* cross.

$\bar{x}_0$ (see Figure 1(a)) is

$$r_n(\bar{x}) \sim e^{-p(\bar{x}-\bar{x}_0)^2/2n} / n^{\frac{1}{2}p} \qquad (3.2)$$

The Corollary of the previous section thus tells us that a reunion at any fixed position, $\bar{x}$, is *certain* if $p = 2$ but is *uncertain* for $p \geq 3$.

If our drunks do not care *where* they meet again we need the *probability of a reunion anywhere*. To this end, we should integrate as follows:

$$R_n^p = \int d\bar{x} r_n(\bar{x}) = n^{\frac{1}{2}} \int d\bar{X} r_n(n^{\frac{1}{2}}\bar{X}) \quad \text{with} \quad \bar{X} = \bar{x}/n^{\frac{1}{2}} \qquad (3.3)$$

where $X \equiv x/n^{\frac{1}{2}}$ is the appropriately scaled distance coordinate for n-step walks: integration thus cancels a factor $n^{\frac{1}{2}}$ in the denominator of (3.2) and yields

$$R_n^p \sim 1/n^{\frac{1}{2}(p-1)} \qquad (3.4)$$

Hence the eventual reunion anywhere of p harmless drunks is *certain* if $p = 2$ or 3 but *uncertain* for $p \geq 4$.

Evidently, it is easy to deal with harmless drunks: in statistical mechanical terms they correspond simply to *non*interacting walkers. As such they are hardly more interesting than an ideal gas. Thus we turn to *interacting walkers*.

4. Vicious Drunks and Their Reunions

The most natural interactions to consider between different walkers are short-range or contact interactions which, in particular, do not allow one walker to pass another. Then, if the p walkers are labelled in linear sequence so that one has

$$x_1 < x_2 < \cdots < x_p \tag{4.1}$$

at time $t = 0$, these same inequalities hold for all subsequent times. This restriction, which we henceforth adopt, is not, however, sufficient to specify the nature of the interactions in full: one must, in addition, say what weights are to be attached to the paths of two colliding walkers after a collision.

One straightforward possibility is to *replace* any forbidden steps, resulting, for example, in two adjacent walkers landing on the same site, by *alternative allowed steps*. The total weight of all walks would then be preserved, the weight of the forbidden collision steps being transferred to the associated allowed steps. Such walkers might be termed *bouncy walkers*: they will bounce off one another, and hence satisfy (4.1), but, statistically, they will tend to remain close to one another because of the increased weight of sequences with many near-collisions. Such bouncing walks can be discussed by the techniques explained below; however, for statistical mechanical applications they seem of less value than the complementary assignment of post-collision weights by which we characterize vicious drunken walkers.

Vicious drunks shoot on sight — we refrain from speculating as to their national origins — but they are short-sighted: thus on arriving at the same site they shoot each other dead: otherwise, they do not interact. In formal terms, the allowed steps carry the same weights, w_1, w_0, or w_{-1}, as before but any forbidden walks with multiple site-occupancy are assigned weight zero. Clearly, the paths of vicious walkers do not cross so that (4.1) is certainly maintained. Three specific models can be handled explicitly in closed form:

(A) *Lock step.* The walkers start either on the even- or the odd-numbered sites of the lattice: at each tick of the clock *each walker moves* either to the right or to the left with weight $w_1 = w_{-1}$ (and $w_0 = 0$) subject to no two walkers occupying the same site at the same time. (See Figure 2(a).)

(B) *Random turns.* The walkers start at distinct but otherwise arbitrary lattice sites: at each tick of the clock a *randomly chosen* walker takes a *random*

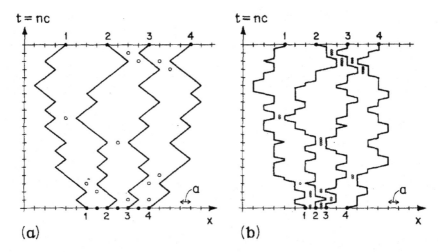

Figure 2. Sample paths for lattice models of vicious walkers for: (a) the lock step model, (A), in which all $p = 4$ walkers start on even-numbered sites and all move at the same moments; and (b) the random model, (B), where walkers can start on any site but only one moves at a time. In this illustration, the *average rate* at which a single walker steps has been chosen to be the same in both models. The open circles mark situations ("close calls") in which, owing to the proximity of another walker, the allowed steps are limited,

step (with weights w_0, $w_1 = w_{-1}$) subject to no two walkers occupying the same site at the same time. (See Figure 2(b).)

(C) *Brownian motion.* The continuum limit in x and t may be taken formally in model (A) or (B). The reason why these particular models are tractable will become clear below.

Now we may ask: "What is the probability that all walkers survive for n steps?" "What is the spatial distribution of the survivors?" "What is the probability of a *reunion* in which all walkers start close to (but not *at*) the origin, say, at spacing a, as illustrated in Figure 1(b) and, after n steps, *all* meet again, not *on* the *same* site but, say, at spacing a at mean position $\bar{x}$?" The basic message is that all these questions can be answered exactly in closed form! To preview some of the principal results, the probability that p walkers survive for n steps decays asymptotically as

$$P_n^{(p)} \sim 1/n^{\frac{1}{4}p(p-1)} \tag{4.2}$$

The probability of a reunion anywhere on the nth step varies as

$$R_n^{(p)} \sim 1/n^{\frac{1}{2}(p^2-1)} \quad \text{as} \quad n \to \infty \tag{4.3}$$

The exponent in this expression,

$$\psi_p = \tfrac{1}{2}(p^2 - 1) = 0, \quad \tfrac{3}{2}, \quad 4, \quad 7\tfrac{1}{2}, \ldots \quad \text{for} \qquad (4.4)$$
$$p = 1, \quad 2, \quad 3, \quad 4, \ldots$$

turns out to be the principal fact needed in most of the applications.

It follows from (4.2) and (4.3) that the (conditional) probability that the survivors enjoy a reunion on the nth step behaves as

$$S_n^{(p)} = R_n^{(p)} / P_n^{(p)} \sim 1/n^{\frac{1}{4}(p-1)(p+2)} \qquad (4.5)$$

as $n \to \infty$. A reunion is thus certain for $p = 2$, which is not so surprising in view of the result for two harmless walkers; however, a reunion of vicious walkers is *not* certain for $p \geq 3$, a fact which seems less than obvious. (Recall that three harmless walkers are certain to have a reunion.)

These results follow by asymptotic analysis of an exact general expression[a] for the distribution of p vicious walkers which is explained in the next section; however, this is a convenient point at which to present some of the more useful exact results.[b] Consider, for simplicity, the situation in which the p walkers are, initially, *equispaced*: with no loss of generality for the asymptotic behavior we may suppose the spacing is a and take the initial coordinates as

$$\vec{x}_0 = (x_{j,0}) \quad \text{with} \quad x_{j,0} = (j-1)a, \quad j = 1, 2, \ldots p \qquad (4.6)$$

Then, as $n \to \infty$, the *asymptotic distribution* for the lattice models is found to be

$$Q_n^{(p)}(\vec{x}) \approx \frac{e^{-p\sigma n} e^{-|x|^2/2n}}{(2\pi n)^{\frac{1}{2}p}} e^{-a^2 s_p/n} \prod_{j>k\geq 1}^{p} (e^{ax_j/n} - e^{ax_k/n}) \qquad (4.7)$$

where $s_p = \tfrac{1}{12} p(p-1)(2p-1)$ and again, we take $b \equiv 1$: the inequalities (4.1) are understood. The leading factor here is identical with the expression for harmless walkers: see (3.1). The second factor yields a negligible correction when $n \to \infty$. The crucial feature is thus the final "death factor" which, itself, is a product of $\tfrac{1}{2} p(p-1)$ factors, one factor, D_{jk}, for each pair of walkers, (j,k). Note also that (4.7) is exact for the continuum model (C).

---

[a] The basic mathematical result was derived and analyzed in connection with a study by Huse and Fisher[2] of fluctuations near a commensurate–incommensurate transition in two-dimensional systems (see also below); although only fairly standard methods of analysis are entailed, the general problem seems not to have been considered previously in the literature.

[b] A reader unconcerned with the details may skim the rest of this section and refer back to particular results only as needed.

Suppose now we are interested in the probability that all p walkers survive to the nth step. This can be obtained from (4.7) by integrating on each of the p final coordinates, x_j, subject, as always, to (4.1). Rescaling with $X_j = x_j/n^{\frac{1}{2}}$, as in (3.3), cancels the factor $n^{-\frac{1}{2}p}$; as $n \to \infty$, the individual death factors then yield

$$D_{jk} = e^{aX_j/\sqrt{n}} - e^{aX_k/\sqrt{n}} \approx a(X_j - X_k)/n^{\frac{1}{2}} \tag{4.8}$$

The overall death factor thus generates a factor $n^{-\frac{1}{4}p(p-1)}$ just as quoted in (4.2). More explicitly, the *probability of survival* may be written

$$P_n^{(p)} = A_p n^{-\frac{1}{4}p(p-1)}[1 + O(p^3 a^2/b^2 n)] \tag{4.9}$$

as $n \to \infty$, with amplitude

$$A_p = \left(\frac{a}{b}\right)^{\frac{1}{2}p(p-1)} \int \frac{d\vec{X}}{p!} \frac{e^{-\frac{1}{2}|X|^2}}{(2\pi)^{\frac{1}{2}p}} \prod_{(j,k)} |X_j - X_k| \tag{4.10}$$

For dimensional consistency the length b has now been exhibited. The integral on $\vec{X}$, which is a pure number, runs over *all* real values of the X_j while the product runs over all $\frac{1}{2}p(p-1)$ distinct pairs (j,k): for fixed values of p is not hard to evaluate the integral more explicitly.

The *probability of a reunion* at $\bar{x}$ also follows directly from (4.7): thus for the equispaced final positions

$$x_j - x_k = (j-k)a' \quad \text{with} \quad \bar{x} = \sum_{j=1}^{p} x_j/p \tag{4.11}$$

each death factor varies, for large n, as aa'/n [see (4.8)] and so one finds

$$r_n^{(p)}(\bar{x}) \approx B_p e^{-p(\bar{x}-\bar{x}_0)^2/2n}/n^{\frac{1}{2}p^2} \tag{4.12}$$

where $\bar{x}_0$ is defined in analogy with $\bar{x}$. If one starts from (4.7) [rather than directly invoking (4.8)] one finds

$$B_p \approx \frac{e^{-(a-a')^2 s'_p/n}}{(2\pi)^{\frac{1}{2}p}} \prod_{k=1}^{p}(1 - e^{-aa'k/n})^{p-k} \tag{4.13}$$

where $s'_p = \frac{1}{24}p(p^2 - 1)$. For large n this simplifies to

$$B_p = (2\pi)^{-\frac{1}{2}p} \left(\frac{aa'}{b^2}\right)^{\frac{1}{2}p(p-1)} \left(\prod_{q=1}^{p-1} q!\right) \left[1 + O\left(\frac{p^3 aa'}{b^2 n}\right)\right] \tag{4.14}$$

The expression (4.12) holds more generally for the probability of a reunion for *arbitrary* fixed initial positions, $x_{j,0}$, and final positions, x_j, if the amplitude is replaced by[c]

$$B_p(\vec{x}, \vec{x}_0) = \frac{\prod_{j>k\geq 1}^{p}(x_j - x_k)(x_{j,0} - x_{k,0})}{(2\pi)^{\frac{1}{2}p} b^{p(p-1)} 1! 2! \cdots (p-1)!} \left[1 + O\left(\frac{p^3 a^2}{b^2 n}\right)\right] \quad (4.15)$$

To find the probability of a reunion *anywhere* we merely need to integrate (4.12) on $\bar{x}$. As in (3.3), this generates a factor $n^{-1/2}$ and so confirms the important conclusion (4.3); more explicitly, one obtains

$$R_n^{(p)} = C_p / n^{\frac{1}{2}(p^2-1)} \left[1 + O\left(\frac{p^3 a^2}{b^2 n}\right)\right] \quad (4.16)$$

with amplitude

$$C_p = \frac{(a/b)^{p(p-1)} \prod_{q=1}^{p-1} q!}{(2\pi)^{\frac{1}{2}p(p-1)} \sqrt{p}} \quad (4.17)$$

It is natural, finally, to enquire if these exact results might not be extended, say, in models (B) and (C), to *dissimilar walkers*, with, in particular, *distinct diffusivities* as measured by their mean square single step lengths, b_i^2. In general it seems difficult to obtain comparable explicit formulae. However, for the simplest nontrivial case of two walkers progress can be made by the methods to be explained. For the continuum model, (C), the exact result can be expressed conveniently as

$$Q_n^{(2)}(\vec{x}) = e^{-(\sigma_1+\sigma_2)n} \frac{e^{-(\bar{x}-\bar{x}_0)^2/\bar{b}^2 n}}{2\pi \bar{b}^2 n} (1 - e^{-aa'/\bar{b}^2 n}) e^{-(a-a')^2/4\bar{b}^2 n} \quad (4.18)$$

where the mean position and diffusivity are defined by

$$\bar{x} = \tfrac{1}{2}\left(\frac{b_2}{b_1}x_1 + \frac{b_1}{b_2}x_2\right) \quad \text{and} \quad \bar{b}^2 = \tfrac{1}{2}(b_1^2 + b_2^2) \quad (4.19)$$

while, with no loss of generality, we have, as before

$$x_2 = x_1 + a' \quad \text{and} \quad x_{2,0} = x_{1,0} + a \quad (4.20)$$

When $b_1 = b_2$ all the previous results for $p = 2$ are easily recaptured.

---

[c]This result follows by analyzing the general result presented in the next section.

5. Absorbing Walls and the Death Factor

5.1. *One walk and a wall: The method of images*

In order to understand the origin and form of the death factor in the interactions of vicious walkers and, thereby, derive the results of the previous section, consider a single random walker who walks near a wall located at the origin. We take the wall to be an *absorbing wall* in the sense that no walk may penetrate the wall and any walk attempting to do so is eliminated. In our anthropomorphic picture, we may visualize an absorbing wall as a *cliff* over which a drunken walker may fall to his death! The interactions between wall and walker are, thus, completely analogous to those between vicious walkers. (We could, equally, discuss a *reflecting wall* which would corresponding to *bouncy* walkers.)

To be concrete, suppose the wall is located at the origin, $x = 0$, and a walker starts from $x_0 = a$. We ask for the weight of all n-step walks to x (> 0) which have never visited the region $x \leq 0$. This is a standard random walk problem[1] which is readily solved by the method of images. To review this technique consider Figure 3, which shows two possible n-step walks from $x_0 = a$ to x. The walk labelled (i) is allowed since it does not touch or cross the wall; by contrast walk (ii) crosses the wall and should be eliminated. To this end suppose walk (ii) *meets* or *hits* the wall for the first time on step n_1 and consider the modified walk, (iii), formed by reflecting in the origin, O, i.e., in the wall, all steps from 0 to n_1. As illustrated in Figure 3 this yields an *image walk* which starts at $x_0 = -a$, the image of the original starting point, meets the wall on step n_1, and reaches x, the original endpoint, on step n. It is evident that each such walk from $x_0 = -a$ to $x > 0$ matches a unique forbidden walk from $x_0 = a$ to $x > 0$. Consequently, if $Q_n^0(x, x_0)$ is the weighted number or partition function for free n-step walks, the corresponding number/partition function for walks in the presence of the absorbing wall is exactly

$$Q_n^W(x, a) = Q_n^0(x, a) - Q_n^0(x, -a) \tag{5.1}$$

where the second term represents the "negative" image walks.

Now, quite generally, if

$$\Phi(\theta) = \sum_l w_l e^{il\theta} \tag{5.2}$$

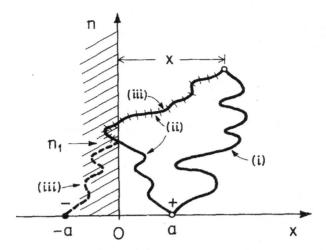

Figure 3. Walks starting from $x_0 = a$ in the presence of an absorbing wall at $x = 0$. Walk (i) is allowed. Walk (ii) is forbidden, since it meets the wall on step n_1, but may be cancelled by the partially reflected, negative walk (iii) which starts at $x_0 = -a$.

is the single-step partition function or generating function for a free walk on a lattice, one has, in the standard way,[1]

$$Q_n^0(la, l_0 a) = \int_{-\pi}^{\pi} \frac{d\theta}{2\pi} e^{-i(l-l_0)\theta} [\Phi(\theta)]^n \quad (5.3)$$

from which, for example, one easily derives the usual explicit binomial expressions for walks with only nearest-neighbor jumps. Likewise, when $b^2 < \infty$, a steepest descent argument yields the basic continuum result, (2.4), as $n \to \infty$. Notice, however, that the argument leading to (5.1) *presupposes* that any walk reaching $x < 0$ must *intersect* the wall at $x = 0$. This tacit hypothesis is valid for Brownian motion, i.e., for a continuum walk, but fails for a lattice walk if, in a single step, the walker can jump more than one lattice spacing. This fact underlies the restrictions to models (A) and (B) in the previous section; but see also below.

Since our primary interest is in asymptotic behavior let us utilize (2.4) or, in order to write *equalities* adopt the continuum model, (C). Then (5.1), with $b = 1$, yields

$$Q_n^W(x, a) = \frac{e^{-\sigma n}}{(2\pi n)^{\frac{1}{2}}} [e^{-(x-a)^2/2n} - e^{-(x+a)^2/2n}]$$

$$= \frac{e^{-\sigma n} e^{-x^2/2n}}{(2\pi n)^{\frac{1}{2}}} e^{-a^2/n} (e^{ax/n} - e^{-ax/n}) \quad (5.4)$$

The last expression here has a close resemblance to (4.7), the basic result for p initially equispaced vicious walkers: evidently the death factor describes simply the interference between the free walk and the negative image walk required to eliminate the forbidden walks which would cross the wall. As we will show, the same mechanism accounts for the death factor in all the other cases.[d]

When a is fixed and $n \to \infty$ the exact (continuum) result simplifies to

$$Q_n^W(x) \approx \frac{e^{-\sigma n}}{\sqrt{(\pi/2)}} \frac{axe^{-x^2/2n}}{n^{3/2}} \tag{5.5}$$

which reveals the distribution of a walk near a wall. Setting $x = a$ then shows that the number of returns of a walk to the wall varies as

$$R_n^W \approx C^W/n^{3/2} \tag{5.6}$$

If one appends a zeroth step from O to $x_0 = a$ and adds an $(n+1)$th step from $x = a$ to O this result equally describes walks which leave the origin on step zero and return *for the first time* on step $n+1$: it is thus reassuring to confirm that the exponent $\frac{3}{2}$ in (5.6) is precisely what follows from the Lemma and Corollary in Section 1! One also notices that this exponent characterizes the reunions of $p = 2$ vicious walkers [see (4.3)]. This is no accident since one may regard a rigid absorbing wall as the limiting case of a vicious walker whose diffusivity vanishes, i.e., $b_1^2 \to 0$, so that he does not move. Indeed, if one calculates the reunions of two dissimilar walkers from (4.18) by integrating (4.18) with respect to $\bar{x}$ and then sets $b_1 = 0$, $b_2 = b = 1$, $\sigma_1 = 0$, $\sigma_2 = \sigma$ and $a' = x$ one reproduces (5.5) precisely for $n \to \infty$.

5.2. Two vicious walkers

Consider now, on its own merits, the case of $p = 2$ identical vicious walkers on the line (and in the absence of a wall).[e] Two coordinates, x_1 and x_2, are needed and they are subject to the restriction $x_1 < x_2$. We may, however, regard x_1 and x_2 alternatively as the coordinates of a *single*, new *compound walker* who walks in a *plane*. Furthermore, the restriction $x_1 < x_2$ then translates into an *exclusion* of our two-dimensional walker from the line $x_1 = x_2$ and from the half-plane lying below the line, as illustrated in

---

[d] A reader accepting this but more interested in the applications will lose little by perusing the next two paragraphs and then skipping to Section 6.
[e] For further details, especially for lattice models, see Huse, Szpilka, and Fisher;[3] Huse and Fisher.[2]

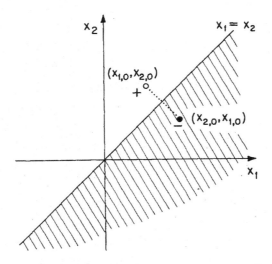

Figure 4. The plane (x_1, x_2) of the compound walker for $p = 2$ vicious walkers showing the starting point $(x_{1,0}, x_{2,0})$, the wall/mirror plane, $x_1 = x_2$, and the corresponding image-walker starting point. The combination of original and image walks satisfies the constraint $x_1 < x_2$.

Figure 4. In other words, we again have a single walker and an absorbing wall but, now, in two dimensions. This may be solved just as before. If the compound walker starts at $\vec{x}_0 \equiv (x_{1,0}, x_{2,0})$ the negative, image walker must start at $\vec{x}_0' = (x_{2,0} x_{1,0})$: see Figure 4; the analog of (5.1) then holds precisely. However, the behavior of the free lattice walks is now given by the generalization of (5.3) to two dimensions. For models (A) and (B) the corresponding single-step generating functions are easily seen to be

$$\Phi_A(\theta_1, \theta_2) = w_1^2 (e^{i\theta_1} + e^{-i\theta_1})(e^{i\theta_2} + e^{-i\theta_2}) \tag{5.7}$$

$$\Phi_B(\theta_1, \theta_2) = w_0^2 + w_0 w_1 (e^{i\theta_1} + e^{-i\theta_1} + e^{i\theta_2} + e^{-i\theta_2}) \tag{5.8}$$

respectively, and one may check[3] that a compound walker starting with $x_{1,0} < x_{2,0}$ cannot reach the half-plane $x_1 > x_2$ without passing through a site *on* the diagonal $x_1 = x_2$, i.e., without intersecting the wall, as required for the validity of the method.[f] Asymptotically, the distribution for a compound walker in d dimensions is simply the product of the distributions for d one-dimensional walkers [and so is given by (3.1) with $p = d$]. On putting everything together we achieve a derivation of (4.7) for the case $p = 2$ (the equispaced condition then being no restriction).

---

[f]Notice that in model A the generating function factorizes, which means that walkers far apart move independently, as desirable for physical applications. This is not the case in model B, other than in a statistical sense, since the second walker may not move on the same tick of the clock as the first walker.

5.3. Many vicious walkers

The problem of $p > 2$ vicious walkers can, with a little more care, be handled similarly.[2] One considers a single compound walker in p dimensions moving subject to the coordinate constraints $x_j < x_k$ for $j < k$, all (j, k). This translates into $\frac{1}{2}p(p-1)$ walls, $x_j = x_k$, in p-space which is thereby divided into $p!$ disjoint segments corresponding to the permutations, $\hat{\pi}$, of labels in the inequality $x_1 < x_2 < \cdots < x_p$. The method of images generates $\frac{1}{2}p(p-1)$ negative walkers from the initial walker by reflection in each of the $\frac{1}{2}p(p-1)$ wall planes. But for $p > 2$ one must go on to consider the positive images of all these negative walkers, and so on, recursively. One discovers that the procedure *closes* after the generation of $p!$ compound walkers, where each compound walker generated by a permutation $\hat{\pi}$ of even parity, $|\hat{\pi}|$, is a normal, positive walker, while each walker corresponding to an odd permutation is a negative walker. Finally, therefore, if $\hat{\pi}\vec{x}$ denotes the vector obtained from $\vec{x}$ by permuting the coordinate labels, the overall distribution of p vicious walkers is given by

$$Q_n^{(p)}(\vec{x}, \vec{x}_0) = \sum_{\hat{\pi}} (-)^{|\hat{\pi}|} Q_n^0(\vec{x}, \hat{\pi}\vec{x}_0) \tag{5.9}$$

where $Q_n^0(\vec{x}, \vec{x}_0)$ denotes the distribution for the appropriate free, compound walker in p dimensions: as indicated above, this is given asymptotically by (3.1) with $\vec{x} \Rightarrow (\vec{x} - \vec{x}_0)$.

From the master formula (5.9) all previously stated results follow. The antisymmetric character of the sum in (5.9) leads to a determinantal expression for the death factor which, in the case of walkers initially (or, finally) equispaced, reduces to a Vandermonde determinant. Such a determinant can be factorized and that yields the product formula (4.7). To obtain the result (4.15) for general initial and final spacing, the determinant elements, $\exp(x_j x_{k,0}/b^2 n)$, are expanded in powers of n^{-1} and a product of two Vandermonde determinants is obtained: details are given in Appendix A of Ref. 2.

5.4. Two walkers and a wall

The effectiveness of the method of images in complex cases depends upon the closure of the set of positive and negative images under all applicable reflections. This, in turn, requires some sort of underlying symmetry in the system. Nevertheless a number of further problems can be discussed. In

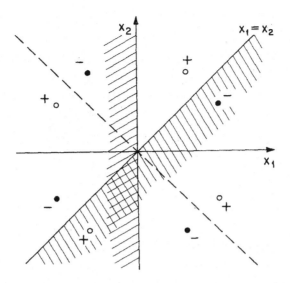

Figure 5. The compound walker plane for two vicious walkers in the presence of an absorbing wall at $x = 0$. The constraints $0 < x_1 < x_2$ yield two mirror planes, $x_1 = 0$ and $x_1 = x_2$, which on reflection generate two more mirror planes, namely, $x_2 = 0$ and $x_1 = -x_2$. As illustrated, a total of eight images, one being the original walker, result from the reflections.

particular, an absorbing wall and two vicious walkers can be handled. As illustrated in Figure 5 reflections in the two basic planes/lines determined by $0 < x_1$ and by $x_1 < x_2$ generate two further effective mirror planes but closure is obtained with eight images as shown. The probability of returns to fixed positions (near) the wall for large n are found to decay as

$$R_n^{W,2} = C_2^W / n^5 [1 + O(a^2/n)] \tag{5.10}$$

where the dependence on the initial and final coordinates is given by

$$C_2^W(\vec{x}, \vec{x}_0) = x_1 x_2 x_{1,0} x_{2,0} (x_2^2 - x_1^2)(x_{2,0}^2 - x_{1,0}^2)/3\pi \tag{5.11}$$

It is interesting that in this case, regarding the wall as just another walker (of low diffusivity) leads to wrong asymptotic behavior since, by (4.3), the probability of a reunion of $p = 3$ vicious walkers varies as $1/n^4$: apparently the positional fluctuations of a third walk allow a slower decay than does a rigid wall.

5.5. *Two dissimilar walkers*

A direct application of the method of images to two *dissimilar* walkers fails because in the first section of a reflected walk the diffusivities, b_1^2 and b_2^2, will necessarily be interchanged: but in the unreflected, second section of the walk interchanged diffusivities will cause the cancellation to fail, in general. This problem may, however, be tackled in the continuum model by separating variables via a change of coordinates to $x = x_2 - x_1$ and $\bar{x}$ as given by (4.19). The "mean" walker, with coordinate $\bar{x}$, diffuses freely; the "internal" walker is subject to $x > 0$ but, as seen, that problem can be handled by the method of images. Combining the results yields (4.18).

5.6. *Other aspects*

The basic expression (5.9) can be manipulated to yield further intriguing expressions describing the asymptotic behavior of p vicious walkers. It would clearly be of interest to obtain comparable exact results in which the limit $p \to \infty$ could be taken maintaining, say, a constant density of walkers. Such results do, in fact, exist and we will make contact with some of them below. However, rather different and more elaborate forms of analysis seem called for. The antisymmetric character of the problem evident in (5.9) can be utilized to construct a representation in which walkers appear as quantum-mechanical particles moving (or "hopping") in one dimension and collectively obeying Fermi statistics. Standard quantum-statistical methods for discussing free fermions then yield valuable exact results. One of the earliest studies along such lines was presented as a model for fibrous structures by de Gennes[4] in 1968. To pursue that theme here would, however, take us too far afield. Instead, let us turn now to some applications of the simple random walk results we have obtained.

6. Phase Transitions in Linear Systems

This section might have been headed "*Phase transitions with short range forces in one dimension (almost).*" Without the final qualifier, "almost," this title would contradict many theorems that prove rigorously that one-dimensional statistical mechanical systems with short range forces (of finitely-many-body character) *cannot* exhibit *any* phase transitions at nonzero temperatures. However, systems which are "almost one-dimensional" in the sense that they have a dominant linear structure but can, in some sense, spread indefinitely in a transverse dimension, may well have phase transitions at positive temperatures even if only short-

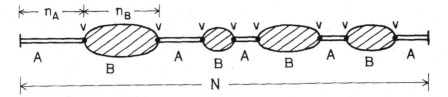

Figure 6. Illustrating a linear system in the form of a *necklace* or *string* of total length N units made up of alternating segments of microstates A and microstates B (constituting "beads" or "bubbles").

range forces are entailed: and, of course, truly one-dimensional systems with forces of sufficiently long range most certainly exhibit phase transitions.[g] Furthermore even one-dimensional systems with interactions of short range can have transitions if many-body forces of indefinitely high order are allowed.[h]

In fact, there is a rather simple but general mathematical mechanism which underlies a broad class of *exactly soluble* one-dimensional models which display phase transitions. This mechanism does not seem to be as well appreciated as it merits and it operates in a number of the applications we wish to discuss. Accordingly it seems appropriate to present a brief exposition here.

Consider the *necklace* or *string* illustrated in Figure 6 which consists of alternating segments, the first containing only a class of microstates labelled A of length n_A discrete units, and the second, only states of a class labelled B (for "bubble" or "bead") of n_B units. If $Z_N(T)$ is the partition function for such a necklace of total length N units, which represents a sum over all possible numbers and lengths of A segments and B segments, we require the *reduced free energy* per unit of length, namely,

$$f(T) = \lim_{N\to\infty} \frac{F_N}{Nk_BT} = -\lim_{N\to\infty} \frac{1}{N} \ln Z_N(T) = \Sigma/k_BT \qquad (6.1)$$

where $\Sigma(T)$ can be interpreted literally as the *tension* of the string.

An effective way to calculate f is to construct the isobaric partition function or generating function

$$G(z, T) = \sum_{N=0}^{\infty} z^N Z_N(T) \qquad (6.2)$$

---

[g] Recall, for example, the renowned work of Dyson[5] on the one-dimensional Ising model with long-range forces; see also the recent results of Fröhlich and Spencer.[6]

[h] This is demonstrated in Refs. 7 and 8, which are of particular relevance in the present context.

in which the activity-like variable z evidently acts as an indicator or counting variable for overall length. Then, if $z_0(T)$ *is the positive real singularity of $G(z)$ lying closest to the origin*, the limiting free energy is given simply by

$$z_0 = e^f \quad \text{or} \quad f(T) = \ln z_0(T) \tag{6.3}$$

This follows since Z_N is real and positive for all N, so that z_0 determines the radius of convergence of the series (6.2). It is important to note that the singularity of $G(z)$ at z_0 may correspond to a divergence such as a simple pole $\sim 1/(z_0 - z)$, but may equally represent a nonanalyticity, such as a square root branch point, at which $G(z)$ remains finite when $z \to z_0-$.

So much is quite general; the power of the generating function approach is that for a necklace such as shown in Figure 6, $G(z)$ can be constructed explicitly in terms of the partial isobaric partition functions

$$G_A(z) = \sum_n Q_n^A z^n \quad \text{and} \quad G_B(z) = \sum_n Q_n^B z^n \tag{6.4}$$

Here $Q_n^A(T)$ and $Q_n^B(T)$ are the (canonical) partition functions for segments of length n of A states and B states, respectively. In fact a little thought shows[i] that one has

$$G(z) = G_A + G_A v G_B v G_A + G_A v G_B v G_A v G_B v G_A + \cdots \tag{6.5}$$

where successive terms correspond to segment sequences, A, ABA, $ABABA$, etc. and, for simplicity we have decreed that a necklace should always start and end with an A segment (a convention of no consequence in the thermodynamic limit if [9]$v \neq 0$). In addition we have introduced a *vertex weight* (or activity) v which can be regarded as a Boltzmann factor associated with each AB or BA junction. Now for sufficiently small z the expression (6.5) is simply a convergent geometric progression: thus we finally obtain

$$G(z) = G_A(z)/[1 - v^2 G_A(z) G_B(z)] \tag{6.6}$$

It is not hard to generalize this formula to describe a necklace with further types of bead $CD \cdots$ arranged in periodic or arbitrary order, etc.

The possible mechanism for a phase transition is now already manifest. Thus, $z_0(T)$, the required singularity of $G(z)$, may be *either* the smallest root of the equation

$$v^2 G_B(z) = 1/G_A(z) \tag{6.7}$$

---

[i]See, e.g., Refs. 2, 3, 7, and 8 and work of Temperley[9] which contains a variety of instructive examples. To include the limit $v \to 0$, it is useful to include also the sequences $B, BAB, \ldots$ so that the numerator in (6.6) becomes $G_A(z) + G_B(z)$.

or the closest real, positive singularity of $G_A(z)$ *or* of $G_B(z)$: then, as the temperature or other parameters change, the condition determining $z_0(T)$ and, thence $f(T)$, may *switch*; such a switch will, in general, be nonanalytic and hence must correspond to a phase transition.

To investigate this in the simplest nontrivial situation suppose that the A states are described merely by

$$A: \quad Q_n^A = u^n \quad \text{with, say,} \quad u = e^{-\epsilon/k_B T} \tag{6.8}$$

so that u increases as T increases, and the generating function is

$$G_A(z) = 1/(1 - uz) \tag{6.9}$$

This has a pole at $z_A = 1/u$; however, it is easy to see from (6.6) that the denominator of $G(z)$ must vanish *before* $G(z)$ diverges at z_A so that this singularity never determines z_0.

For the B states we suppose that the canonical partition function for large n, which will be all that matters as regards the nature of any phase transitions, behaves as

$$B: \quad Q_n^B \approx q_0 w^n / n^\psi \quad \text{with} \quad w = e^{-\sigma_0(T)} \tag{6.10}$$

This is a rather natural generalization of (6.8) but its appropriateness will become quite clear when we consider various applications. Notice that the reduced free energy per unit length of a long bead or bubble is given simply by $f = \sigma_0(T)$. Now the behavior of the corresponding generating function as $wz \to 1-$ is easily estimated with the aid of the binomial theorem which yields

$$\begin{aligned} G_B(z) &\approx G_s/(1-wz)^{1-\psi}, & \text{for } \psi < 1 \\ &\approx G_s \ln(1-wz)^{-1}, & \text{for } \psi = 1 \\ &\approx G_c - G_s(1-wz)^{\psi-1} + G_1(1-wz) + \cdots, & \text{for } \psi > 1 \end{aligned} \tag{6.11}$$

where G_s, the amplitude of the leading singularity, which occurs at $z_B = w^{-1}$, is positive, as is

$$G_c = \sum_n Q_n^B / w^n < \infty \quad (\psi > 1) \tag{6.12}$$

When ψ is an integer (≥ 2) the term $G_s(1-wz)^{\psi-1}$ gains a factor $\ln(1-wz)^{-1}$. (Notice, also, that for $\psi > 3$ further analytic terms dominate.)

Now notice that $G_B(z)$ *diverges* at z_B when (a) $\psi \leq 1$. Then, as explained for G_A, the denominator of G must vanish *before* z reaches z_B. Consequently,

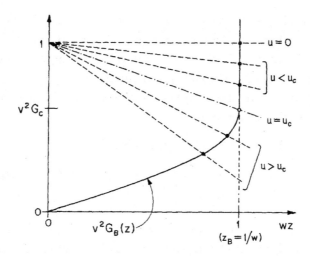

Figure 7. Plot versus wz of $v^2 G_B(z)$, which remains finite but becomes singular at $z = z_B = 1/w$ when $\psi > 1$. The dashed lines represent $1/G_A(z) = 1 - uz$. By (6.7) the circles locate the singularity, $z_0(T)$, which thus sticks at $z = z_B$ for $u < u_c$.

the free energy is *always* given by the smallest root of (6.7): this varies analytically with u, v etc. so there can be *no phase transition*. On the other hand, when (b) $\psi > 1$ it is not hard to see from Figure 7 that the root $z_0(T)$ will stick at $z_0 = z_B = 1/w$ for small u, provided $v^2 G_c < 1$, but will switch to the smallest root of (6.7) when u exceeds the critical value

$$u_c = w(1 - v^2 G_c) \tag{6.13}$$

The free energy *below* u_c is thus given simply by

$$f = \sigma_0(T) \quad (\text{all } \psi > 1) \tag{6.14}$$

which describes a "frozen" phase consisting, essentially, of one infinitely long bubble, i.e., only B states are realized in the thermodynamic limit. Above u_c, if one measures the deviation from the transition by

$$t = (T - T_c)/T_c \sim u - u_c \tag{6.15}$$

one finds *critical behavior*,

$$f = \sigma_0(T) - A_s t^{1/(\psi-1)} + \cdots, \quad \text{for} \quad 1 < \psi < 2 \tag{6.16}$$

with a logarithmic factor at $\psi = 2$, but a *first-order transition* plus singular corrections,

$$f = \sigma_0(T) - A_1 t + A_s t^{\psi-1} \cdots, \quad \text{for} \quad \psi > 2 \tag{6.17}$$

(Logarithmic factors in t appear in the corrections when ψ is integral and further terms, $A_2 t^2, \ldots$ dominate when $\psi > 3$.) The critical behavior (6.16) corresponds to a specific heat singularity with exponent[10]

$$\alpha = (2\psi - 3)/(\psi - 1) \tag{6.18}$$

The mean length of a given bubble in the "melted" or "disordered" phase can be calculated from

$$\bar{n}_B = z \frac{\partial}{\partial z} \ln G_B(z) \tag{6.19}$$

which, on using (6.16) for $1 < \psi < 2$, yields

$$\bar{n}_B \sim \frac{1}{(1 - wz)^{2-\psi}} \sim \frac{1}{(\sigma_0 - f)^{2-\psi}} \sim \frac{1}{t^{(2-\psi)/(\psi-1)}} \tag{6.20}$$

Not surprisingly, this diverges as $t \to 0$. On the other hand, $\bar{n}_B$ remains finite at the first-order transition when $\psi > 2$. When the transition is critical one should expect to find a divergent, *longitudinal correlation length*, $\xi_{\|}(T)$. Its magnitude will be set by the average bubble length, $\langle \bar{n}_B \rangle$; however, longer bubbles occupy more space on the necklace and so must be weighted accordingly in computing the average. Thus, if c is the length of a unit, we have

$$\xi_\| = c\langle \bar{n}_B \rangle = c\overline{n_B^2}/\bar{n}_B = (c/\bar{n}_B)\left(z\frac{\partial}{\partial z}\right)^2 \ln G_B(z)$$

$$\sim \frac{c}{1 - wz} \sim \frac{c}{\sigma_0 - f} \sim \frac{c}{t^{\nu_\|}} \tag{6.21}$$

where the correlation length exponent is seen to be

$$\nu_\| = 1/(\psi - 1) \tag{6.22}$$

Comparison with (6.16) and (6.18) shows that this satisfies the standard hyperscaling relation, $d\nu_\| = 2 - \alpha$, with $d = 1$, in accord with normal expectations.[10,11]

Armed with this general formalism and the results for vicious walkers, let us, finally, turn to some concrete applications.

7. Wetting of a Boundary Wall in Two Dimensions

Consider a system in thermodynamic equilibrium in a phase A which is on the point of a bulk first-order transition at which coexistence of the phase A with a second phase, B, becomes possible. Now if a wall of the container enclosing the bulk A phase becomes sufficiently attractive to the complementary phase, B, as, say, T increases, we may expect to see a *wetting transition*.[12,13] At such a transition the amount of B-like material adsorbed on the wall, or the mean thickness, l_W of the adsorbed, B-like film becomes *infinite* so that, the wall is, in fact, covered by a *macro*scopically thick *wetting layer* of bulk phase B; this is separated from the bulk A phase by a normal $A|B$ interface with surface tension

$$\Sigma_{AB}(T) = k_B T \sigma \tag{7.1}$$

To model such a system we suppose, following Abraham,[14] that the wall affects the bulk interactions locally so that an interface located close to the wall has a modified interfacial tension, $\Sigma_{AB}^{(1)} = k_B T \sigma_1$. If, as we suppose, $\Sigma_{AB}^{(1)} < \Sigma_{AB}$ holds for low T, the interface tends to be pinned to the wall. One may similarly represent the specific attraction of the wall for the phase B by a local field or (reduced) chemical potential, h_1, although this parameter plays no essential role in the present case. Abraham[14] solved this problem[j] *exactly* for a ($d = 2$)-dimensional Ising model and, indeed, discovered a wetting transition at a temperature T_W, below the bulk critical point, T_c. We will solve a simplified version of Abraham's model or, to express it differently, solve his model approximately by heuristic arguments which prove, however, good enough to capture the principal results exactly![k]

Our model or "picture" of a wall in a two-dimensional system below the wetting transition is embodied in Figure 8. The interface is pinned to the wall along segments of varying length but escapes from the wall over alternate segments to form bubbles or droplets of B phase bounded by sections of $A|B$ interface. The fluctuations of the interface normal to the wall, which we take to be the x direction, are modeled by a random walk for which the y coordinate, parallel to the wall, represents time, i.e., we take $y \equiv t = nc$. In this picture "overhangs" or "double-backs" are disregarded as in the original Temperley[19] or solid-on-solid (SOS) model[16] of a $d = 2$ Ising model interface. In a long, free, statistically straight interface, overhangs can certainly be absorbed into a fully renormalized value of the interfacial tension:

---

[j]Abraham[14] described the phenomenon as an interface *roughening transition*[15] since the unbound interface above T_W is, in fact, rough: however, since a *free* interface in two dimensions is *always* rough it was recognized later that the transition in Abraham's model is primarily a wetting or *interface delocalization* transition: see also Chui and Weeks.[16]

[k]Other treatments of versions of the simplified model have been given by Chui and Weeks,[16] by Burkhardt,[17] and by van Leeuwen and Hilhorst.[18] However, these authors have used somewhat more elaborate approaches based on the Schrödinger-like diffusion equation for a walk in a potential which is discussed below.

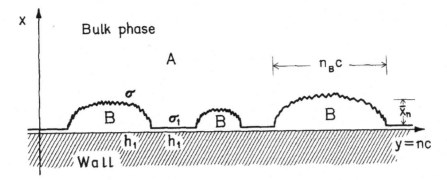

Figure 8. Model of a prewetting layer on a wall in a two-dimensional system close to coexistence between bulk phases A and B. The reduced tensions σ and σ_1 characterize a free interface and one pinned to the wall, respectively, while h_1 is a local field or chemical potential favoring the B phase.

however, it is not easy to assess directly whether the neglect of overhangs is of any greater significance in curved and restricted interfaces of finite, albeit long length such as entailed in bubbles on a wall.

As clear from Figure 8, our model is just a special type of *necklace* of the sort discussed in the previous section. The interaction parameters are to be identified as

$$u = e^{-\sigma_1 + h_1} \quad \text{and} \quad w = e^{-\sigma + h_1} \tag{7.2}$$

The vertex weight may, for present purposes, be absorbed into the prefactor q_0 in (6.10). The all-important exponent ψ follows from our analysis of a walk near an absorbing wall: to this end, recall from (5.6) that the number of n-step returns to the wall, R_n^W, decays as $1/n^{3/2}$. Hence we have

$$\psi = \tfrac{3}{2} \tag{7.3}$$

The results can now be read off from (6.14) to (6.20). As the tension, $\Sigma_{AB}(T)$, of a free interface decreases with increasing T bubbles of greater and greater length appear and the wall free energy or, equivalently, the modified interfacial tension finally exhibits a transition where it varies as[2,20]

$$\begin{aligned} f \equiv \Sigma_W/k_BT &= \sigma(T) - A_0(T - T_{cW})^2 + \cdots, & \text{for } T \leq T_{cW} \\ &= \sigma(T), & \text{for } T \geq T_{cW} \end{aligned} \tag{7.4}$$

Evidently this wetting transition is critical and, indeed, corresponds precisely to a classical second-order transition with a simple discontinuity in the specific heat ($\alpha = 0$); see also Figure 9. This result agrees precisely with the exact Ising model calculations of Abraham![14] For T above the transition

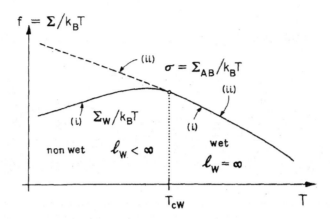

Figure 9. Schematic variation (i) of the wall free energy, $\Sigma_W(T)$, (or wall-modified tension) through a critical wetting transition in a two-dimensional system. The tension Σ_{AB} of a free $A|B$ interface, (ii), is approached quadratically as T approaches T_{cW} from below, while the thickness, l_W, of the adsorbed film, diverges as $1/(T_{cW} - T)$.

the equality $f = \sigma$ shows that one has a free interface uninfluenced by the wall so that $l_W(T) = \infty$; i.e., the wall is *completely wet*.

The longitudinal correlation length, $\xi_\parallel(T)$, describing density or composition correlations parallel to the wall below T_{cW}, now diverges with exponent $\nu_\parallel = 2$ [see (6.22)]. Physically, it is also interesting to investigate the *transverse correlation length*, $\xi_\perp(T)$, describing correlations normal to the wall, and to determine the rate of divergence of the adsorption or, equivalently, the mean thickness l_W of the microscopic film of B-like phase. To do this let us examine the characteristic:

7.1. Size and shape of a droplet on a wall

Accordingly, in a bubble of length nc consider the displacement, x, of the interface from the wall at a distance mc from one end of the bubble. The interfacial segments of m steps and $(n - m)$ steps, respectively, may be described by *independent* walks from a point at a small distance, a, from the wall to the point x. From (5.5) the distribution of x for large m and n is thus proportional to

$$Q_m^W(x) Q_{n-m}^W(x) \propto a^2 x^2 \exp\left(-\frac{x^2}{2b^2 m} - \frac{x^2}{2b^2(n-m)}\right) \quad (7.5)$$

From this the mean value of x, the dispersion in x, etc. are readily calculated. These all scale with the most probable value of x. If the origin for y is taken at the midpoint of the bubble, so that $m = \frac{1}{2}n + (y/c)$, this is readily found to obey

$$x^2 + (2b^2/nc^2)y^2 = \tfrac{1}{2}nb^2 \tag{7.6}$$

Hence the bubble is, on average, *elliptical in shape*! Furthermore the mean area and width are given by

$$\bar{A}_n \approx \frac{\pi bc}{4\sqrt{2}} n^{3/2} \quad \text{and} \quad \bar{x}_n \approx \frac{\pi b}{4\sqrt{2}} n^{\frac{1}{2}} \tag{7.7}$$

Of course the exponents here could have been guessed easily since the spread of a random walk of n steps is always of order $bn^{\frac{1}{2}}$.

With this information we can estimate the transverse correlation length and the adsorption.[2,3] For the former we have

$$\xi_\perp \approx \langle \bar{x}_n \rangle \sim b \langle \overline{n_B^{1/2}} \rangle \sim b \overline{n_B^{3/2}} / \bar{n}_B \sim b/(1-wz)^{\frac{1}{2}}$$
$$\sim b(\xi_\parallel/c)^{\frac{1}{2}} \sim b/t^{\nu_\perp} \quad \text{or} \quad \nu_\perp = \tfrac{1}{2}\nu_\parallel \tag{7.8}$$

so that $\nu_\perp = 1$. The adsorption follows similarly[2] as

$$\rho_B \propto \langle \bar{A}_n \rangle / \langle \bar{n}_B \rangle \sim \overline{n_B^{5/2}} / \overline{n_B^2} \sim (1-wz)^{-\frac{1}{2}} \sim \xi_\perp \tag{7.9}$$

and hence diverges in the same way as the transverse correlation length, as is physically reasonable. The mean *thickness*, $l_W(T)$, of the prewetting film can equally be regarded as measured by $\xi_\perp$.

7.2. Interface pinning in the bulk

We have seen above that if an interface is attracted to a wall it may be pinned at the wall but can then break loose, or "delocalize," at a wetting transition at some $T = T_{cW}$. [To check that the transition does actually occur in a given model it is necessary to pay a little more attention to the vertex weight, v, than we did above and, in particular, to verify that the transition point given by (6.13) actually falls within the physical region; e.g., T_{cW} should be less than the bulk critical point T_c.] It is natural to enquire further whether an interface in a *bulk* two-dimensional system can be similarly pinned by a *linear imperfection*, (away from any walls) and then undergo a *depinning* or *delocalization transition*.[16–18] To be concrete, consider a nearest-neighbor ferromagnetic square lattice Ising model with a row parallel to the y axis (i.e., "horizontal" in the presentation of Figure 8) of weakened x bonds ("vertical" bonds in Figure 8). An interface parallel to the y axis will, clearly, be pinned on the row of weakened bonds at $T = 0$, since the interfacial tension will be lowest there. As T increases, however, segments of the interface may break loose, wander on *either side* of the attractive row of bonds, and then return to form a pinned segment. Since the interface may wander *across* the attractive row without actually experiencing the pinning potential its behavior may, in essence, be described just by the returns of a free walker. By (2.7) we thus have $\psi = \frac{1}{2}$ and hence, by the analysis following (6.12), there is *no transition*: in other words the interface *always remains pinned* however weak the attractive row potential! This conclusion is confirmed, again, by the more elaborate arguments.[16–18]

8. Multiphase Systems: Interfacial Wetting

8.1. *Two-dimensional fluids*

Consider a multiphase two-dimensional system in which three or more different phases $A, B, C, \ldots$ can coexist. Between these phases various distinct interfaces may be formed. Then, as one varies the temperature or composition, etc., one of these interfaces, say $A\|C$, may undergo a wetting transition, becoming wet by an "intermediate phase," say, B, and decomposing into an $A|B$ interface and a separate $B|C$ interface, as illustrated schematically in Figure 10. Indeed current experiments on three-dimensional systems are performed on such multiphase fluid systems in which, typically, A is vapor phase.[21–24] The analogous experiments on multicomponent adsorbed fluid films should be possible.

Now if, as before, we model the various interfaces by random walks we see from Figure 10 that an intact $A\|C$ interface may be regarded as a necklace of segments, say, "threads," in which the local composition profile changes rapidly with x from bulk A-like to bulk C-like, and "bubbles" describing regions where a sliver of intermediate, B-like phase penetrates as a fluctuation. The former segments may be described by a reduced "bare" or "local" tension, σ_{AC}, while the latter, consisting essentially of two distinct, more-or-less parallel interfaces can be described by a total reduced tension $\sigma_{AB} + \sigma_{BC}$ where

$$\sigma_{AB} = \Sigma_{AB}/k_B T \quad \text{and} \quad \sigma_{BC} = \Sigma_{BC}/k_B T \tag{8.1}$$

in which $\Sigma_{AB}(T)$ and $\Sigma_{BC}(T)$ are the tensions of free, well-separated $A|B$

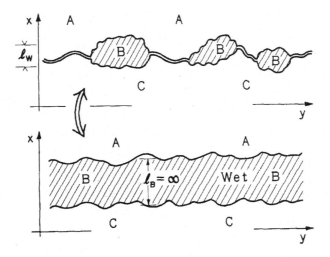

Figure 10. Illustration of an interfacial wetting transition in a two-dimensional three-phase system. The free $A\|C$ interface has fluctuations in which microscopic droplets of B-like phase appear on the interface. When these grow and coalesce the $A\|C$ interface can decompose into distinct $A|B$ and $B|C$ interfaces separated by a macroscopic wetting layer of B phase.

and $B|C$ interfaces. The necklace parameters may then be taken as

$$u = e^{-\sigma_{AC}} \quad \text{and} \quad w = e^{-\sigma_{AB}-\sigma_{BC}} \tag{8.2}$$

while the vertex activity, v can be used to allow for the deviations of the total bubble tension from $\sigma_{AB} + \sigma_{BC}$ induced by the two sides of the bubble coming together at the vertex.

Finally, the exponent ψ now required is that associated with the reunion of $p = 2$ vicious walkers so that by (4.4) we have

$$\psi = \tfrac{3}{2} \tag{8.3}$$

This is the *same* as for the wetting of a rigid wall in a two-phase system. Thus the wetting transition will again be critical with classical, second-order exponents. In particular, the surface tension of the intact, nonwet $A\|C$ interface varies when T approaches T_{cW}, say, from below, as

$$\Sigma_{AC}(T) = \Sigma_{AB}(T) + \Sigma_{BC}(T) - A_0(T - T_{cW})^2 + \cdots \tag{8.4}$$

while in the wet region, $T > T_{cW}$, one has simply

$$\Sigma_{AC}(T) = \Sigma_{AB}(T) + \Sigma_{BC}(T) \tag{8.5}$$

This represents Antonow's rule[25] for the situation in hand and, clearly, describes two well-separated interfaces with a wetting layer of macroscopic thickness between them, i.e., $l_B = \infty$ (see Figure 10). Graphically, Figure 9 still applies with the curve (i) representing the tension $\Sigma_{AC}(T)$, while curve (ii) represents the sum $\Sigma_{AB} + \Sigma_{BC}$. The longitudinal and transverse correlation length exponents and the exponent for the adsorption of B-like material on the $A\|C$ interface must all be the same as found in the previous section.

Notice that in quoting (4.4) as authority for $\psi = \frac{3}{2}$ we were implicitly assuming that the two distinct interfaces $A|B$ and $B|C$ could be modeled by *similar* random walks. In general, the $A|B$ and $B|C$ interfaces will be physically *dissimilar* so this is not a satisfactory approximation. Nevertheless, as we have seen in Section 5, the reunions of two dissimilar walkers are, equally, described by $\psi = \frac{3}{2}$ so that (8.3) remains valid. This fact also makes it clear why the wetting of a rigid wall exhibits the same critical behavior.

8.2. *Denaturation of a biopolymer*

The upper part of Figure 10 may be regarded as a portrayal of a double-stranded biopolymer molecule, such as DNA, in which, over some segments, the two strands have become unwound from one another. Once the unwinding encompasses the whole length of the polymer molecule it dissociates into two separate chains, corresponding schematically to the "wet" arrangement of interfaces in the lower half of Figure 10. The accompanying transition in a long biopolymer molecule may be very sharp:[1] it is generally referred to as the helix-coil ($A \equiv$ "helix," $B \equiv$ "coil" in the necklace picture), denaturation, or, simply, "melting" transition.

This transition can be discussed on the basis of the necklace theory[26] and one discovers that the sharpness of the transition is directly related to the smallness of the vertex activity, v, which in this context is sometimes called the "initiation factor." The exponent ψ enables one to account for the self-avoiding character of the two, partially dissociated, "coiled" strands of polymer which constitute a necklace bubble.[28] The necklace theory does not, however, take account of the self-avoiding or "excluded volume"

---

[1]See the review and reprint collection by Poland and Scheraga[26] and, for example, Stevens and Felsenfeld.[27]

requirement between different bubbles or between bubbles and bound, "helical" segments. To correct this deficiency represents a very difficult problem but one may hope it is not too serious in the case that the helical sections are comparatively rigid, as is so in most real situations.

8.3. *Commensurate adsorbed phases*

When submonolayers of an atomic or simple molecular gas are adsorbed on smooth crystalline substrates one frequently observes (by x-ray and electron scattering techniques) the formation of *commensurate surface phases* in which the adsorbate atoms are, predominantly, ordered in registry with the underlying substrate lattice. Such phases may be formed and studied under conditions of thermodynamic equilibrium. Then, as the temperature, T, and chemical potential, μ (which controls the overall adsorbate coverage), are varied the degree of order changes and the commensurate phase may melt into a disordered, fluid phase, or undergo a transition into a different commensurate phase or into an incommensurate phase.[m]

The simplest type of commensurate phases to study are the uniaxial $p \times 1$ phases in which, on a substrate of rectangular symmetry, the adsorbate atoms define a superlattice in which the x-axis lattice constant is an integer, p, times the corresponding substrate lattice constant, say, a. In the ideal, fully ordered situation the adsorbate atoms (or molecules) will thus form uniformly spaced chains parallel to the y axis and at distance pa apart. A physical realization of such $p \times 1$ rectangular phases for $p = 1, 2$, and 3 is formed by dissociated, and hence atomic, hydrogen on the (110) face of crystalline iron[30] (although in the $p = 3$ phase it is chains of vacancies rather than of atoms which are separated by three lattice spacings).

Now it is evident that a $p \times 1$ phase gives rise to p physically distinct but equivalent types of domain $A, B, C, \ldots$. If different domains coexist on the substrate they must be separated by domain walls or interfaces where there is a mismatch in the ordering. However, even though all domains are completely equivalent, there will, as has recently been stressed,[2,3,20] be $(p-1)$-*distinct types* of physically *non*equivalent *domain walls*. This is evident from Figure 11, which illustrates the case $p = 4$ schematically. The types of wall or interface may be labeled by q, the discrete phase shift measured in units of the x lattice spacing which is generated on crossing the interface

---

[m]See the collection of lectures edited by Dash and Ruvalds.[29]

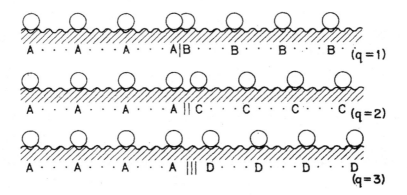

Figure 11. Schematic cross-sections showing two contiguous domains of a $p \times 1$ rectangular commensurate phase on a substrate when $p = 4$ and demonstrating the existence of three physically distinct types of domain wall or interface with phase shifts $q = 1, 2,$ and 3. (See also Huse and Fisher, Refs. 2 and 20.)

from one domain to its neighbor. Thus for $p = 4$ we have the three types:

$$[+1] \equiv [-3]: \quad A|B, \quad B|C, \quad C|D, \quad \text{and} \quad D|A$$
$$[+2] \equiv [-2]: \quad A\|C, \quad B\|D, \quad C\|A, \quad \text{and} \quad D\|B \qquad (8.6)$$
$$[+3] \equiv [-1]: \quad A\|\|D, \quad B\|\|A, \quad C\|\|B, \quad \text{and} \quad D\|\|C$$

Each type of wall will have its own distinct free energy or tension, $\Sigma_q(T, \mu)$.

Consider a $p = 3$ phase. There are just two wall types: $[+] \equiv [+1]$ and $[-] \equiv [-1] \equiv [2]$. As T and μ vary, their tensions, $\Sigma_+(T, \mu)$ and $\Sigma_-(T, \mu)$, will change and thus we may anticipate wetting transitions. Figure 10 would then represent the transition

$$[-] \Rightarrow 2[+], \quad \text{i.e.,} \quad A\|C \Rightarrow A|B|C \qquad (8.7)$$

which will occur if $2\Sigma_+(T, \mu)$ approaches $\Sigma_-(T, \mu)$. Our theory indicates that the transition should be continuous with classical, second-order thermodynamic exponents, $\alpha = 0$, etc., and correlation length exponents $\nu_\| = 2$ and $\nu_\perp = 1$.

More generally, for $p \geq 4$ one can have[2]

$$[-] \equiv [p-1] \Rightarrow (p-1)[+], \quad \text{e.g.,} \quad A\|\|D \Rightarrow A|B|C|D \qquad (8.8)$$

To discuss this on the necklace picture we need ψ for $q = (p-1)$-stranded bubbles, which is described by the reunions of q vicious walkers. For $q \geq 3$ we have $\psi_q \geq 4$ and hence, by (6.17), we expect a *first-order transition* (although singular corrections will be present on the coherent, or intact, nonwet side of the transition). However, there are other possibilities for $p \geq 4$. Thus it may well happen that in one region of the phase diagram only [+] walls are stable, all other $[q \neq 1]$ walls being wet, i.e., dissociated into $(p-q)$ [−] walls. This leads one to consider a necklace model in which *both* A and B segments are characterized by generating functions like (6.10) with (6.11) so that one has two exponents, $\psi_A \equiv \psi_{p-q}$ and $\psi_B \equiv \psi_q$. The general theory still applies but it transpires that the vertex weight, v, plays a more important role. If v is less than a critical value

$$v_c = [G_c^{(p-q)} G_c^{(q)}]^{1/2} \qquad (8.9)$$

[extending the notation of (6.11) and (6.12) in an obvious fashion] one always has a first-order, wet-to-wet transition, such as $2[-] \Rightarrow 2[+]$ for $p = 4$. This is illustrated in Figure 12, where the ordinate is v while the abscissa, labelled by the chemical potential, corresponds, precisely, to $\mu = \sigma_- - \sigma_+$ where, in the general theory, $w_A = e^{-\sigma_-}$ and $w_B = e^{-\sigma_+}$. As shown in the figure, the wetting transitions above v_c are normal (i.e., an intact, coherent wall dissociates) and may be either first order or second order in character, depending on the values of p and q. Consequently, the phase diagrams can exhibit a *bicritical point* from which spring two critical lines, a *critical endpoint*, or a *triple point*; see Figure 12.

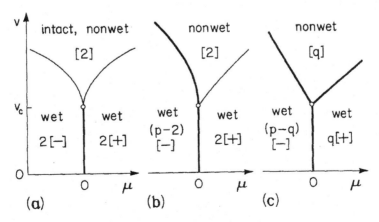

Figure 12. Phase diagram for wet-to-wet domain wall transitions in $p \times 1$ commensurate adsorbed phases for a [2] wall when (a) $p = 4$ and (b) $p \geq 5$, and (c) for a $[q \geq 3]$ wall with $p - q \geq 3$. First order transitions are denoted by bold lines; the light lines denote classical second-order transitions. An intact or coherent wall exists only for $v > v_c$. (After Huse and Fisher, Ref. 2.)

8.4. *Chiral melting*

There is good evidence, both theoretically and experimentally,[n] that the melting of a two-dimensional commensurate phase may be *continuous*. The question then arises as to the universality class of the transition or, more concretely, as to the values of the critical exponents. This issue has been reopened recently in light of the observations above concerning different types of domain wall.[2,20] Let us consider, specifically, a $p \times 1$ commensurate adsorbed phase with $p = 3$: then there are just two types of wall, namely,

$$
\begin{array}{llll}
[+]: & A|B, & B|C, & \text{or} \quad C|A \\
[-]: & A\|C, & B\|A, & \text{or} \quad C\|B
\end{array}
\tag{8.10}
$$

Now a continuous melting transition implies the existence of large-scale fluctuations: but the important fluctuations in an initially fully ordered domain, say, an A domain, are those which can be represented as small subdomains of contrasting character, i.e., B and C. Near criticality such subdomains will become large and, thus, their perimeters will resemble the domain walls [+] and [−]. The four simplest types of subdomain fluctuations are illustrated schematically in the upper left part of Figure 13.

This picture supposes, however, that both [+] and [−] walls remain intact, i.e., are *non*wet. In this situation the critical fluctuations will resemble those in a two-dimensional ($p = 3$)-state *Potts model*. For nearest-neighbor coupling on a rectangular lattice this model can be described by the *clock model* Hamiltonian

$$
\mathcal{H} = -J_x \sum_{\langle i,j \rangle}^{\perp} \cos \frac{2\pi}{p} (n_i - n_j) - J_y \sum_{\langle i,j \rangle}^{\|} \cos \frac{2\pi}{p} (n_i - n_j)
\tag{8.11}
$$

with $p = 3$. Here the first sum, labelled $\perp$, runs over nearest-neighbor pairs of transverse bonds, parallel to the x axis, while the second sum runs over longitudinal bonds, parallel to the y axis: the clock variables take the values $n_i = 0, 1, 2, \ldots (p-1)$, [mod p]. In the ordered state of the clock and Potts models p physically equivalent types of domain exist just as in the $p \times 1$ commensurate phases. Thus, in agreement with earlier considerations,[33,34] we are led to conclude[2,20] that if, as the phase boundary is approached, both [+] and [−] walls remain intact, the continuous melting transition should be in the same universality class as the three-state Potts model. By this

---
[n]See, e.g., Refs. 2, 20, and, experimentally, Bretz[31] and Moncton *et al.*[32]

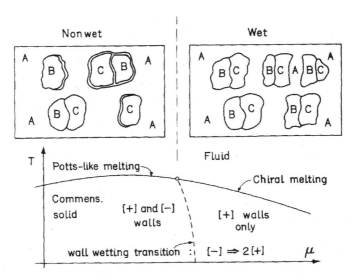

Figure 13. Schematic depiction of the predominant heterophase subdomain fluctuations near a continuous commensurate melting transition when [−] walls are not wet, as on the left, *or* are wet, by the intermediate domain, as on the right. The corresponding (schematic) phase diagram exhibits a wall-wetting transition locus within the commensurate solid phase and a continuous melting line which changes from Potts-like to chiral character. Such a phase diagram may apply to $\sqrt{3} \times \sqrt{3}$ commensurate phases as found in krypton on graphite; see Ref. 20.

token the specific heat exponent should be $\alpha = \frac{1}{3}$, and so on. (In this connection we should recall Baxter's exact calculations for the melting of hard hexagons on a triangular lattice.[35,36] This transition is also expected to be in the $p = 3$ Potts class and Baxter found $\alpha = \frac{1}{3}$.)

On the other hand, suppose that as the chemical pressure increases the [−] wall undergo a wetting transition and so decompose entirely into pairs of [+] walls. The phase diagram, see Figure 13, now has a wetting line (shown dashed) and a region where any long [−] walls are unstable. Thus the previously identified set of simplest critical fluctuations will change character drastically as illustrated in the upper right part of Figure 13: the most probable large fluctuations are now only $A(B|C)A$ subdomains, the complementary or reflected $A(C\|B)A$ subdomains being suppressed. The configurational combinations of these fluctuations are unlike those of the three-state Potts model and thus, in contrast to the earlier considerations,[33,34] (which overlooked the lack of symmetry which allows different types of wall), we conclude[2,3,20] that continuous melting should be in a *new*, *chiral melting* universality class when only one type of wall remains intact as the phase boundary is approached.

The name "chiral" has been given to this new type of melting[2,20] to highlight the reflection noninvariance of the true adsorbate Hamiltonian that results in the two-domain configurations $A|B|C$ and $C\|B\|A$ having different free energies. In the standard Potts or clock models these two configurations are precisely equivalent and have the same free energy. At this point we do not know the details of chiral melting; in particular, the critical exponents have not been estimated reliably. However, the nature of the critical scattering in the disordered, fluid phase does provide an observable hallmark which distinguishes chiral from nonchiral continuous melting.[2,20]

8.5. *Chiral clock models*

To this point our discussion of domain wall wetting and commensurate melting has been quite phenomenological. One may reasonably doubt whether such wetting transitions will ever arise naturally in real systems or in sensible models and, even if they do, whether the difference in domain wall free energies, which is what ultimately drives the wetting transitions, will represent a *relevant perturbation*, in the renormalization group sense,º at the continuous melting transition.

The first issue can be answered physically by noticing from Figure 11 that the different types of domain wall differ primarily in terms of the local density of the adsorbate atoms in the region of the wall:[20] if the adsorbate particles repel on close approach, a "heavy" or denser-than-average wall will have a higher free energy (or tension) than a "light" wall. In reaching this conclusion, however, the overall surface pressure, which is controlled by the chemical potential, μ, has been overlooked. As μ increases the coverage will increase above the ideal dictated by perfect registry with the substrate. The consequent "crowding" of the adsorbate will favor heavy walks reducing their free energy, say, Σ_+, relative to that of the light wall, say, Σ_-. Ultimately, this process can lead to an *incommensurate phase* (see further below): before this happens, however, the wetting condition $2\Sigma_+ = \Sigma_-$ is likely to be encountered and the light, say, [−] walls will disappear.

A more concrete theoretical answer is provided by studying the so-called *chiral clock models* devised by Ostlund,[38] with this specific issue in mind, and, independently, by Huse.[39] These models allow for the chiral character of real adsorbed phases by, in the uniaxial case, generalizing the

---

ºSee, e.g., Refs. 70 and 11 and references therein.

first term in the clock Hamiltonian (8.11) by supposing j always denotes the right-hand neighbor of i and taking

$$-J_x \sum_{\langle i,j \rangle}^{\perp} \cos \frac{2\pi}{p}(n_i - n_j + \Delta) \qquad (8.12)$$

The parameter Δ measures the degree of chirality; thus for $p = 3$ the energy of the configurations $\{n_i\} = \cdots 012012 \cdots$ of "spins" along the x axis, deviates from the energy of $\{n_i\} = \cdots 210210 \cdots$ by a term proportional to Δ. One may hence associate Δ with the chemical potential μ.

The chiral clock models can be studied by systematic series expansions at low and high temperatures.[3] In this way one discovers, first, that there are two distinct wall tensions obeying

$$\Sigma_+(T,\Delta) - \Sigma_-(T,\Delta) \approx -D(T)\Delta, \quad \text{as} \quad \Delta \to 0 \qquad (8.13)$$

and second, that domain wall wetting transition lines, $T_W(\Delta)$, do, indeed, lie within the commensurate phase region,[p] resembling the behavior illustrated in Figure 13 when $T_W \to 0$. Furthermore, one finds[3] for $p = 3$ that at the pure Potts critical point, which is realized when $\Delta \equiv 0$, the chirality is a *relevant* perturbation so that for $\Delta \neq 0$ the melting transition should cross over from Potts-like behavior and exhibit the anticipated new chiral character.

9. The Forces Between Walks and Walls

Studies of incommensurate phases in two and three dimensions have led various workers to the notion that parallel interfaces at separation l in a multiphase system act as if there were a well-defined force of interaction between them, at least for separations $l \gg b$, where b measures the intrinsic width of the interface.[q] This viewpoint has been particularly stressed by Pokrovsky,[40] by Villain,[41] and by Halperin and coworkers.[42] It can be extended to discuss the dependence of critical exponents on dimensionality regarded as a continuous parameter[43,44] and, as will be explained below, it can be used to discuss interfaces and wetting transitions in the presence of an external field as well as other related problems.[r] Let us approach this

---

[p] As $T \to 0$ the wetting lines for $[-] \Rightarrow 2[+]$ and for $[+] \Rightarrow 2[-]$ approach $\Delta_W = \pm\frac{1}{4}$, respectively, whereas the commensurate phase, represented at $T = 0$ by $n_i = 0$ (or 1 or 2) for all i, extends to $\Delta_m = \pm\frac{1}{2}$; see Ref. 3.
[q] The interface width b should be comparable to the transverse correlation length, $\xi_\perp$, in the bulk phases.
[r] Attention may be drawn to Natterman's further developments of interface phenomenology to include dipolar couplings, random field effects, etc.[44]

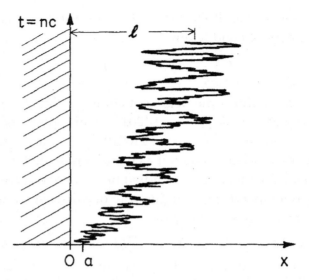

Figure 14. Portrayal of a walk near an adsorbing wall showing the drift of the most probable position of the walker, $l = bn^{1/2}$, away from the wall under the influence of the effective wall–walk force of repulsion.

topic by asking for the effective force between a rigid, absorbing wall and a nearby random walk as illustrated in Figure 14.

We learn from (5.5) that the partial partition function for a walk which starts at a small distance, a, from a wall at the origin and, after n steps, reaches the point x is given by

$$Q_n^W(x) \sim ae^{-\sigma n} x e^{-x^2/2nb^2}/n^{3/2} \tag{9.1}$$

The most probable distance from the wall after n steps is thus

$$l = bn^{\frac{1}{2}} \tag{9.2}$$

The mean distance, of course, varies similarly with n. The total partition function for n step-walks near the wall is

$$Z_n^W = \int dx\, Q_n^W(x) \sim e^{-\sigma n}/n^{\frac{1}{2}} \tag{9.3}$$

and so the reduced *free energy per step* at the nth step is given by

$$f_n = -\ln(Z_{n+1}/Z_n) = \sigma + \tfrac{1}{2}\ln(1+n^{-1}) \tag{9.4}$$

Now re-express this result in terms of l and expand for $l^2/b^2 > 1$. This tells us that the free energy per step [i.e., per unit of length, c, of walk parallel to the wall in the (x,y) plane] varies with l, the (most probable or mean) distance from the wall, as

$$f(l) = \sigma + \tfrac{1}{2}\frac{b^2}{l^2} + O\left(\frac{b^4}{l^4}\right) \qquad (9.5)$$

The first term, σ, represents simply the reduced free energy per step of a free walk which, naturally, is realized when $l \to \infty$. The further terms are thus due to the interactions between the walk and the wall; indeed they correspond to an *effective force* with a repulsive, i.e., positive *potential*

$$W(l) \approx \tfrac{1}{2} k_B T \frac{b^2}{l^2} \qquad (l/b \gg 1) \qquad (9.6)$$

per step [or per length of walk c in the (x,t) plane]. The dependence of this potential on temperature serves as a reminder that the origin of the force lies in the *loss of entropy* a walk incurs when the walker approaches but cannot penetrate the absorbing wall.

How literally can we accept the result (9.6)? Does it apply in other circumstances? To test the interpretation, consider a walk which is confined between *two* absorbing walls at spacing $2L$: see Figure 15 which, in part (b), also shows the total effective potential, or free energy increment due to the walls,

$$\Phi(l; L) = W(l) + W(2L - l) \qquad (9.7)$$

constructed by supposing that each wall may be regarded as acting independently on the walker. By symmetry, the most probable position of the walker is at $l_0 = L$; but this location can, alternatively, be regarded as that which minimizes $\Phi(l)$. Thus we anticipate that the appropriate walk free energy can be written in the form

$$f_2(L) \approx \sigma + 2c_2 \tfrac{1}{2}(b^2/L^2), \quad \text{for} \quad L^2/b^2 \gg 1 \qquad (9.8)$$

If (9.6) is taken literally the coefficient here has the value $c_2 = 1$.

On the other hand, let us return to the original lattice model of the walk described in Section 1. It is easy to see that the partial partition functions satisfy the *difference equation*

$$Q_{n+1}(x) = w_{-1} Q_n(x+a) + w_0 Q_n(x) + w_1 Q_n(x-a) \qquad (9.9)$$

Two absorbing walls at $x = \pm L$ are then described by the boundary conditions

$$Q_n(L) = Q_n(-L) = 0 \quad (\text{all } n) \qquad (9.10)$$

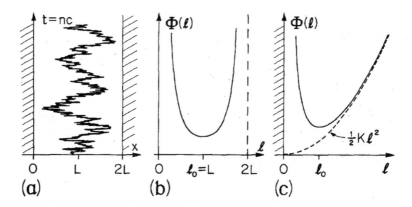

Figure 15. (a) Depiction of a random walk between two absorbing walls at $x = 0$ and $x = 2L$; (b) the total effective walk–wall potential, $\Phi(l)$, for a walk between two walls; (c) the effective potential for a walk in a harmonic well with an absorbing wall at $x = 0$.

A linear difference equation with constant coefficients is easily solved: the form (9.8) turns out to be quite correct but one discovers that the coefficient has the exact value[43]

$$c_2 = \pi^2/8 \simeq 1.2337 \tag{9.11}$$

The description in terms of two superposable wall forces is thus qualitatively exact; quantitatively it is in error only by about 19%! Incidentally, the value of c_2 remains unchanged if the walk is allowed to make *arbitrarily large* weighted jumps at each step (provided only that b^2 remains finite).[43]

Further, more challenging tests are possible. Suppose that the walker moves in an external potential described by

$$U(x) = V(x)/k_B T \tag{9.12}$$

so that a Boltzmann factor $e^{-U(x)}$ is to be associated with each step on which a walker arrives at or remains on the site x. The recursion relation (9.9) thus becomes

$$Q_{n+1}(x) = e^{-U(x)}[w_{-1}Q_n(x+a) + w_0 Q_n(x) + w_1 Q_n(x-a)] \tag{9.13}$$

This equation (which is well known in studies of the solid-on-solid model of an interface in an external field[15–18]) is, in general, hard to solve exactly. However, if $U(x)$ varies smoothly, one may go to the continuum limit in x and t and obtain the *Schrödinger-like diffusion equation*

$$\frac{\partial Q}{\partial t} = \tfrac{1}{2}b^2 \frac{\partial^2 Q}{\partial x^2} - [\sigma + U(x)]Q \tag{9.14}$$

for the continuum partial partition function, $Q(x;t)$. [In taking the limit the weights w_i must be written in appropriate form and the free energies f and σ, and the potential, $U(x)$, must be redefined to refer to unit time: then the Boltzmann factor in (9.13) becomes $e^{-cU(x)}$ with $c \to 0$, and so on.] This generalized diffusion equation is somewhat more tractable than the discrete equation as we shall see.

As a second test problem consider a walker moving in a harmonic potential well,

$$U(x) = \tfrac{1}{2}Kx^2 \qquad (9.15)$$

which acts outside an absorbing wall at the origin; see Figure 15(c). If K is not too large $U(x)$ varies slowly on the scale of b and, heuristically, we can write down the total free energy for a walk at distance l from the wall

$$f_3(K;l) = \sigma + \Phi(l) \approx \sigma + \frac{b^2}{2l^2} + \tfrac{1}{2}Kl^2 \qquad (9.16)$$

Minimization yields the most probable walk location as

$$l_0 = b^{\frac{1}{2}}/K^{\frac{1}{4}} \qquad (9.17)$$

and substitution in (9.16) then shows that the free energy should vary with K as

$$f_3(K) = \sigma + c_3 b\sqrt{K} \qquad (9.18)$$

with $c_3 = 1$.

To check this we must solve the diffusion equation (9.14) subject to the boundary condition $Q(0;t) = 0$ (all t). Since, as one learns in elementary quantum mechanics, harmonic potentials go well with Schrödinger's equation, one can actually find an exact solution for all t with the general initial condition $Q(x,0) = \delta(x - x_0)$. However, since only the asymptotic free energy is required, it is easiest to ask merely for the long time behavior and hence to solve the corresponding ground-state Schrödinger eigenvalue problem. In this particular case, however, it is simplest just to guess a trial solution of the form

$$Q(x,t) = e^{-ft} x e^{-x^2/2l_0^2} \qquad (9.19)$$

One readily checks that this solves (9.14) only if l_0 is given by (9.17). Thus the most probable walk location is given *exactly* by the heuristic arguments! From (9.19), (9.17), and (9.14) one discovers that the form (9.18) for the free energy is, again, quite correct. This time, however, the coefficient c_3 is in error by about 33%, the exact value being $c_3 = \frac{3}{2}$; but this reduced accuracy is not really surprising since the direct effects of the harmonic potential on the walker were ignored in writing (9.16). Clearly a walker is restricted even by a pure harmonic potential and so loses some entropy: the effect corresponds precisely to the zero-point motion in a pure harmonic well which yields a free energy increment

$$\Delta f_0 = \tfrac{1}{2}b\sqrt{K} \equiv \tfrac{1}{2}b[(d^2U/dx^2)]^{\frac{1}{2}} \tag{9.20}$$

(corresponding to the usual zero point energy $\tfrac{1}{2}h\nu$). If this is added to (9.18) with $c_3 = 1$ the exact answer is recaptured! The second form written here for Δf_0 suggests that the heuristic ansatz leading to (9.7) and (9.16) could be improved for slowly varying potentials, $U(x)$, by adding a local term, $\Delta f_0(l)$, evaluated from the second derivative in accord with (9.20).

The tests presented above should convince one that the effective wall-walk potential, $W(l)$, found in (9.6) will give a good general account of the effects of walls and other external potentials on an otherwise random walker. What about interactions between two or more walks? Consider two dissimilar walkers at separation l: from the viewpoint of the first walker the second may be approximated by a wall at distance l, and vice-versa for the second walker. Thus we would guess that the total entropy loss should be described by the *walk–walk interaction potential*

$$W(l) \approx k_B T \frac{b_1^2 + b_2^2}{2l^2} \tag{9.21}$$

Certainly, this reduces correctly to (9.6) when either b_1 or b_2 vanish, a non-diffusing walk corresponding, as before, to a stationary wall. More generally, however, one can check the validity of (9.21) by recalling the result (4.18) for two dissimilar vicious walkers and repeating the general reasoning that led to (9.6).

A more stringent test is provided by considering $p = 3$ similar vicious walkers for which the total free energy follows from (4.2) or (4.9) [the coefficient $\tfrac{1}{2}$ in (9.4) being replaced, in general, by $\tfrac{1}{4}p(p-1)$]. By symmetry, the three walkers will on average be equally spaced and from (4.7) and (4.8) one finds that the most probable spacing is $l = \sqrt{(3/2)}bn^{\frac{1}{2}}$. If this is used in (9.21) to evaluate the sum $2W(l) + W(2l)$, which allows for *all* pairs of walk interactions, the exact result for the free energy is again reproduced.

Since the expression (9.21) for the effective forces felt by a walk passes its trials with flying colors, let us try it in some applications.

10. Critical Prewetting

Consider the phase diagram of a fluid system, as sketched in Figure 16, which exhibits a first-order transition from phase A, say, "vapor," to phase B, say, "liquid" on a phase boundary $\mu = \mu_\sigma(T)$.[45,46] It is convenient to define the external bulk field by

$$h = \mu - \mu_\sigma(T) \tag{10.1}$$

Then our previous discussion of the two-dimensional wetting transition of a wall attractive to phase B but in contact with bulk phase A was confined to the phase boundary itself, i.e., $h = 0$. Somewhat more precisely, we required $h = 0-$ since an infinitesimal negative field was needed to ensure the presence of bulk A phase, as illustrated in Figure 8. When T increased a critical wetting transition was reached at $T = T_{cW}$; this is indicated by the open circle in Figure 16. For $T \geq T_{cW}$ a wetting layer of macroscopic thickness, $l_W = \infty$, covers the wall; the dots along the phase boundary in Figure 16 denote this wet state.

Now a wetting film of *finite* thickness, l_W, will persist even in the presence of a negative field, $h = -|h|$. The wetting transition can then be described by saying that $l_W(T,h)$ remains finite as $h \to 0-$ for $T < T_{cW}$ but *diverges to infinity when* $h \to 0-$ if $T \geq T_{cW}$. But what is the form of this divergence of $l_W(T,h)$ as h vanishes? And, how do the thickness, $l_W(T,h)$, and the wall free energy, $f(T,h)$, vary with T at nonzero h near the wetting transition? These questions have been discussed by Abraham and Smith[47] on the basis of the Schrödinger-like diffusion equation (9.14). Here we present a heuristic treatment which yields an understanding of their results and goes slightly further in some respects.

Consider, first, the wetting film in a field $h = -|h|$ well above the transition at T_{cW}. Adopting the previous picture (see Figure 8) we construct the free energy for a film of thickness l as

$$f(T,h;l) = \sigma - h_1 + |h|\frac{cl}{\bar{a}^2} + \frac{1}{2}\frac{b^2}{l^2} \tag{10.2}$$

The first term on the right is the (reduced) tension of the free interface; the second represents the net short range attraction of the wall for phase B; in the third term cl represents the area of the interface per step (parallel to the y axis), while one has

$$\bar{a}^2 = k_B T / \Delta\rho \tag{10.3}$$

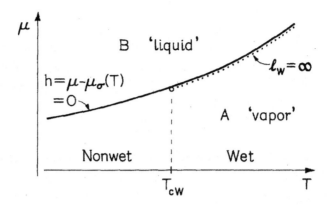

Figure 16. Bulk phase diagram showing a critical wetting transition at $T = T_{cW}$. (See also Refs. 45 and 46.)

in which $\Delta \rho$ is the jump in density across the bulk phase boundary. (Recall that μ couples to the density, ρ.) Finally, the last term in (10.2) represents the interface–wall interaction in accord with (9.6). Minimization on l yields a (most probable) wetting layer thickness diverging as

$$l_W \approx \bar{b}/|h|^{\frac{1}{3}}, \quad \text{with} \quad \bar{b} = (b^2 \bar{a}^2 / c)^{\frac{1}{3}} \tag{10.4}$$

that is with an exponent $\nu_W = \frac{1}{3}$. The free energy then varies for small fields as

$$f(T, h) \approx \sigma - h_1 + \tfrac{3}{2} B |h|^{\frac{2}{3}} \tag{10.5}$$

with $B = (bc/\bar{a}^2)^{2/3}$. As in the tests of the wall–walk force law we should anticipate that the factor 3/2 is not exact although it should be accurate to 20% or so.

Now to discuss the transition region we must allow for the situation, illustrated in Figure 8, in which, statistically, the wetting film is composed of a distribution of droplets of B-like phase adhering to the wall. The crucial new feature that arises is the need to include the external field, h, in the bubble generating function; thus the partition function Q_n^B, for a bubble of length n (in units of c), should have a factor

$$\exp[-n(\sigma - h_1) - \bar{A}_n |h|/\bar{a}^2] \tag{10.6}$$

where $\bar{A}_n$ is the mean area of a bubble. If the external field can be neglected, the area is given by (7.7) which can be written

$$\bar{A}_n = \bar{a}^2 n^{3/2} \quad \text{with} \quad \bar{a}^2 = (\pi/4\sqrt{2})bc \tag{10.7}$$

One is therefore tempted merely to insert a factor $\exp(-k_0|h|n^{3/2})$ with

$$k_0 = \tilde{a}^2/\bar{a}^2 \tag{10.8}$$

[in accord with (10.2)] into the original form (6.10) for Q_n^B. To do this, however, would overlook the effects of the field itself on the shape and size of a typical bubble.

In order to account for the field properly one should solve the recursion relation (9.13) or the Schrödinger diffusion equation (9.14) with a linear potential $U(x) \propto |h|x$. In the latter case, closed-form but complex results could be anticipated in terms of Airy functions.[47] However, we can embody the physically important features in a simple approximation which can be explained with the aid of Figure 17. For a sufficiently small droplet the field has negligible effect on the shape and we may use (10.7) in (10.6) to obtain a satisfactory representation of Q_n^B. But once the maximum width for a typical free bubble, namely, $x_n \approx b(\frac{1}{2}n)^{\frac{1}{2}}$ [see (7.6)] exceeds the wetting film thickness for the field in question, namely, $l_W(h) \approx \bar{b}/|h|^{1/3}$ [see (10.4)], the width of a bubble will be effectively limited by the field. Equating x_n to l_W shows that this crossover in bubble shape occurs at

$$n_0(h) \approx 2(\bar{b}/b)^2|h|^{-2/3} \tag{10.9}$$

For $n > n_0$ the bubble area will, except for end effects decreasing as (n_0/n), be given simply by $\bar{A}_n = ncl_W$ so that $\bar{A}_n|h| \sim n|h|^{2/3}$ which correctly reflects the result (10.5). Thus the simplest approximation for Q_n^B is constructed by using this form in (10.6) for $n \geq n_0(h)$ but to use (10.7) for $n \leq n_0(h)$. A better approach is to utilize a smooth changeover function chosen to preserve the analyticity at small h which the exact generating function must enjoy. A simple but satisfactory option is to replace n in $(\bar{A}_n/n)$ and in the prefactor for Q_n^B by $\tilde{n}(h) \equiv n/(1 + k_1^3 n^3 h^2)^{1/3}$ which becomes proportional to $n_0(h)$ when $n \to \infty$. This yields the physically reasonable approximation

$$Q_n^B z^n \approx q_0 \frac{e^{-n\Delta f}}{n^{3/2}}(1 + k_1^3 n^3 h^2)^{\frac{1}{2}} \exp\left[\frac{-k_0 n^{3/2}|h|}{(1 + k_1^3 n^3 h^2)^{1/6}}\right] \tag{10.10}$$

in which

$$\Delta f = \sigma - h_1 - f \tag{10.11}$$

The changeover parameter, k_1, should now be chosen so that

$$k_0/k_1^{\frac{1}{2}} \equiv \tilde{a}^2/\bar{a}^2 k_1^{\frac{1}{2}} = \tfrac{3}{2}B \equiv \tfrac{3}{2}(bc/\bar{a}^2)^{2/3} \tag{10.12}$$

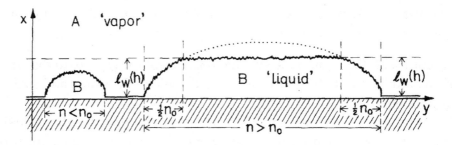

Figure 17. Droplets or bubbles of "liquid," B, adhering to a wall in the presence of an external field, h, which favors the bulk "vapor" phase, A. Droplets of length less than $n_0(h)$ (in units of c) are effectively uninfluenced by h but when n exceeds n_0 the width of a droplet is limited by $l_W(h)$.

This serves to ensure that (10.5) is reproduced well above the transition. Other limits are readily checked.

Whichever approximation is adopted, it is not hard to see that, near its singularity, the bubble generating function approaches the scaling behavior

$$G_B(z,h) - G_c \approx -|h|^{\frac{1}{3}} \Gamma\left(\frac{\Delta f}{|h|^{\frac{2}{3}}}\right) \tag{10.13}$$

Indeed this must also be a consequence of the exact analysis. The approximation (10.10) yields the closed expression

$$\Gamma(s) = q_0 k_1^{\frac{1}{2}} \int_0^\infty \frac{dr}{r^{3/2}} \{1 - (1+r^3)^{\frac{1}{2}} \exp[-sr/k_1 - k_2 r^{3/2}/(1+r^3)^{1/6}]\} \tag{10.14}$$

where $k_2 = k_0/k_1^{3/2}$; although not exact, this reproduces the correct results in the various limits and should be useful for practical purposes. From (10.13) the scaling behavior of the wall free energy follows straightforwardly by the methods of Section 6. One obtains

$$f \equiv \frac{\Sigma_W(T,h)}{k_B T} \approx \sigma - h_1 - |h|^{\frac{2}{3}} \Upsilon\left(\frac{T - T_{cW}}{|h|^{\frac{1}{3}}}\right) \tag{10.15}$$

which, again, is necessarily a consequence of the more exact theory. In particular, the exponents entering are exact.[47] The scaling function $\Upsilon(u)$ can be expressed in terms of $\Gamma(s)$. If one accepts the approximation (10.14) as adequate, or considers the possible behavior of $G_B(z;h)$ more generally, one discovers that there is *no transition* in the presence of a nonzero field.[47] Thus the phase diagram in Figure 16 correctly represents the situation.

The general nature of the scaling function $\Upsilon(u)$ follows from the known limiting behavior. Thus, $\Upsilon(u)$ approaches $-\frac{3}{2}B$ as $u \to +\infty$ so reproducing

(10.5). On the other hand when $|h| \to 0$ at fixed $T < T_{cW}$ so that $u \to -\infty$ the scaling function behaves as $A_0 u^2$ which yields the original critical wetting expression (7.4). One may hope that it will eventually prove possible to check these predictions for two-dimensional wetting in the laboratory.

11. Ising Model Correlations in Two Dimensions: Anomalous Decay Law

Consider a two-dimensional Ising model on, say, a rectangular lattice with a spin variable $s_{xy} = \pm 1$ at each lattice site (x,y). In accord with our previous conventions we will take the x and y lattice spacings as $a_\perp \equiv a$ and $a_\parallel \equiv c$, respectively. The net, two-point correlation function is defined, in the standard way, by

$$C(x,y) = \langle s_{00} s_{xy} \rangle - \langle s_{00} \rangle \langle s_{xy} \rangle \tag{11.1}$$

The decay of the correlations with distance is a matter of perennial interest! Away from the immediate critical region the Ornstein–Zernike theory should apply to the two-point function; this predicts

$$C(x,y) \approx D_\theta e^{-\kappa_\theta r}/r^{\frac{1}{2}}, \quad \text{as} \quad r \to \infty, \quad \text{for} \quad d=2 \tag{11.2}$$

where $\kappa_\theta \equiv 1/\xi(T,\theta)$ denotes the inverse correlation length for the direction θ, measured, say, from the y axis, while r is the radial distance. Of course, κ_θ, and the amplitude, D_θ, depend on the temperature and the (reduced) magnetic field, h; the exponent $\frac{1}{2}$ of the power in the decay law is, however, universal (away from criticality);[48] in fact, it reflects a dominant simple pole at $k = \pm i \kappa_\theta$ in the Fourier transformed correlation function, $\hat{C}(\vec{k})$.

The Ornstein–Zernike prediction (11.2) is verified by the exact calculations of the two-point correlations for planar Ising models in zero field ($h=0$) above T_c.[49–52] Perturbation expansions allowing for the field h suggest that (11.2) remains valid in the presence of a nonzero field.[48,53,54] However, the exact calculations for zero field *below* T_c show that (11.2) is violated;[49,52] instead, the two-point correlation function[s] obeys the *anomalous decay law*

$$C(x,y) \approx D_\theta e^{-\kappa_\theta r}/r^2 \quad (d=2, h=0, T<T_c) \tag{11.3}$$

---

[s]Strictly, one must take the limit $h \to 0\pm$ or replace the product $\langle s_{00} \rangle \langle s_{xy} \rangle$ by the square of the spontaneous magnetization, $m_0(T)$, or by the long range order, $\langle s_{00} s_\infty \rangle$.

that is, the exponent $\frac{1}{2}$ becomes 2. This corresponds to a branch point and associated cut in the complex k plane in place of the Ornstein–Zernike pole.

The analysis leading to the exact solutions throws little light on the reason for this anomalous decay; nor does it indicate whether or not it should persist in the presence of a small field below T_c. This issue has been addressed in a general context using the *transfer matrix approach*[48,54] and it has been shown[55] that the anomalous behavior below T_c is, in fact, special *both* to $d = 2$ dimensions *and* to zero field. To check this conclusion, McCoy and Wu[56] have tackled the problem of including a small field in the exact calculations for the two-dimensional Ising model; they discovered the mathematical mechanism by which the behavior (11.3) goes over to (11.2) as the field h is switched on[†] but, again, rather little physical insight can be gleaned! More recently, Abraham[57] has treated the problem approximately and shown how the exact results can be understood heuristically on the basis of a bubble picture. As we will show, simplifying Abraham's analysis still further, the issue becomes isomorphic to the behavior of a wetting layer in the presence of a bulk field: essentially, then, we will merely recast the arguments of the previous section in a new language!

The starting point is explained in Figure 18. Low-temperature expansions for Ising models are constructed by turning over spins from the fully aligned or + state ($s_{xy} = +1$ all x, y). One thus discovers that contributions to the two-point correlation function, $C(x,y)$, arise only from configurations in which both sites $(0, 0)$ and (x, y) are *linked* by a chain of neighboring overturned spins so that they reside in the same island of − spins. The energy of such a configuration in *zero field* is determined solely by the total length of *perimeter* of the islands of − spins: in the presence of a field there is an additional contribution proportional to h and to the total *area* of the islands. At *large separations*, $r \gg a, c$, the dominant contributions will thus come from those islands which just reach from $(0, 0)$ to (x, y); however, there is entropy to be gained if the perimeters of the islands are allowed to wander. In *zero field*, therefore, we anticipate that the behavior will be

---

[†]The McCoy–Wu analysis[56] is based on a formal expansion in powers of the field, h. In the complex k plane the anomalous branch cut present in zero field breaks up into a sequence of poles. McCoy and Wu calculate the locations and residues of the poles near the tip of the cut for small h. The location of these poles is also discussed from a perspective somewhat closer to that developed here (and in Ref. 55) by Stone.[37] However, as pointed out by Abraham[57] the approximate real-space expression presented for $C(x,y)$ by McCoy and Wu is actually inadequate for large (x, y) as can be seen by calculating the fluctuation sum, which should yield a finite value for the susceptibility, $\chi(T,h)$, but which, in fact, diverges. Discussions with Barry M. McCoy on these issues are appreciated.

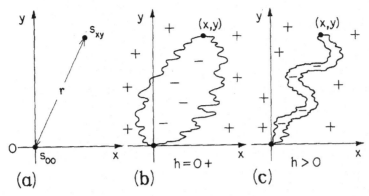

Figure 18. Configurational interpretation of the net spin–spin correlation function, $C(x,y)$, of a two-dimensional Ising model below T_c [see (11.1)], (b) in the case of zero field (in the "up" or "+" phase) and (c) in a small positive field showing the narrowing of a typical subdomain of overturned spins which contributes to the dominant behavior of $C(x,y)$ for large r.

determined by a "bubble" of — spins which constitute a small subdomain as illustrated in Figure 18(b). The free energy to be associated with the perimeter is thus just the interfacial tension, $\Sigma(T)$, times the total length.

Let us now suppose that the angle θ is relatively small ($\theta^2 \approx x^2/y^2 \ll 1$): then, as previously, we may regard the two sides of the subdomain or bubble stretching from $(0, 0)$ to (x, y), as *two strings* representing the paths of two vicious random walkers on the (x, y) plane with $y \equiv t = nc$ denoting time. The statistical weight of the subdomain is then proportional to the partial walk partition function and, in turn, yields the correlation function in leading order. From the expression (4.12) for the reunion of vicious walkers we thus obtain

$$C(x,y) \propto e^{-2\sigma n - \frac{1}{2}(2x^2/b^2 n)}/n^2 \tag{11.4}$$

the crucial exponent coming merely from $\frac{1}{2}p^2$ at $p = 2$. The free energy per step, σ, required here is that for an interface running parallel to the y axis so that $\sigma = c\Sigma_\parallel/k_B T$. Now it is not hard to show that the diffusivity of a walker representing an interface should be related to the interfacial tension via[43]

$$b^2 \approx ck_B T/\tilde{\Sigma}(T) \quad \text{with} \quad \tilde{\Sigma} = \Sigma_0 + \Sigma'' \tag{11.5}$$

Here we have used the fact that the interface is rough so that $\Sigma(T;\theta)$ is a differentiable function of the angle θ; then one has[43]

$$\Sigma_0 = \Sigma(T;0) \equiv \Sigma_\parallel \quad \text{and} \quad \Sigma'' = \left(\frac{\partial^2 \Sigma}{\partial \theta^2}\right)_{\theta=0} \tag{11.6}$$

Finally a little algebra shows that (11.4) can be rewritten, up to errors of order $\theta^4 \approx x^4/y^4$ in the exponent, as

$$C(x,y) \propto e^{-2\Sigma(T;\theta)r/k_BT}/r^2 \tag{11.7}$$

where $r^2 = x^2 + y^2$. This agrees with the exact anomalous decay law (11.3) and implies a definite relation between the inverse *correlation length* and the *surface tension* at angle θ, namely,

$$1/\xi(T;\theta) \equiv \kappa_\theta = 2\Sigma(T;\theta)/k_BT \tag{11.8}$$

But this relation is, in fact, an *exact* result for the two-dimensional Ising model for all $T < T_c$![52,58,59] It is also exact for the ordered regions of Baxter's hard square/hexagon model for directions parallel to the square lattice axes.[36] Thus the bubble "picture" for the large distance behavior of the correlations in zero field has proved surprisingly accurate. Notice that the factor 2 in (11.8) corresponds to the *two* strings defining the perimeter of the dominant bubble, while the anomalous power $1/r^2$ arises directly from the diffusive repulsion between the two strings.

What will happen in a nonzero magnetic field? As explained, the statistical weight of a bubble will be reduced by a factor representing the action of the field on the *area* of the bubble. This is precisely the same effect as in the wall wetting analyzed in the previous section. For θ not too large and y less than $cn_0(h)$, where $n_0 \sim |h|^{-2/3}$ may be taken from (10.9), the field will have negligible effect on $C(x,y)$. For larger values of r, however, the width of the bubble will saturate at $\Delta x \approx b(\frac{1}{2}n_0)^{\frac{1}{2}} \sim |h|^{-\frac{1}{3}}$ and, instead of two freely diffusing strings, we will be left with a *single* double string.[57] As suggested in Figure 18(c), this double string will diffuse but it is unrestricted by collisions. Thus we can put $p = 1$ in the general walker expressions and obtain

$$C(x,y) \propto e^{-\sigma_2 n - \frac{1}{2}x^2/b_2^2 n}/n^{\frac{1}{2}} \tag{11.9}$$

Thus the original $1/r^{\frac{1}{2}}$ decay law is restored by the field, as anticipated. The walk tension, σ_2, now denotes the free energy per step of a double walk. But this must depend on the magnetic field in the same way as the surface free energy in the wall wetting problem. From (10.5) we can thus read off the dependence of the correlation length on the field as

$$c/\xi_\|(T,h) - c/\xi_\|(T,0) \approx B|h|^{2/3} \tag{11.10}$$

This singular variation with h is confirmed by the exact calculations and by the approximate but more detailed analyses.[56,57] The diffusivity b_2^2 appearing in (11.9) must reduce to $\frac{1}{2}b^2$ as $h \to 0$ but may have a nontrivial field dependence which should also be reflected in the angular dependence of $\xi(T,h;\theta)$.

An approximate crossover form for the correlation function may be constructed along the lines used in writing down (10.10). The essential, and exact feature is a scaling behavior in terms of the combination $r|h|^{2/3}$; this leads to a prefactor $|h|$ in the long-range behavior in a field in agreement with the exact results.[56] Although an approximate formula analogous to (10.10) will probably be fairly accurate, its Fourier transform will not exhibit the correct mathematical structure for $h \neq 0$. In fact, the exact analysis of Wu and McCoy[56] shows that the anomalous cut in the k plane breaks up into a sequence of simple poles at spacing of order $|h|^{2/3}$. However, Abraham's more detailed discussion of the bubble approximation on the basis of the Schrödinger-like diffusion equation,[57] does reproduce this feature correctly. The essence of the matter is thus once more captured by the heuristic walk/interface/string picture.

12. Commensurate–Incommensurate Transitions

Under variation of temperature and chemical potential (or pressure, etc.) a commensurate phase may undergo a transition into an *incommensurate phase*; a typical phase diagram is sketched in Figure 19(a). Now, from the work of many authors[u] we know that a uniaxial incommensurate phase can, not too far from the transition, be viewed as a sequence of domains of *commensurate* phase separated by parallel sheets of domain walls or interfaces at some mean spacing l. This characterization of an incommensurate phase derived from a two-dimensional $p \times 1$ commensurate phase is illustrated in Figure 20. If there are N walls, say, parallel to the y axis, in a length L along the x axis the linear density of walls, $N/L = 1/l$, may be regarded as a type of order parameter for the incommensurate phase; in practice one usually defines the *incommensurability*

$$\bar{q}(T,\mu) \equiv 2\pi/pl \qquad (12.1)$$

which can be observed in scattering experiments as the displacement of an adsorbate Bragg peak from the associated commensurate Bragg peak (or reciprocal lattice vector) of the underlying substrate lattice. If

$$t \propto T - T_c(\mu) \qquad (12.2)$$

---

[u] See the review by Bak[60] and references therein.

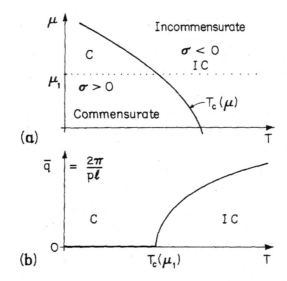

Figure 19. (a) Phase diagram showing a commensurate–incommensurate (or C–IC) transition in the (T,μ) plane; (b) variation of the incommensurability, $\bar{q}(T,\mu)$ on the locus $\mu = \mu_1$.

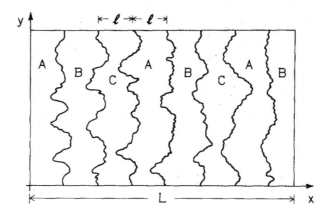

Figure 20. Schematic representations of a 3×1 incommensurate phase showing domain walls at mean spacing l.

measures the distance from the commensurate–incommensurate (C–IC) phase boundary, $T_c(\mu)$, one may expect a power law variation

$$\bar{q} \approx Q_0 t^{\bar{\beta}} \quad (t \to 0+) \tag{12.3}$$

where the value of the exponent $\bar{\beta}$ is of prime interest (see Figure 19(b)).

This problem has been approached by a variety of fairly sophisticated mathematical techniques,[40,60–63] in particular by representing the walls as one-dimensional *fermions* moving on the x axis with, just as in the interpretation in terms of walkers, the y axis representing time.[v] But, as pointed out by Pokrovsky and Talapov[40] who were the first to calculate the value of $\bar{\beta}$, the correct answer is readily obtained and understood by thinking about the *interactions* between the domain walls.[43] If $W(l)$ denotes the interaction potential of one domain wall with a neighbor at mean separation l, the singular part of the free energy density can be constructed as

$$f_s(T, \mu; l) \approx L^{-1}[N\sigma + NW(l)/k_B T] \tag{12.4}$$

where $\sigma = \sigma(T, \mu)$ is the interfacial tension of an isolated domain wall parallel to the y axis. Now, in the commensurate phase the tension σ is positive and an isolated wall is stable. However, as the C–IC transition is approached σ vanishes and becomes negative in the commensurate phase: the resulting instability leads to the appearance of many walls. To leading order we may thus write

$$\sigma \approx -\sigma_1 t \quad \text{with} \quad \sigma_1 > 0 \tag{12.5}$$

But, as we have seen in Section 9, we should anticipate repulsive forces acting between walls which will stabilize the incommensurate phase as an array of walls. At large separations, l, which are relevant near the transition, we may use the basic result (9.21) for the interface interactions and thus conclude[40,43]

$$f_s(T, \mu; l) \approx -\sigma_1 t/l + b_0^2/l^3 \tag{12.6}$$

where the modified wall diffusivity, $b_0^2 \propto b^2$, allows for the additional interactions between further neighboring walls. Note that the value of the isolated wall diffusivity, b^2, is given in terms of the interfacial tension as in (11.5).[43]

Minimizing on l, as previously, yields the mean value

$$1/l = (\sigma_1/3b_0^2)^{\frac{1}{2}} t^{\frac{1}{2}} \quad \text{for} \quad t \geq 0 \tag{12.7}$$

whence, in agreement with the more sophisticated theories,[40] we conclude $\bar{\beta} = \frac{1}{2}$ (for $d = 2$).[w] Likewise the free energy is given by[2]

$$\begin{aligned} f_s(T, \mu) &= 0, & \text{for} \quad T \leq T_c(\mu) \\ &\approx -A_0 t^{3/2}, & \text{for} \quad T \geq T_c(\mu) \end{aligned} \tag{12.8}$$

---

[v] As mentioned in Section 5, an interesting early paper using fermions to represent strings or interacting walkers is by de Gennes.[4]
[w] The generalization of this result to other values of d is considered in Refs. 43 and 44.

where $A_0 = 2(\sigma_1^3/3b_0^2)^{1/2}$. Thus the transition is critical but the exponents are *non*classical: in fact the specific heat diverges when T approaches $T_c(\mu)$ from above as $t^{-\alpha}$ with exponent

$$\alpha = \tfrac{1}{2} \tag{12.9}$$

However, there is *no divergence* on the commensurate side of the transition. The transverse correlation length in the incommensurate phase must be simply proportional to l and hence may be written as[2]

$$\xi_\perp \approx b_1/t^{\nu_\perp} \quad \text{with} \quad \nu_\perp = \bar{\beta} = \tfrac{1}{2} \tag{12.10}$$

and, thence, via the diffusive character of the walk as seen previously,

$$\xi_\| \approx c(l/b)^2 \approx c_1/t^{\nu_\|} \quad \text{with} \quad \nu_\| = 2\nu_\perp = 1 \tag{12.11}$$

These results are in agreement with the fermion type of analysis.[62] Evidently the anisotropic hyperscaling relation

$$2 - \alpha = \nu_\| + \nu_\perp \tag{12.12}$$

is satisfied. One can generalize these arguments by including an ordering field, h, which favors one type of commensurate domain: this is found to scale with an exponent $\Delta = 1\tfrac{1}{2}$.[2]

It is interesting that the result for the specific heat, including its characteristic asymmetry with no divergence below T_c, can be verified by exact calculations for what is, on the face of it, a completely different type of model. Consider the *dimer problem* in which "hard dimers" occupy the bonds of a lattice (and the two associated terminal sites). If a planar lattice is filled with dimers the partition function can be calculated exactly by Pfaffian methods.[64–66] In particular, Kasteleyn,[67] has considered a brick (or, equivalently, honeycomb) lattice, as illustrated in Figure 21, in which dimers on vertical or y bonds have different statistical weights or a chemical potential excess, say, $\Delta\mu$. This dimer model has a transition when $\Delta\mu/k_B T_c = \ln 2$. The low-temperature phase is "frozen," all dimers pointing "up," i.e., parallel to the y axis, and the specific heat vanishes; above T_c, however, horizontal dimers are present and the specific heat diverges as $t^{-\frac{1}{2}}$.[67]

Why do these results confirm the heuristic analysis leading to (12.8) and (12.9)? The answer is contained in Figure 21 which serves to demonstrate that horizontal dimers can appear *only* in nearest-neighbor strings that run,

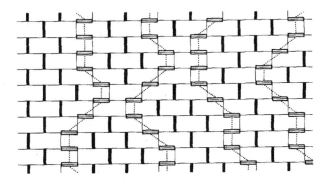

Figure 21. Dimer problem on a brick lattice in which vertical dimers (bold) have a different chemical potential than horizontal dimers. The dotted lines indicate that a configuration of horizontal dimers can always be regarded as composed of nonintersecting strings or walks running, on average, vertically.

on average, vertically through the lattice. An isolated string can be regarded as a domain wall in a $p \times 1$ commensurate phase with $p = 1$; above T_c the state of many walls clearly corresponds to the associated incommensurate phase. (Note that the heuristics gives $\alpha = \frac{1}{2}$ and $\bar{\beta} = \frac{1}{2}$ independently of p.) Once again, then, the simple picture of domain walls in two dimensions which interact through a $1/l^2$ potential reproduces the exact asymptotic behavior!

13. Dislocations and their Effects

The account just presented of the commensurate–incommensurate transition neglected completely the possibility that the array of domain walls might exhibit any type of topological defect formed by some of the walls merging. In a $p \times 1$ phase the elementary defects consist of *dislocations* (or *vortices*) at which p walls come together and terminate. Allowing for such dislocations, say, with an activity v, alters the picture of both the commensurate and the incommensurate phases. In the former case, as illustrated in Figure 22(a) for $p = 3$, heterodomain fluctuations arise which prefigure the nature of the new, incommensurate phase.[2,x] On the other hand, as illustrated in Figure 22(b), dislocations can appear in various ways in the incommensurate phase[42,63] and should be considered in giving a full account of its properties. Indeed, one must ask if the presence of disloca-

---

[x] It is instructive to compare Figure 22(a) with Figure 13 and to note that, as seems physically reasonable, we consider only one type of wall near the incommensurate phase transition, i.e., we suppose the transition occurs from a fully *wet* region of the commensurate phase.

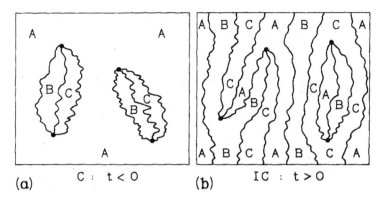

Figure 22. Representation of (a) fluctuations arising in a 3 × 1 commensurate phase close to the C–IC transition when dislocations are allowed and (b) configurations of domain walls in the corresponding incommensurate phase when dislocations are present.

tions for $v > 0$ might not change the nature of the singularities at the C–IC transition,[2] or even destroy the incommensurate phase altogether![42,63]

Physically, dislocations must always be allowed for even if, in practice, v turns out to be small. Ideally, however, one may consider $v = 0$; indeed, the brick lattice dimer model corresponds to this situation since one can easily check from Figure 21 that *no* dislocations can occur there. For $v = 0$ we may accept the previous theory of C–IC transition (as checked by the exact model results[40,67] and then ask[2] "Does the dislocation activity, v, represent a *relevant* or *irrelevant* perturbation of the transition in the renormalization group sense?"[11,68–71] More concretely, with t defined as in (12.2) the scaling form

$$f_s(t,v) \approx |t|^{3/2}\Upsilon_\pm(v/|t|^{\phi_p}) \tag{13.1}$$

should hold for small v. Here $\Upsilon_\pm(z)$ represents the scaling function for $t \to 0\pm$ while ϕ_p is the appropriate *crossover scaling exponent*.[11,68,70] If ϕ_p is negative v is irrelevant and only higher-order corrections to the transition behavior should arise; positive ϕ_p, on the other hand, implies that v is relevant which implies that the transition is destroyed or, at least, changed in character.

Now,[2] from (13.1) we have

$$f_0'' \equiv \left(\frac{\partial^2 f_s}{\partial v^2}\right)_{v=0} \approx \Upsilon_-''(0)|t|^{3/2-2\phi_p} \tag{13.2}$$

for $t < 0$. On the other hand,[2] if we expand the full partition function for the commensurate phase to quadratic order in v we see that f_0'' is proportional to the total partition function, Z_2^p for a single "bubble" of p strands terminated by two dislocations, as illustrated in Figure 22(a). But the partition function, $\Theta_p(x,y)$, for a bubble with one dislocation at $(0,0)$ and one at (x,y) can be described by the reunions of p walkers where, as before, the step weight σ represents the reduced surface tension for a wall parallel to the y axis, which, by (12.5), vanishes like $|t|$ when the transition is approached. Thus, for $t < 0$, we have

$$Z_2^p(T) = \int dx \int dy \Theta_p(x,y) \approx \sum_n e^{-p\sigma n} R_n^{(p)}$$
$$\approx C_p \sum_n e^{+p\sigma_1 in}/n^\psi \quad \text{with} \quad \psi = \tfrac{1}{2}(p^2 - 1) \quad (13.3)$$

where we have appealed to the walk result (4.16). The leading singular behavior follows as in (6.11) and is proportional to $|t|^{\psi-1}$ but with a factor $\ln|t|$ when ψ is an integer. Comparing with (13.2) finally yields the crossover exponent[2]

$$\phi_p = \tfrac{1}{4}(6 - p^2) \quad (13.4)$$

Now for $p = 1$ or 2 we see that ϕ_p is positive so that v is *relevant*. In fact, dislocations then destroy the incommensurate phase altogether as one may anticipate from the explicit results for Ising models which describe the cases $p = 1$ or 2: only a commensurate ordered phase and disordered fluid phases are seen.[2] Conversely, for $p \geq 3$ the dislocations are irrelevant, and the incommensurate phase should remain stable for $v > 0$. These conclusions agree with the analysis of Coppersmith et al.[42] and Villain and Bak[63] (who, however, primarily address the question of the stability of the incommensurate phase not too close to the transition itself).

Even though the dislocations are technically irrelevant for $p \geq 3$ one sees from (13.1) that they lead to correction-to-scaling factors of the form

$$[a_0^\pm + a_1^\pm |t|^{\theta_p} + a_2^\pm |t|^{2\theta_p} + \cdots] \quad \text{with} \quad \theta_p = \tfrac{1}{4}(p^2 - 6) \quad (13.5)$$

In the commensurate phase ($t < 0$), however, only even powers of $|t|^{\theta_p}$ appear, since dislocations must occur in bound pairs; furthermore, when $2\theta_p$ is integral a factor $\ln|t|$ enters. Since $2\theta_p = 1\tfrac{1}{2}$ and 5 for $p = 3$ and 4, respectively, the corrections are of relatively high order. Nevertheless, they should be visible for $p = 3$ since, as seen in the previous section, the leading singularities vanish identically on the commensurate side of the transition so that the "corrections" actually play the dominant role! Thus

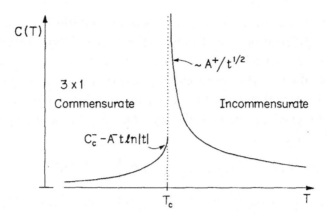

Figure 23. Sketch showing the variation of the specific heat through a 3×1 commensurate–incommensurate transition when dislocations can be generated. In the absence of dislocations the specific heat below T_c should be smooth and analytic.

one finds that the specific heat should exhibit[2] a $|t|\ln|t|$ behavior below T_c, which, as sketched in Figure 23, represents a cusp with vertical slope in contrast to the $|t|^{-1/2}$ behavior above T_c. Likewise, one can show[2] that the ordering susceptibility, which diverges like $|t|^{-3/2}$ in the incommensurate phase, should diverge as $\ln|t|^{-1}$ below the transition.

In principal, one might hope to observe the dislocation–dislocation correlation function, $C_2^p(x,y)$. In the commensurate phase near the transition this follows from the reunion distribution (4.12). If one scales the x and y coordinates in accord with (12.10) and (12.11) by writing

$$X = x/\xi_\perp \approx x|t|^{\frac{1}{2}}/b_1, \quad Y = y/\xi_\parallel \approx y|t|/c_1 \qquad (13.6)$$

one finds a scaling form

$$C_2^p(x,y) \approx c_p |t|^{\frac{1}{2}p^2} \Gamma_\pm(X,Y) \qquad (13.7)$$

where the walk distribution yields[2]

$$\Gamma_-(X,Y) = e^{-pR}/R^{\frac{1}{2}p^2} \quad \text{with} \quad R^2 = X^2 + Y^2 \qquad (13.8)$$

for $t < 0$ and $X^2 = O(Y)$. (The coefficient c_p is a nonuniversal amplitude.) Schultz, Halperin, and Henley[72] have used fermion methods for $t > 0$ to obtain

$$\Gamma_+(X,Y) \sim 1/R^{\frac{1}{2}p^2} \quad \text{as} \quad R \to \infty; \qquad (13.9)$$

but note that this applies only for $r \gg \xi_\parallel \sim l^2/b^2$. More generally, the behavior of the dislocation–dislocation correlation function in the incommensurate phase requires an analysis of domain wall configurations such as shown in Figure 22(b). This is beyond the present scope of our random walk formulation but one may hope that the approach can be extended to yield similarly transparent results for the reunions of walkers moving within a sea of other walkers!

14. More Than Two Dimensions? Lipid Membranes

The applications of random walks and their interactions which we have studied so far have been restricted to physical systems of two or fewer dimensions. To close our study let us ask if one can obtain interesting results by similar means for systems of higher dimensions? Two obvious lines of development present themselves. A domain wall separating d-dimensional bulk phases is a curve for $d = 2$, and hence representable as a trajectory of a walker, but a *surface* for $d = 3$. One would, thus, like to study self-avoiding surfaces and the interactions between such surfaces and a wall or between two or more surfaces which may not intersect one another. This is a difficult problem even if it is simplified, in analogy with our treatment for $d = 2$, by regarding one dimension as time-like and examining the motion of a self-avoiding chain which lies in a plane. In fact, some heuristic arguments applicable to the commensurate–incommensurate transition have been developed.[43,44] In addition, some foundations for a rigorous theory have been laid out recently.[73] However, no fully convincing explicit results seem available.

On the other hand, one may enquire after the properties of one-dimensional interacting strings or chains which are embedded in three or more dimensions. If, as illustrated in Figure 24(a), a three-dimensional system is uniaxial with the strings lying, on average, parallel to one (time-like) axis, they may be regarded as the trajectories of vicious walkers moving in a plane. A model for which this picture is relevant has recently been proposed by Izuyama and Akutsu,[74] who studied it using approximate methods. Following earlier work by Nagle,[75] who adapted Kasteleyn's dimers-on-a-brick-lattice[67] to describe interacting lipid and polymer molecules in two dimensions, Izuyama and Akutsu introduced a similar dimer model for bulk, three-dimensional lipid membranes which undergo a thermal melting transition. In their model, square lattice layers are stacked vertically and connected in brick-like fashion (generalizing Figure 21).[74] This lattice is filled with dimers but vertical dimers again

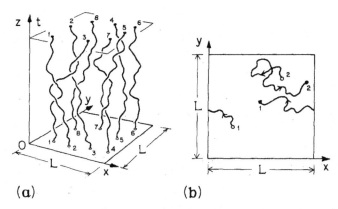

Figure 24. A configuration of nonintersecting strings which lie predominantly parallel to the z axis but wander in the (x, y) directions. The strings may be interpreted as dislocation lines in the Izuyama–Akutsu model of a lipid membrane. (b) Paths of two vicious walkers on a two-dimensional $L \times L$ torus. Although the walkers cannot *meet*, their trails may cross in the (x, y) plane; however, their trajectories in $(x, y, t \equiv z)$ space will not intersect.

enjoy an excess chemical potential, say, $\Delta\mu(> 0)$. As in two dimensions, it is easy to see that the low-temperature phase is completely frozen, with all dimers in a vertical orientation, but that a melting transition occurs when $\Delta\mu/k_B T = \ln 4$. Above T_c horizontal dimers appear but, as in Figure 21, their only possible configurations are such that they can be regarded as defining strings or "dislocation lines;" as suggested in Figure 24(a), these strings must run vertically through the lattice but may wander sideways in the layers provided they do not intersect one another. The nature of this membrane melting transition has not, as yet, been convincingly determined. Inspired by the success of simple heuristic ideas in two dimensions, we may, however, attack the problem as follows.[76]

Suppose the cross-section of the system transverse to the principal axis has dimensions $L \times L$, as indicated in Figure 24(a), and that there are N strings in the system which repel one another via a pair potential $W(r)$ depending on their mutual separations, r. If $l = L/N^{\frac{1}{2}}$ is the mean interstring spacing and the reduced tension of a single string is $\sigma \approx -\sigma_1 t$, with $t \propto (T - T_c)$, the free energy density, as in Section 12, may be expressed phenomenologically by

$$f(T;l) \approx \frac{1}{L^2}\left[N\sigma + jN\frac{W(l)}{k_B T}\right] \approx -\frac{\sigma_1 t}{l^2} + \frac{jW(l)}{l^2 k_B T} \quad (14.1)$$

where j is a geometric factor accounting for the mean coordination number and further neighbor interactions.

So far so good; but how are we to determine the interaction potential $W(r)$? This task may be approached in two steps:[76] first, in analogy with our arguments for walkers on a line, consider the case $N = 2$, illustrated in Figure 24(b). Two vicious walkers, who must not meet, move in two-dimensions within a square of side L. To simplify the problem let us take periodic boundary conditions, so that the "square" is really a torus, as suggested in the figure. To determine $W(r)$ we note that the mean separation of the walkers is $l = L/\sqrt{2}$ and ask for the incremental free energy of the system over the value 2σ for two widely separated strings ($L \to \infty$). The problem of two vicious walkers on a torus can be simplified by going to center-of-mass coordinates (as used to discuss two dissimilar walkers on a line): this yields one free walker on a torus and one "internal" walker who must not visit the origin of the torus. This latter problem can be solved exactly along the lines used in proving (2.9). With due care the requisite asymptotic analysis can be performed for large L and yields[76]

$$W(L/\sqrt{2}) \approx \pi b^2/L^2 \ln(L/b) \quad \text{for} \quad N = 2 \qquad (14.2)$$

Now this result, although exact, applies only for $N = 2$ walkers and $L \to \infty$, i.e., to a situation of asymptotically zero density; however, we really require $W(l)$ for small but nonzero overall density, $\rho = N/L^2 = 1/l^2$. In order to leap this gap we may introduce what is essentially a strong finite-size scaling ansatz,[57,77] namely, we suppose that (14.2) remains true for large but finite l, at least up to some proportionality factor, if L is replaced by $\sqrt{2}l$. Thereby we conclude that the potential varies for large l/b as

$$W(l) \approx \pi b_0^2/l^2 \ln(l/b_0) \qquad (14.3)$$

where $b_0 \simeq b/\sqrt{2}$.

If one accepts this reasoning, one can proceed to minimize $f(T;l)$ with respect to l and thence determine the critical behavior of the free energy. One finds that the specific heat above the transition varies as[76]

$$C(T) \propto \frac{\partial}{\partial T}\left(\frac{b_0^2}{l^2}\right) \approx A_0 \ln t^{-1} + \cdots . \qquad (14.4)$$

In other words, the transition is critical with $\alpha = 0$ and a logarithmically divergent specific heat![76]

More generally, if one carries through the analogous argument for a d-dimensional system, regarding d as a continuous variable, one finds $\alpha = \frac{1}{2}(3-d)$ for $1 < d < 3$. Note that this reproduces correctly the original result $\alpha = \frac{1}{2}$ for $d = 2$ [see (12.9)]. On the other hand, for all $d > 3$ one obtains $\alpha = 0$ but with $C(T)$ bounded as $T \to T_c+$, i.e., classical, second-order critical behavior. Thus $d = 3$ represents the upper *borderline dimension* for the lipid membrane melting transition. As such one is not surprised to see logarithmic factors appearing.[69,70] Nevertheless, some caution is in order since experience with systematic renormalization group analyses[69,70] indicates that nontrivial powers of logarithms may appear. Consequently, although (14.4) is most appealing and suggestive it cannot be regarded as definitive.

Clearly, then, the problem of interacting walkers moving in more than one dimension demands further study: the walkers' paths may tangle seriously but one may hope that diligence will be rewarded by the unravelling of some further problems!

Acknowledgments

Much of the research work reviewed above was performed in close association with David A. Huse whose collaboration was much appreciated. I am grateful to Daniel S. Fisher for rekindling my interests in random walks and alerting me to new developments. Some of my first lessons in random walk theory came from Cyril Domb, whose tutelage and early encouragement I would like to acknowledge particularly on this occasion. The stimulation provided by many other colleagues through personal contacts and communications and through the media of preprint and reprint has been important to me: I regret I cannot acknowledge all by name. Anthony M. Szpilka participated in some of the researches reported and kindly commented in detail on the manuscript. I am grateful to the Institut des Hautes Études Scientifiques for hospitality during a visit when the lecture on which this article is based was prepared. Last but not least, I am indebted to the National Science Foundation for ongoing support, provided in part through the Materials Science Center at Cornell University.

References

1. W. Feller, *An Introduction to Probability Theory and its Applications*, Vol. 1 (John Wiley & Sons, New York, 1950).
2. D. A. Huse and M. E. Fisher, *Phys. Rev. B* **29**, 239 (1984).
3. D. A. Huse, A. M. Szpilka and M. E. Fisher, *Physica* **121 A**, 363 (1983).

4. P.-G. de Gennes, *J. Chem. Phys.* **48**, 2257 (1968).
5. F. J. Dyson, *Commun. Math. Phys.* **21**, 269 (1971).
6. J. Fröhlich and T. Spencer, *Commun. Math. Phys.* **84**, 87 (1982).
7. M. E. Fisher, *Physics* **3**, 255 (1967).
8. M. E. Fisher and B. U. Felderhof, *Ann. Phys.* (N.Y.) **58**, 176, 217, 268, 281 (1970).
9. H. N. V. Temperley, *Phys. Rev.* **103**, 1 (1956).
10. M. E. Fisher, *Rept. Prog. Phys.* **30**, 615 (1967).
11. M. E. Fisher, *Proc. Nobel Symp.* **24**, *Collective Properties of Physical Systems*, B. Lundqvist and S. Lundqvist, eds. (Academic Press, New York, 1974), p. 16.
12. J. W. Cahn, *J. Chem. Phys.* **66**, 3667 (1977).
13. C. Ebner and W. F. Saam, *Phys. Rev. Lett.* **38**, 1486 (1977).
14. D. B. Abraham, *Phys. Rev. Lett.* **44**, 1165 (1980).
15. J. D. Weeks, in *Ordering in Strongly Fluctuating Condensed Matter Systems*, T. Riste, ed. (Plenum, New York, 1980), p. 293.
16. S. T. Chui and J. D. Weeks, *Phys. Rev. B* **23**, 2438 (1981).
17. T. W. Burkhardt, *J. Phys. A* **14**, L63 (1981).
18. J. M. J. van Leeuwen and H. J. Hilhorst, *Physica* **107A**, 319 (1981).
19. H. N. V. Temperley, *Proc. Cambridge Philos. Soc.* **48**, 683 (1952).
20. D. A. Huse and M. E. Fisher, *Phys. Rev. Lett.* **49**, 793 (1982).
21. M. R. Moldover and J. W. Cahn, *Science* **207**, 1073 (1980).
22. O'D. Kwon, D. Beaglehole, W. W. Webb, B. Widom, J. W. Schmidt, J. W. Cahn, M. R. Moldover and B. Stephenson, *Phys. Rev. Lett.* **48**, 185 (1982).
23. D. W. Pohl and W. I. Goldburg, *Phys. Rev. Lett.* **48**, 1111 (1982).
24. J. W. Schmidt and M. R. Moldover, *J. Chem. Phys.* **79**, 379 (1983).
25. J. S. Rowlinson and B. Widom, *Molecular Theory of Capillarity* (Clarendon Press, Oxford, 1982).
26. D. Poland and H. A. Scheraga, *Theory of Helix-Coil Transitions in Biopolymers* (Academic Press, New York, 1970).
27. C. L. Stevens and G. Felsenfeld, *Biopolymers* **2**, 293 (1964).
28. M. E. Fisher, *J. Chem. Phys.* **45**, 1469 (1966).
29. J. G. Dash and J. Ruvalds, eds., *Phase Transitions in Surface Films* (Plenum Press, New York, 1980).
30. R. Imbihl, R. J. Behm, K. Christmann, G. Ertl and T. Matsushima, *Surf. Sci.* **117**, 257 (1982).
31. M. Bretz, *Phys. Rev. Lett.* **38**, 501 (1977).
32. D. E. Moncton, P. W. Stephens, R. J. Birgeneau, P. M. Horn and G. S. Brown, *Phys. Rev. Lett.* **46**, 1533 (1981); **49**, 1679 (1982).
33. S. Alexander, *Phys. Lett.* **54 A**, 353 (1975).
34. E. Domany, M. Schick, J. S. Walker and R. B. Griffiths, *Phys. Rev. B* **18**, 2209 (1978).
35. R. J. Baxter, *J. Stat. Phys.* **26**, 427 (1981).
36. R. J. Baxter and P. A. Pearce, *J. Phys. A* **15**, 897 (1982); **16**, 2239 (1983).
37. M. Stone, *Phys. Rev. D* **23**, 1862 (1981).
38. S. Ostlund, *Phys. Rev. B* **24**, 398 (1981); **23**, 2235 (1981).
39. D. A. Huse, *Phys. Rev. B* **24**, 5180 (1981).
40. V. L. Pokrovsky and A. L. Talapov, *Zh. Eksp. Teor. Fiz.* **78**, 269 (1980) [*Sov. Phys. JETP* **51**, 134 (1980)].
41. J. Villain and M. Gordon, *J. Phys. C* **13**, 3117 (1980); F. D. M. Haldane and J. Villain, *J. Phys. (Paris)* **42**, 1673 (1981).
42. S. N. Coppersmith, D. S. Fisher, B. I. Halperin, P. A. Lee and W. F. Brinkman, *Phys. Rev. Lett.* **46**, 549 (1981); *Phys. Rev. B* **25**, 349 (1982).

43. M. E. Fisher and D. S. Fisher, *Phys. Rev. B* **25**, 3192 (1982).
44. T. Nattermann, *J. Phys. (Paris)* **43**, 631 (1982); *J. Phys. C* **16**, 4125 (1983).
45. R. Pandit and M. Wortis, *Phys. Rev. B* **25**, 3226 (1982).
46. H. Nakanishi and M. E. Fisher, *Phys. Rev. Lett.* **49**, 1565 (1982).
47. D. B. Abraham and E. R. Smith, *Phys. Rev. B* **26**, 1480 (1982).
48. W. J. Camp and M. E. Fisher, *Phys. Rev. Lett.* **26**, 73 (1971); *Phys. Rev. B* **5**, 946 (1972).
49. T. T. Wu, *Phys. Rev.* **149**, 380 (1966).
50. L. P. Kadanoff, *Nuovo Cimento* **44B**, 276 (1966).
51. R. Hartwig and M. E. Fisher, *Adv. Chem. Phys.* **15**, 333 (1969).
52. B. M. McCoy and T. T. Wu, *The Two-dimensional Ising Model* (Harvard University Press, Cambridge, Massachusetts, 1973).
53. H. B. Tarko and M. E. Fisher, *Phys. Rev. B* **11**, 1217 (1975).
54. W. J. Camp, *Phys. Rev. B* **6**, 960 (1972).
55. M. E. Fisher and W. J. Camp, *Phys. Rev. Lett.* **26**, 565 (1971).
56. B. M. McCoy and T. T. Wu, *Phys. Rev. D* **18**, 1259, 1243, 1253 (1978).
57. D. B. Abraham, *Phys. Rev. Lett.* **50**, 291 (1983).
58. M. E. Fisher and A. E. Ferdinand, *Phys. Rev. Lett.* **19**, 169 (1967).
59. D. B. Abraham, G. Gallavotti and A. Martin-Löf, *Physica* **65**, 73 (1973).
60. P. Bak, *Rept. Prog. Phys.* **45**, 587 (1982).
61. P. Bak and V. J. Emery, *Phys. Rev. Lett.* **36**, 978 (1976).
62. H. J. Schulz, *Phys. Rev. B* **22**, 5274 (1980); *Phys. Rev. Lett.* **46**, 1685 (1981).
63. J. Villain and P. Bak, *J. Phys. (Paris)* **42**, 657 (1981).
64. P. W. Kasteleyn, *Physica* **27**, 1209 (1961).
65. H. N. V. Temperley and M. E. Fisher, *Philos. Mag.* **6**, 1061 (1961).
66. M. E. Fisher, *Phys. Rev.* **124**, 1664 (1961).
67. P. W. Kasteleyn, *J. Math. Phys.* **4**, 287 (1964).
68. F. J. Wegner, *Phys. Rev. B* **5**, 4529 (1972).
69. K. G. Wilson and J. Kogut, *Phys. Rept.* **12C**, 75 (1974).
70. M. E. Fisher, *Rev. Mod. Phys.* **46**, 597 (1974).
71. K. G. Wilson, *Rev. Mod. Phys.* **47**, 773 (1975); **55**, 583 (1983).
72. H. J. Schulz, B. I. Halperin and C. L. Henley, *Phys. Rev. B* **26**, 3797 (1982).
73. B. Durhuus, J. Frohlich and T. Jonsson (preprint, 1983).
74. T. Izuyama and Y. Akutsu, *J. Phys. Soc. Japan* **51**, 50, 730 (1982).
75. J. F. Nagle, *Ann. Rev. Phys. Chem.* **31**, 157 (1980); *Proc. R. Soc. London Ser. A* **337**, 569 (1974).
76. S. M. Bhattacharjee, J. F. Nagle, D. A. Huse and M. E. Fisher, *J. Stat. Phys.* **32**, 361 (1983).
77. M. E. Fisher, in *Critical Phenomena*, Proceedings of the 1970 Enrico Fermi International School of Physics, Course No. 51, Varenna, M. S. Green, ed. (Academic Press, New York, 1971), p. 1.

5

Condensed Matter Physics: Does Quantum Mechanics Matter?*

Michael E. Fisher

Cornell University[†]

HERMAN Feshbach, the organizer of this Symposium in honor of Niels Bohr, asked me, in his original invitation, for a review of the present state of condensed matter physics, with emphasis on major unsolved problems and comments on any overlap with Bohr's ideas regarding the fundamentals of quantum mechanics. That is surely a difficult assignment and, indeed, goes well beyond what is attempted here; nevertheless, I will take the liberty of raising one issue of a philosophical or metaphysical flavor.

Does Quantum Mechanics Matter?

Condensed matter physics is a multifaceted subject — a beautiful diamond. The aim of this lecture is, in part, to hold this diamond up to view and to turn it around, alas rather rapidly, so that a few flashes may catch the eye. In doing this I will, however, concurrently address the question, "Does quantum mechanics matter for condensed matter physics?" This is not meant facetiously — it is not a rhetorical question. Some of my colleagues at Cornell were, I believe, a bit horrified on learning that I could consider this question seriously or even pose it! One of them indeed, said, "That seems

---

*Reprinted with permission from *"Niels Bohr: Physics and the World,"* Eds. H. Feshbach, T. Matsui and A. Oleson (Harwood Academic Publisher, Chur, 1988) pp. 65–115.
[†]Now at Institute for Physical Science and Technology, University of Maryland, College Park, Maryland 20742.

to me as crazy as discussing planetary motion and asking 'Does Newtonian mechanics matter?'" However, I want to suggest that the question for condensed matter physics is not really quite like that. Consider the analogous but hopefully less controversial question, "Doe human biology matter to sociology?" Obviously there *is* some connection. But suppose we go further towards more fundamental levels of science, asking the same question:

> *Sociology* — Does *human biology* matter?
> Does *cell biology* matter?
> Does *biochemistry* matter?
> Does *molecular physics* matter?
> Does *nuclear physics* matter?
> Does *quantum chromodynamics* matter?

Surely no one can seriously maintain that quantum chromodynamics matters to sociology!

For every area in science, there are other areas of direct relevance, further areas of less relevance, and also relevant interconnections between what may properly be regarded as *different levels* of discourse. I believe that this remark reflects, to some extent, Bohr's idea of *complementarity*; the issue is, however, interesting and subtle. I would like to explain my view of it in further detail.

To a very good degree of approximation, many disciplines in the sciences are "essentially" self-contained in the sense that there is a natural language which "fits" the phenomena and a corresponding "calculus", which, according to the case at hand, may be more-or-less mathematical. It is perhaps easiest to grasp this within the discipline of physics itself. It has long been granted, for instance, that thermodynamics has a wide domain of validity within which one does not have to worry about the atomic constitution of matter or about statistical mechanics. The situation as regards most applications of Maxwell's equations is really very similar, i.e., they are used to discuss electrical circuits on appropriate scales, to design antennae, etc., without worrying about their more fundamental basis in quantum electrodynamics or condensed matter physics (the latter as regards any magnetic, dielectric or conducting properties). This situation seems even clearer as regards the discipline of chemistry where, unfortunately, a mathematically complete calculus is, in practise, absent in most cases: Schrödinger's equation really does have *almost no* relevance to what one actually does in vast areas of chemistry.

The issue is truly the same in condensed matter physics. The basic problem which underlies the subject is to understand the many, varied manifestations of ordinary matter in its condensed states and to elucidate the ways in which the properties of the "units" affect the overall, many-variable systems. In that enterprise it has become increasingly clear that an important role is and, indeed, should be played by various special "models." It is, nowadays, a triviality that the behavior of dilute gases is universal and in most regards quite independent of the chemical constitution of the molecules or atoms. We no longer regard the universal ideal gas laws as peculiar or mysterious. One should not, therefore, be so surprised that other theoretical models that have been abstracted from complicated physical systems, like the basic Ising and Heisenberg models originally devised for magnetism, take on a life of their own quite regardless of the underlying atomic and molecular physics (for which quantum mechanics does matter).

This situation serves to illustrate another point, namely, the picture of *levels* or *strata* in science. One might, as a theorist, argue that more attention should be paid to the connection between models, like the ideal gas or the Ising and Heisenberg models, and fundamental physical principles as embodied in quantum mechanics. The various models so important in modern condensed matter physics, often appear to the outsider as inspired guesses, whose success is judged solely by their ability to explain the phenomena which motivated them. Would it not be an equally great success for theory to *derive* any of these models from atomic theory? Is that not where quantum mechanics enters modern condensed matter physics?

Certainly, the connection between levels of description deserves clarification in all science. However, the mere fact that such an 'artificial,' 'nonfundamental' model as the Ising model provides insight into a wide range of *contrasting* examples of condensed matter such as anisotropic ferromagnets, gas-liquid condensation, binary alloys, structural phase transitions, etc., shows that the question of "connecting the models with fundamental principles" is *not* a very relevant issue or central enterprise. I stress again, that in any science worth the name, the important point is to gain *understanding*. The language in which the understanding is best expressed cannot be dictated ahead of time, but must, rather, be determined by the subject as it develops. Accordingly, it really would *not* be a "great success" to derive the Ising model from atomic theory or quantum mechanics. Indeed, for which physical system would one be "deriving" it? For a ferromagnet where the electronic spins are crucial? For a liquid-gas system where the atoms or molecules are the basic unit? For an alloy where the metallic ions are the "variables"? Or for a ferroelectric material in which the rotation of some large unit of correlated atoms in a lattice is the basic dynamical variable? Of course, in each of these cases, one *does* want to understand what the important variables actually are; but, in many ways, it is more important to understand how these variables interact together and what their "cooperative" results will be.

On the other hand, it is certain that all the sciences are linked together. Furthermore, some disciplines are more "fundamental" than others. However, it is not so easy to identify "fundamental principles" *per se*. Does Schrödinger's equation constitute a truly "fundamental principle"? Surely not! It must merely be some sort of low-energy approximation to a more basic theory. The same goes for quantum chromodynamics: evidently that is also only a limiting theory! And so on. Personally, I find the elucidation of the connections between various disciplinary levels a topic of great interest. Typically, delicate matters of understanding appropriate asymptotic limits, both physical and mathematical, are entailed. The issues involved are often very subtle; nevertheless, one must admit, they seldom add significantly to one's understanding of either the "fundamental" starting theory or of the target discipline. Sad but true! Every now and then, when one is lucky, one can uncover some special corner of the "higher" discipline where more basic or fundamental principles reveal some limitations that have not been suspected before or where "natural assumptions" can be prove false; but that is a bonus not to be expected generally. Nevertheles, I am myself interested in and will continue to encourage those braver souls who really try to understand the interconnections between different disciplines, or subdisciplines, or between layers within a discipline. Many, however, who would claim to be working from fundamental principles do not, in my view, really approach the task appropriately. As indicated, the aim should be, for instance, to justify the Ising model for some specific system on the basis of, say, Schrödinger's equation. To often, however, individual theorists have, under what one may fairly characterize as the prevailing absolutist philosophy in physics, attempted to derive *all* the properties of a real system from "fundamental principles," often using gross approximations, rather than attempting to justify the intermediate language that one has discovered is appropriate.

Clearly there are elements of Bohr's "complementarity" in these points of view. However, I think it is a mistake to view complementarity as merely a two-terminal black box. In real science there are many layers and interleavings, some of which we may hope to see disappear, but many of which will remain, both complementary and supplementary. In a word, a fully reductionist philosophy, while tenable purely as philosophy, is the wrong way to practice real science!

The Task of Condensed Matter Physics

With that over-lengthy preamble let me return to the main subject. My aim is, first of all, to explain what I think condensed matter physics does, or should be doing, and what defines condensed matter physics, and thence to approach the question, "Does quantum mechanics matter?"

The key fact about condensed matter physics is the existence of many states of matter — and we keep discovering more! In fact, one of the things that makes condensed matter physics exciting is that every year or two, something new turns up.[a] We will consider some of these new things.

So the first question is, "What are the states of matter?" Can we elucidate them? When we have discovered them, can we characterize them? What is their nature? More specifically, this usually means determining the type of order, and the particular spatial and temporal correlations that build up. Then, we may ask, "How do the various states transform into one another? How are they interrelated?" These latter questions demonstrate that the study of phase transitions, my own love, is one of the fundamental aspects of condensed matter physics.

The Stability of Matter

Now there is one most basic way in which quantum mechanics truly matters. We would not even have condensed matter if we did not have quantum mechanics! But even here it is worth enquiring as to which aspects of quantum mechanics are actually important. The first issue to appreciate

---

[a] Since this lecture was presented verbally the startling discovery of "high T_c" superconductors,[1] which lose their resistance already above 90°K, has borne out this assertion!

is that the very term "condensed matter" implies a large number of elementary units, atoms, electrons and nuclei, spins, etc., say N in number. Then the so-called "thermodynamic limit" in statistical mechanics, namely $N \to \infty$, is an important and basic idealization. (Nevertheless, both theoretically and experimentally, one is increasingly able to study large but still finite systems. And 'large', amusingly enough, sometimes turns out to be not much bigger than 10, 20 or 30. From the viewpoint of interconnections with wave mechanics and the more basic theories, this is quite a puzzle: Why is 'large' so small?)

There is a second, related issue which underpins all statistical mechanics. Statistical mechanics relates microscopic mechanics to macroscopic thermodynamics. Thermodynamics postulates — or embodies the deep physical insight — that there is a bulk free energy function which is proportional to the amount of matter or, one says, is an "extensive" quantity. At low temperatures the free energy becomes just the energy. Statistical mechanics must yield, or explain, the extensive nature of the free energy and energy and thus that matter is *stable*. More concretely if $E_0(N)$ is the ground state energy of a system of size N, say N electrons and N protons, stability means that

$$E_0(N) \geq -\varepsilon N \quad \text{for all } N, \tag{1}$$

where ε, a finite constant independent of N, is essentially the limiting binding energy per degree of freedom (one might also well work on a per unit volume basis).

Now one of the fundamental things Bohr did with his theory of the atom was to explain why it was stable — why it did not collapse, the electrons falling into the nucleus to yield a ground state energy unbounded below. However, the finite energy of the Bohr atom, which is naturally measured in terms of the Rydberg[b]

$$\varepsilon_0 = me^4/2\hbar^2, \tag{2}$$

is *not* enough to explain why, granted Bohr mechanics and Coulomb interactions, bulk matter is stable with, as one observes, ε not so different from ε_0.

Indeed, some years back, F. J. Dyson and A. Lenard[2] proved the converse! Suppose one has the standard Schrödinger wave equation — kinetic energy in the usual form — and Coulomb forces: What is the ground state

---
[b]Here m and e denote, as usual, the mass and charge of an electron.

energy of a neutral system of N such particles? The answer is provided by the *upper* bound

$$E_0(N) < -\varepsilon_1 N^{7/5} \quad \text{for some fixed} \quad \varepsilon_1 > 0; \tag{3}$$

note the exponent $\frac{7}{5}$ which exceeds unity. So although one single hydrogen atom is stable, according to Bohr, many hydrogen atoms would seem to collapse; and, indeed, to collapse quite catastrophically!

It should be mentioned here, that we leave gravity aside. We will choose to work on scales where it is unimportant — of course, all real matter will eventually collapse into a black hole if N gets large enough!

But what then is the answer to the puzzle posed by Eq. (2)? Why is matter stable? The missing piece, is in fact, provided by Fermi-Dirac statistics. Dyson and Lenard in their very beautiful paper,[2] proved that if the negative particles, i.e., the 'electrons', are fermions, and if there are no more than q species of them (with masses $m_j^- \leq m$ and charges $|e_j| \leq e$ for $j = 1, 2, \ldots, q$), then there is, in fact, a lower bound of the required form, namely,

$$E_0(N) > -\varepsilon_2 q^{2/3} N, \tag{4}$$

with ε_2 proportional to ε_0. Thus wave mechanics *with* Fermi statistics is enough to stabilize matter.[c]

This result echoes a theme running through modern condensed matter physics: "Fermions are vital." They are hard to deal with; we do not really understand them well yet despite a half-century's acquaintance. But we know they are vital, and in this sense quantum mechanics certainly matters.

On the other hand, there is "another end" to the question of the existence of the free energy for Coulomb forces — namely, what happens to the N-particle interactions at long distances. One must prove that the forces are not too repulsive. More concretely the free energy, $F(N)$, must satisfy a "tempering condition",[3] $F(N) < \varepsilon_\infty N$. How does this come about? Coulomb forces can be strongly repulsive at long distances; but Coulomb forces screen! This is a crucially important property, but one that is completely classical! The delicate proof[4] that screening tempers the repulsions uses only the ideas of Newton, specifically the fact that when there are

---

[c]In retrospect one recalls that the ground state of the unrestricted many-particle Schrödinger wave equation will have a nodeless wave function symmetric under the interchange of particle coordinates. Such a wave function is acceptable for bosons but, of course, is unacceptable for fermions which must, therefore, have a higher energy.

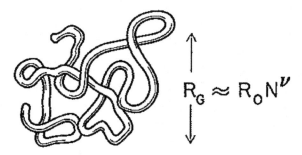

Figure 1. Depiction of a polymer molecule of N units and its mean size or "radius of gyration" R_G.

spherical distributions of charges, only the total charge outside a given region matters — a reflection of the fact that Coulomb's law is a true inverse square law (or, more appositely, an inverse $(d-1)$-power law where d is the spatial dimensionality).

Now I will discuss several interesting forms of matter, describing in each case the experimental situation and then giving some idea of our theoretical understanding — and concurrently pointing out the degree to which quantum mechanics enters.

Polymeric Matter

The first sort of condensed matter I would like to consider is polymeric matter. Polymers are long molecules with some finite thickness: see Figure 1, the importance of their thickness has been stressed by many researchers, most notably P. J. Flory, and is realized to be a crucial feature determining their behavior. It means that a polymer chain of length N units (or monomers) cannot be thought of as a free random walk: such a walk may revisit any region of space. But a real polymer chain interacts with itself and different segments cannot approach too closely. This poses the so-called *excluded volume problem*.

The mean size of a polymer molecule is properly described by its *radius of gyration*, R_G, which is the square root of the average squared distance between all the monomers that compose it. Because of the excluded volume, R_G grows in a nontrivial way with polymer length, namely, according to the law

$$R_G \approx R_0 N^\nu, \tag{5}$$

where the exponent, ν, differs from the value $\nu_0 = \frac{1}{2}$ which characterizes a free random walk. Indeed, we know that $\nu > \frac{1}{2}$ (for spatial dimensionalities $d < 4$) so that the rate of growth exceeds that for an unimpeded random walk; in fact, one has $\nu \simeq 0.588$ in $d = 3$ dimensions.

Now, thanks to renormalization group ideas and other theoretical ways of looking at the problem, the exponent ν is fairly well understood: but other challenging problems remain beyond the size of a single isolated molecule. One might say, "Well, aren't polymers for chemists?" The answer is that many of their properties have proved too difficult for traditionally trained chemists! Consider a solution of polymer molecules of overall concentration c. It has been the condensed matter physicists who have finally had to think hard about this problem and to solve it. To give a feel for the theory, consider the simplest thermodynamic quantity of such a solution, namely, the osmotic pressure, π, regarded as a function of the concentration, c. This must also depend strongly on the parameter, N, the length of the polymer chains: but how? It turns out that one can construct a universal theory — actually a "scaling theory"[5] — for the osmotic pressure and other solution quantities.

To be more explicit, suppose one has n polymeric molecules in a total volume V so that the concentration is

$$c = n/V = \rho_1/N, \qquad (6)$$

where ρ_1 is the overall density of the *monomers*. Consider the *reduced* osmotic pressure

$$\tilde{\pi} = \pi(c, N)/ck_BT; \qquad (7)$$

in the limit of extreme dilution, $c \to 0$, one expects this to approach the ideal gas value, $\tilde{\pi} = 1$; but interactions between different molecules must change $\tilde{\pi}$ drastically already at low polymer concentrations, especially when N is large. One concludes theoretically[6] that the combined dependence on c and $N \gg 1$ can be written

$$\tilde{\pi} = \frac{\pi(c, N)}{ck_BT} \approx Z(X), \qquad (8)$$

where $Z(X)$ is a *universal function* of the *scaled concentration*

$$X = c/c^* \quad \text{with} \quad c^* = c_0/N^{d\nu}. \qquad (9)$$

Note the appearance here of the exponent ν which describes the excluded volume of a single chain via Eq. (5). The parameter c_0 depends only on the nature of the individual monomers.

At low concentrations one can expand in powers of X and so finds

$$\tilde{\pi} = 1 + B_2(N)c + O(c^2) \tag{10}$$

where the second virial coefficient obeys $B_2 = b_2 N^{d\nu}/c_0$, in which b_2 is a numerical coefficient. As anticipated, B_2 depends strongly on N. However, for high concentrations the theory shows[6] that

$$Z(X) \approx Z_\infty X^{1/(d\nu-1)} \quad (X \to \infty) \tag{11}$$

where Z_∞ is a constant. This, in turn, implies the surprising result

$$\pi/k_B T \approx C_0 \rho_1^\delta \quad \text{with} \quad \delta = d\nu/(d\nu - 1), \tag{12}$$

in which C_0 depends only on the properties of a monomer. Thus at high concentrations the osmotic pressure becomes *independent* of the length, N, of the polymer chains! Furthermore π then increases with monomer density according to a universal power law with exponent $\delta \simeq 2.31$ (for $d = 3$).

How well does all this theory work? As an example, regard Figure 2: this shows the osmotic compressibility, $\partial \pi/\partial c$ vs. X, which, by Eq. (8), just depends on the scaling function Z and its derivative. The scaled concentration, $X = c/c^*(N)$, runs over four decades, the compressibility over two — and the data come from solutions in different solvents and for a range of N. The only physical parameter used to draw the theoretical line in Figure 2 is c_0 which is fitted to the second virial coefficient, B_2. Equally good agreement with theory[6] is found for the pressure itself, for the monomer density correlation length, etc.

Thus we see in polymeric matter new, subtle and universal behavior which we have succeeded in understanding theoretically. But quantum mechanics has had essentially nothing to say about the problem! Indeed, one feels that if some of the giants of the past, like Boltzmann or Gibbs or Rayleigh, were able to rejoin us today, they would be able to engage in research at the cutting edges of condensed matter physics without taking time off to study quantum mechanics first! No doubt they would *want* to learn about quantum mechanics — who would not? But while learning quantum mechanics, they could all be doing front-line research using their knowledge of classical physics, Newtonian mechanics, statistical mechanics and electrodynamics, and their mathematical and modelling skills.

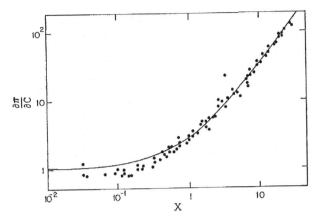

Figure 2. Logarithmic plot of the osmotic compressibility, $\partial\pi/\partial c$, versus the scaled concentration, $X = c/c^*$, for polystyrene solutions in toluene and in methylethylketone verifying the universal scaling form (solid line) with asymptotic slope $1/(d\nu - 1) \simeq 1.31$. [After Y. Oono, Ref. 6.]

Magnetic Matter

Let us now consider magnetic matter, which will yield further insights into modern condensed matter physics. The simplest example of an interesting magnetic material is a ferromagnet, something that can point north and enable us to navigate. Ferromagnetism, of course, ultimately rests on the fact that electrons have spin. An electronic spin is governed by quantal laws, so quantum mechanics is surely relevant. However, we must enquire further.

Historically, the first microscopic model of magnetism that was addressed theoretically was the so-called Ising model: here the individual spins are supposed just to point 'up' or 'down' — the simplest possibility. More concretely, the spins are pictured as sitting at the sites of a space lattice so that at site i, a classical spin variable, s_i, takes on only the two values $s_i = \pm 1$.

Now the fundamental question about the model was "At a given temperature, will all the spins tend to point the same way?" In other words, "Is there any ferromagnetism?" At high enough temperatures the spins will surely be disordered (yielding a paramagnet). But suppose one couples nearest-neighbor spins on the lattice together with an interaction energy $-Js_is_j$ ($J > 0$), so there is a tendency to align;[d] what then happens? This

---

[d]That is, the pair configurations $s_i = s_j = 1$ and $s_i = s_j = -1$ have energy $-J$, while the configurations in which $s_i = -s_j$ have the higher energy $+J$.

was analyzed sixty years ago by E. Ising in his thesis. In an exact calculation he found there was *no ferromagnetism* for a linear chain of spins, i.e., for a one-dimensional system. There was *no* transition from paramagnetism. One could *not* describe real ferromagnets by such a model!

Now one can argue (and, as far as I understand the history, I believe it was argued): "Well, this Ising model is so crude, it does such injustice to the true quantum-mechanical state of affairs, that it is hardly surprising that it fails to describe ferromagnetism. One should improve matters by using proper quantum-mechanical spin variables, namely, Heisenberg spins. These can point in any direction; they are vectors, $\mathbf{S} = \frac{1}{2}\hbar(\sigma_x, \sigma_y, \sigma_z)$; and, furthermore, they are operators, σ_x, σ_y and σ_z obeying the Pauli commutation relations."

Accordingly, let us examine the Heisenberg model of ferromagnetism with couplings $-J\mathbf{S}_{ij} \cdot \mathbf{S}_j$. It turns out, however, that the corresponding chain of Heisenberg spins (again a one-dimensional system) also has no phase transition! It never becomes ferromagnetic while $T > 0$. So quantum mechanics was no help on this problem. Indeed, it turns out to have been an actual hinderance.

We know now that it is the *one-dimensionality* of the spin chain, *not* classical mechanics, that was the reason why Ising found no phase transition in his model. Neither for Ising spins nor for Heisenberg spins does one have spontaneous magnetization (or spontaneous "order") in one dimension.[7] But for Heisenberg spins matters are worse: there is even no ferromagnetism in two dimensions.[7] On the other hand for Ising spins — not withstanding that they represent crude approximations — there is ferromagnetism in two dimensions; and this accords with the observation of ferromagnetism even in films of a few monolayers thickness and in other isolated magnetic layers.

Finally, in three dimensions there is ferromagnetism for both Heisenberg and Ising spins. So, again, it is not quantum mechanics that determines whether spontaneous magnetization occurs. Well then, what does? The answer lies in the interplay of entropy, fluctuations, and symmetry. What was "wrong" with the Heisenberg model was its symmetry. Not all real magnets have the full rotational spin symmetry of the Heisenberg model. That symmetry is broken, at the microscopic level, by further interactions, some of which turn on the subtleties of quantum mechanics. Ising spins

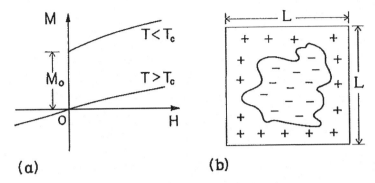

Figure 3. (a) Plot of magnetization, M, vs. magnetic field, H, at fixed temperature, T, for a ferromagnetic material, illustrating the existence of a spontaneous magnetization, M_0, below the Curie point, T_c. (b) Example of a counter-domain or droplet of overturned spins (of linear dimensions L) in a ferromagnetic domain of 'up' or + spins.

embody only a discrete, 'up-down' reflection symmetry: but for condensed matter physics, that is not the prime issue.

The Importance of Dimensionality

The real issue underlying the existence of ferromagnetic order was pointed out by R. E. Peierls. Let me remind you of the arguments that Peierls and L. D. Landau developed which give us an understanding of what matters as regards such a phase transition. Consider Figure 3(a): at high temperatures, any magnetic material has a smooth magnetization curve, M versus H, which does nothing very interesting; but at low enough temperatures — specifically below the ferromagnetic Curie temperature or critical point, T_c — a jump, measured by the spontaneous magnetization, $M_0(T)$, appears at zero field. What allows or prevents such a spontaneous magnetization?

The way to understanding is found by considering the thermodynamic stability of a counter-domain of, say, 'down' or '$-$' spins embedded in a ferromagnetic domain of 'up' or '+' spins, as illustrated in Figure 3(b). The spins around the outer boundary are held fixed, pointing 'up', and we ask whether overturning a cluster or droplet of spins in the middle yields a state of lower or higher free energy; in other words, is the ferromagnetic 'up' state stable or not? The change in free energy for a counter-domain with linear dimensions of order L may be estimated, neglecting all constants of portionality, as follows:

$$\Delta F_L = \Delta E - T\Delta S,$$
(i) $\sim M_0 H L^d$,   from the bulk change of magnetic energy,
(ii) $+JL^{d-1}$,   the surface/interface energy of the droplet,
(iii) $-k_B T L^{d-1}$,   the corresponding surface entropy,
(iv) $-k_B T \ln L^d$,   the positional entropy of the droplet center,
(13)

where d is the dimensionality, J the strength of the ferromagnetic couplings, M_0 the magnetization and H the external field. In zero field we can ignore the first term (i). What Peierls showed then matters is the free energy associated with the droplet surface or interface. There is an energetic term (ii) and also an entropic term (iii). In leading approximation both are proportional to the size (or 'area') of the droplet surface. Finally, there is also a positional entropy, (iv), since the counter-domain can occupy many distinct spatial locations.

Now, if the ferromagnetic state is to be stable, the free energy change, ΔF_L must be positive for all L however large: the presence of a large counter-domain (which could reduce the mean magnetization to a small value) must not lower the free energy. By comparing terms, we see that one first needs $T < J/k_B$, so that the temperature must be sufficiently low; secondly, one must have $L^{d-1} > \ln L^d$, which implies $d > 1$. The spatial dimensionality, d, is thus crucial. If $d = 1$ the last term, (iv), dominates and the spontaneous magnetization is destroyed; for $d > 1$ an Ising-like system will display magnetic order for $T < T_c$ (with $T_c > 0$).

To make these arguments rigorous one must allow for "bubbles" within the droplets and for bubbles within the bubbles and so on. The full analysis was first accomplished by R. B. Griffiths[8] but the physical basis of his rigorous proof is just the principle of control by the interface identified by Peierls.

If one looks similarly at a Heisenberg system, it turns out[7(c)] that, because of the continuous symmetry, there is no sharp interface bounding an overturned droplet. Rather, there is a diffuse transition region — an infinitely thick Bloch wall: that reduces the corresponding free energy increment by a factor of L. Consequently, one then needs $L^{d-2} > \ln L^d$ which pushes up the borderline dimensionality by 1, so that $d > 2$ is required for spontaneous magnetization. Evidently the only feature of the quantum

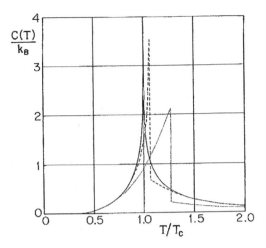

Figure 4. The exact specific heat of the square lattice Ising model (solid curve) and two approximate calculations (dashed and dotted plots). [After L. Onsager, Ref. 9.]

mechanics of Heisenberg spins used is the existence of the full rotational symmetry; but that enters equally in the classical limit ($\hbar \to 0$ or $S \to \infty$).

Criticality

Now the actual nature of the transition to spontaneous order can be studied in much greater detail. The two-dimensional Ising model was solved in a wonderful manner by Lars Onsager, now over forty years ago.[9] He discovered that the specific heat should have a divergent singularity of characteristic logarithmic form, namely,

$$C(T) = A \ln |t| + B + \cdots , \qquad (14)$$

where $t = (T - T_c)/T_c$ measures the deviation from the critical point T_c. Figure 4, from Onsager's original paper,[9] shows the exact form of $C(T)$ for the square lattice compared with the best that approximate theories could then do. One sees that the approximations are not only quantitatively wrong; they are *qualitatively wrong* as well, the functional form being simply incorrect — a somewhat shocking outcome!

Now experimentalists these days can perform most precise and delicate thermodynamic measurements. Figure 5 shows the corresponding curve for a real two-dimensional magnetic material.[10] Actually, it is a layered antiferromagnetic crystal, Rb_2CoF_4 (but Onsager's results apply equally to

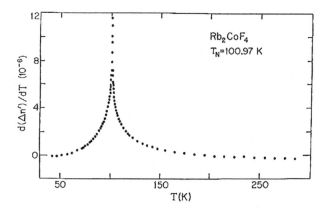

Figure 5. The specific heat of the layered antiferromagnet Rb_2CoF_4 (as measured via the linear birefringence). [After P. Nordblad et al., Ref. 10.]

the square lattice antiferromagnet, with $J < 0$). The real electronic spins act like Ising spins here because there are anisotropic interactions that keep them aligned predominantly parallel (or antiparallel) to a crystalline axis. The interactions between layers are very weak (for special structural reasons) and matter only for T extremely close to T_c. The experimental results in Figure 5 can be superimposed upon the Onsager solution of Figure 4 with essentially no adjustments! So again it is the two dimensions and Ising-like symmetry that matters; quantum mechanics does not.

Now there is yet another great discovery by Onsager for two-dimensional Ising models. This concerns the spontaneous magnetization, $M_0(T)$, and, by analogy, the liquid-gas coexistence curve, $\Delta\rho(T)$, for fluids: these both vanish when $T \to T_c-$ as

$$M_0(T), \quad \Delta\rho(T) \approx B|t|^\beta, \quad t = (T-T_c)/T_c, \qquad (15)$$

where the exponent takes the universal value $\beta = \frac{1}{8}$ for $d=2$ dimensions. For a long time we did not have any well controlled two-dimensional lattice gases on which to experiment! But recently it has become possible to work with submonolayers of simple gases adsorbed on almost perfect crystalline substrates. Figure 6 shows the data of Kim and Chan[11] for the gas-liquid coexistence curve of methane adsorbed on graphite. One sees a very steep, bullet-nosed curve, implying a rather small value of the exponent β. Now, methane is a fascinating chemical compound. As explained by R. N. Zare, earlier in this Symposium, quantum mechanics has quite a job describing its detailed molecular properties. But we do not need all that here. Only the

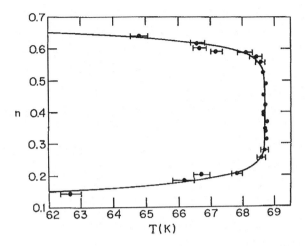

Figure 6. The gas-liquid coexistence curve for methane, CH_4, adsorbed on graphite at submonolayer coverages, n, vs. temperature, T. [After H. K. Kim and M. H. W. Chan, Ref. 11.]

uniform two-dimensional character of the substrate and the facts that the methane molecules sit adsorbed on the graphite, and feel mutual, short-range attractions are essential. Indeed, the experimental data yield an exponent $\beta = 0.127 \pm 0.020$; despite the relatively large uncertainty — which is not surprising given the difficulties of the experiment — the agreement with Onsager's prediction $\beta = 0.125$ is remarkable.

Another interesting question which I would have liked to have gone into is "Why are these two-dimensional critical exponents all rational numbers?". There have been wonderful developments in the last few years coming out of conformal field theory and string theory.[12,13] Perhaps, they are best characterized as mathematical physics rather than condensed matter physics, but, in a nutshell, conformal covariance in two dimensions tells us why two-dimensional critical exponents — not just the few I have mentioned but many others — will, in fact, be rational numbers.[13,14] As an extra bonus, the analysis can also generate detailed information about the correlation functions precisely at two-dimensional critical points.[12,14]

The Irrelevance of $\hbar$ to Critical Phenomena

Allow me to enter a little more technically into the question of quantum mechanics and critical phenomena. One can show that Planck's constant, $\hbar$, is an "irrelevant parameter" for criticality (at nonzero temperatures)! In saying this one refers to calculations performed within a renormalization group framework and the associated, now standard, terminology. One starts by considering the behavior of, say, a spin system for very large values of the spin quantum number, S. Now when $S \to \infty$ the quantal spin variables simply become classical vectors. Then one can ask, "Does the actual value of S matter to the critical behavior? Is finite S different from $S = \infty$?"

Consider also the transition to superfluidity at the lambda point in liquid helium. For describing this, one is inclined, at first, to say that surely quantum mechanics is essential. However, it transpires that for this normal-to-superfluid transition there is really only one feature of quantum mechanics that is essential, namely, that to describe superfluid order one needs, in essence, a wave function. Now quantum mechanics tells us something important about wave functions which we tend to hide from beginning students: specifically, a wave function is a complex number that has both a magnitude *and* a phase. Furthermore, the actual value of the phase cannot be observed owing to the so-called gauge symmetry. Gauge symmetry is an $O(2)$ symmetry corresponding simply to rotations in a plane, say, the complex plane or an (x, y) Cartesian plane.

But the same $O(2)$ or XY symmetry of the basic "order parameter" applies equally to many magnetic systems: their spin vectors are free to rotate into any direction in some crystalline plane; and the analogous result applies to other sorts of materials as well. This XY symmetry, together with the appropriate spatial dimensionality, serves to determine the nature of the critical behavior. For example, the superfluid density, $\rho_s(T)$ of liquid helium, vanishes according to

$$\rho_s(T) \approx D|t|^\zeta, \tag{16}$$

when T approaches the lambda point, T_c, with a characteristic exponent ζ. The same happens for the magnetic analog, the so-called helicity modulus, which measures what happens when one twists the alignment of the spins at the two ends of a ferromagnetic system. For the Bose–Einstein condensation of an ideal Bose gas one has $\zeta_0 = 1$. However, the exponent for real helium, an interacting Bose gas, takes a value $\zeta \simeq \frac{2}{3}$; the same goes for XY magnets. Again the value of ζ is the same whether the system is classical or quantal.

It may be instructive for some to see how this insensitivity to quantum mechanics comes about mathematically within a now traditional perturbation-theoretic formulation. For a classical system, say an $S = \infty$ magnet, there will be a 'bare' or unperturbed Green's function behaving like

$$G_0(k, T) \propto \frac{1}{k^2 + t_0}. \tag{17}$$

The square of the momentum or wave vector, k, derives from short-range spin-spin couplings re-expressed in Fourier space, while t_0 depends on T and becomes small as the transition is approached. For a quantum mechanical system matters are, of course, more complicated: e.g., for bosons of mass m the unperturbed Green's function becomes

$$G_0(k, \bar{\omega}_n, T) \propto \frac{1}{k^2 + i\bar{\omega}_n + t_0} \tag{18}$$

where the (rescaled) Matsubara frequencies are given by

$$\bar{\omega}_n = 4\pi(mk_BT/\hbar^2)n \quad \text{with} \quad n = 0, 1, 2, \ldots .$$

The effect of quantum mechanics is, thus, to introduce the infinite set of frequencies $\bar{\omega}_n$. But if the temperature is *nonzero* these are discrete nonvanishing frequencies for $n \geq 1$. Thus as t_0 becomes small near the transition only the $n = 0$ component of G_0 can become large when $k^2 \to 0$. All the other components remain finite and harmless. Hence the two expressions (17) and (18) become virtually identical. The terms with $n = 1, 2, \ldots$ serve only to fix details like the precise value of T_c, the magnitude of certain amplitudes, such as D in (16), and the size of the asymptotic corrections to a pure power law as criticality is approached.

Helium in Vycor — Crossover to Ideal-Bose-like Behavior

My colleague at Cornell, John Reppy, has in recent years been measuring very precisely the variation of the superfluid density of helium adsorbed in a porous glass — a sponge-like material called "Vycor." The point of studying helium in Vycor is that by controlling the overall density one can effectively modulate the interactions between helium atoms. Thus the superfluid transition temperature can be reduced far below the usual bulk lambda point, $T_c = 2.17°$K. Figure 7 shows the data for $\rho_s(T)$ obtained by Reppy and collaborators[15] for fillings of the Vycor which yield transition temperatures of 80 mK or less. The figure also illustrates a $\zeta \simeq \frac{2}{3}$ power law of ρ_s, characteristic of bulk, fully interacting helium, and the ideal-Bose

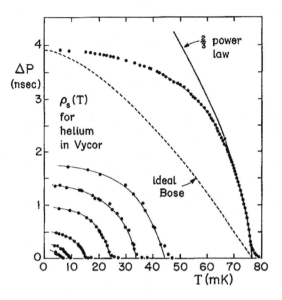

Figure 7. Measurements of the superfluid density of helium-four adsorbed in Vycor glass at various overall densities showing the crossover from a critical behavior characterized by an exponent $\zeta \simeq \frac{2}{3}$, to ideal-Bose-like behavior (see dashed curve) as the critical temperatures, T_c, fall below 25 mK. [After Crooker *et al.*, Ref. 15.]

behavior which corresponds to an exponent $\zeta_0 = 1$, i.e., a straight line near the transition. Now, by varying the amount of helium in the Vycor one can control the critical temperature, T_c. For the higher T_c values, one finds good agreement with the bulk $\frac{2}{3}$ law (except very near T_c where imperfections, probably miscellaneous spatial inhomogeneities near the boundary etc., result in a small tail). But notice that by the time one pushes T_c down to 20 mK or less, the plots change shape and approach the transition in rather linear fashion. In fact one seems to be observing a tuning of the system away from bulk helium criticality over towards ideal-Bose-gas behavior; lowering T_c is apparently serving to switch off the interactions.

Now it turns out that scaling and renormalization group theory serve to give an understanding of this crossover in critical behavior. The data of Figure 7 (and more) can be rearranged and thence reduced to a single experimental curve — see the dots, circles and triangles in Figure 8. Furthermore, this plot can be well described by a theoretically calculated curve (solid line) which actually answers an old theoretical question, namely, "How does ideal-Bose criticality change when interactions between the bosons are switched on?" To be more concrete, we regard T_c for a particular filling of

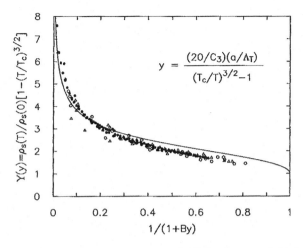

Figure 8. Scaling plot of the data of Crooker et al., (Ref. 15) for the superfluid density of helium adsorbed in Vycor glass, including eleven values of T_c: from 6 to 12 mK (open circles); from 15 to 35 mK (triangles); and from 37 to 77 mK (solid circles). In the definition of the scaling variables (see also the text) $C_3 \simeq 4.2$ is a numerical constant and $\Delta_T = (\hbar^2/2\pi m^* k_B T_c)^{1/2}$. [After Rasolt et al., Ref. 16(a).]

the Vycor as the control parameter. Then theory predicts[16] that the ratio

$$Z = \frac{\rho_s(T;T_c)}{\rho_s^{\text{ideal}}(T/T_c)} \qquad (19)$$

should, as $T_c \to 0$, be a function only of a scaled variable y given, apart from unimportant details, by

$$y^2 \propto T_c/|t|^{\phi_T} \quad \text{with} \quad t = (T-T_c)/T_c. \qquad (20)$$

The so-called crossover exponent, ϕ_T, is found generally to be[16]

$$\phi_T = 2(4-d)/(d-2)^2, \quad (2 \leq d \leq 4) \qquad (21)$$

so that for the real three-dimensional system one has $\phi_T = 2$. Figure 8 represents, in somewhat different form, a plot of the scaling function $Z(y)$ and a check that the data do, indeed, obey scaling with $\phi_T = 2$. There is only a single fitting parameter entering the figure: this is $\mathscr{R} = (a^*/a)(m^*/m)^{1/2} \simeq 1.8$ in which a and m refer to the scattering length and mass of isolated helium atoms while the stars denote the effective values resulting from the motion of the atoms in the porous Vycor glass. The calculation of the scaling function, $Z(y)$, is an exercise in purely classical statistical mechanics (and has been carried out to leading order in an $\varepsilon = 4-d$ expansion):[16] its form depends only on the dimensionality and on the XY symmetry. Quantum mechanics enters in the scale factor required in Eq. (20) to make y dimensionless. This is also where the ratios m^*/m and a^*/a are needed: their calculation certainly represents a quantum-mechanical problem, indeed a rather hard one! But the outcome should serve only to refine the rather reasonable fitted value, 1.8, of the product $\mathscr{R}$.[e]

Planck's Constant is Worth Only One Extra Dimension

Now consider a situation in which it is established that Planck's constant $\hbar$ enters from the outset: I have already emphasized that, in modern condensed matter physics, the dimensionality, d, is important to both theorists and to experimentalists. Experiments on two- and three-dimensional systems have been described above; and one-dimensional systems are likewise realizable in the laboratory. Indeed, one is almost not a condensed matter physicist these days if the dimensionality is not one of your parameters, and, often, one of the most important. My thesis, however, is that in order to calculate the partition function, and other system properties, the cost of including Planck's constant amounts to no more than allowing for an extra spatial dimension! In some ways this is a well known, almost trivial point; but it is still worth emphasizing and I include it for those who have not seen it before. So here is one way of demonstrating the fact.

Suppose the Hamiltonian operator for a system in d dimensions can be written $\mathscr{H} = \mathscr{H}_0 + \mathscr{H}_1$ where the eigenvalues, $U(\sigma)$, of $\mathscr{H}_0$ are known, i.e., one can write

$$\mathscr{H}_0|\sigma\rangle = U(\sigma)|\sigma\rangle, \qquad (22)$$

where the eigenfunctions $|\sigma\rangle$ are labelled by the set of quantum numbers, σ. Quantum numbers are just classical variables so that evaluation of the partition function, $Z(T)$, for $\mathscr{H}_0$ alone, reduces to a problem in ordinary, classical, d-dimensional statistical physics. But the full partition function is given by

$$Z(T) = \text{Tr}\{e^{-\beta(\mathscr{H}_0 + \mathscr{H}_1)}\}, \quad (\beta = 1/k_B T) \qquad (23)$$

where, in general, $\mathscr{H}_0$ and $\mathscr{H}_1$ *do not commute*: that is, of course, precisely where $\hbar$ enters in a crucial way! One can calculate the matrix elements

$$\langle\sigma|\mathscr{H}_1|\sigma'\rangle \equiv V(\sigma, \sigma'); \qquad (24)$$

but how can one deal with the noncommutativity in the partition function?

---

[e]This necessarily brief, summary of the experiments and theory overlooks the random character of Vycor and does not address the *onset* of superfluidity at a finite filling value at $T = 0$. The latter problem certainly requires quantum mechanics; but see the next section.

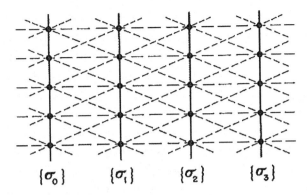

Figure 9. Schematic representation of the reduction of a quantal statistical mechanical problem in d dimensions which are represented vertically, to a classical statistical mechanical problem in one extra dimension represented horizontally. The sets of quantum numbers $\{\sigma_k\}$ for $\mathcal{H}_0$ in the kth vertical layer can be regarded as spin-like variables coupled, between layers, via the noncommuting part, $\mathcal{H}_1$, of the original Hamiltonian, $\mathcal{H} = \mathcal{H}_0 + \mathcal{H}_1$.

The answer is to say the magic word "Trotterize"! This means that one uses Trotter's formula for the exponential of two non-commuting matrices (or more general operators) which can be written

$$e^{\mathbf{A}+\mathbf{B}} \approx \underbrace{e^{\mathbf{A}/n}e^{\mathbf{B}/n}e^{\mathbf{A}/n}\cdots e^{\mathbf{B}/n}}_{2n \text{ factors}}, \quad (25)$$

as $n \to \infty$. This is a generalization of the familiar definition of the standard exponential function via $e^x = \lim_{n\to\infty}[1 + (x/n)]^n$. Substituting this formulae in Eq. (23) gives rise to

$$Z(T) = \sum_{\{\sigma_0\}}\sum_{\{\sigma_1\}}\cdots\sum_{\{\sigma_n\}} e^{-(\beta/n)[U(\sigma_0)+V(\sigma_0\cdot\sigma_1)+U(\sigma_1)+V(\sigma_1\cdot\sigma_2)+U(\sigma_2)+\cdots]} \quad (26)$$

At first inspection, this is a miserable-looking expression for the partition function, perhaps worse than what we started with; but when all is said and done, one finds that what has arisen corresponds merely to a classical system in one more dimension. In other words, d has become $d+1$!

To understand this in a little more detail consider Figure 9. Each vertical layer represents a d-dimensional classical system resembling the original one with $\mathcal{H}_0$ only but at a temperature nT; the kth layer introduces a factor $\exp[-\beta U(\sigma_k)/n]$. Between adjacent layers, however, there appear interactions in the extra direction, indicated by the dashed lines which represent the factors $\exp[-\beta V(\sigma_k, \sigma_{k+1})]$ in Eq. (26). Note that all the quantities now appearing are ordinary, classical, commuting variables. Of course, the extra dimension and further couplings represent a complication. But who is scared of anisotropic systems? One often deals with them. Incidentally, when the analogous steps are taken in quantum field theory, the extra dimension represents time and then Lorentz invariance comes in to help. The time-like dimension finishes up looking just like the other dimensions so full spatial isotropy is actually restored! Quantum field theory, from this point of view, can be regarded as classical statistical mechanics, in four-dimensional space! These connections are now quite well explored and have proved mutually fruitful. Nevertheless, if one wants to deal with quantum-mechanical systems more generally — as one certainly does in condensed matter physics — one is basically faced with a problem of classical mechanics in one more dimension. As we have seen, varying the number of dimensions is of great interest, in any case, so quantum mechanics does not present an intrinsically new challenge.[f]

Disordered Matter — Ferromagnets in Random Fields

Now I would like to convey the flavor of some recent advances in condensed matter physics. As a first example, I will describe some theoretical progress tied to some challenging experiments. One form of matter which is always there to plague experimentalists is matter that is a bit "mucky" or "dirty": it is disordered by, for example, impurities that are randomly frozen in. Condensed matter physicists (with the aid of chemists and expert crystal growers) have recently gained the ability to prepare materials with controlled disorder and to perform careful experiments on them. There are many different, fascinating aspects to this; I shall focus on just one, which I hope will be found both entertaining and instructive.

Take a ferromagnet — for instance an Ising-like magnet — and imagine that there is disorder in the form of random local magnetic fields. To be specific suppose the total field at site i is

$$H_i = \bar{H} + h_i, \qquad (27)$$

where the random piece, h_i, varies from site to site with a zero mean, $\bar{h}_i = 0$, but has a nonzero variance $\overline{h_i^2} \neq 0$: the simplest example is $h_i = \pm h_R$. Now the central question is, do the random fields, h_i, destroy the ferromagnetism? The current theoretical view is that in high enough dimensionali-

---

[f]It should, perhaps, be added that dynamics presents a distinct challenge both classically and quantum-mechanically: the former is now fairly well understood; the latter still presents profound technical and conceptual questions.

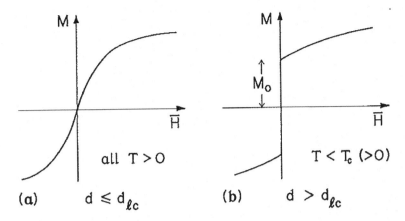

Figure 10. Illustrations of the effects to be expected when random local magnetic fields (of zero mean) are imposed on a ferromagnet: (a) at and below the lower critical dimension, d_{lc} the spontaneous magnetization for $T > 0$ is destroyed by the randomness; (b) above d_{lc} the ferromagnetic order survives below some nonzero critical point T_c.

ties, typically $d \gtrsim 4$, nothing drastic will happen; the ferromagnetic order will remain intact. So the question becomes one of determining that special dimensionality above which the original phase transition in the pure system is not spoiled by the disorder. This dimensionality is usually called d_{1c} the *lower critical dimension*. Above d_{lc} as illustrated in Figure 10(b), the system exhibits a first-order magnetic transition; but *below d_{lc}* something new can happen: as indicated in Figure 10(a), one expects the original transition to be destroyed in some way by the randomness when $d \leq d_{lc}$, however low the temperature.

Now, theorists came to hold contrasting views as to the correct answer to this question; one can read the different opinions in the literature. Some said d_{lc} should be 2; others said 3. Both of these conclusions are experimentally significant: two-dimensional magnets have already been mentioned, and one can certainly study three-dimensional examples: as indicated, the randomness is controlled by careful addition of impurities. The first theoretical argument was presented by Imry and Ma[17] in 1975: as we will see, that led to $d_{lc} = 2$. Later others supported this conclusion. But in the meanwhile a number of reputable theorists started to argue for $d_{lc} = 3$. And, indeed — although it is, perhaps, a little unkind of me to record it — some individuals backed both horses! Table 1 shows the situation reached by 1983–84. It lists the opinions expressed by theorists from various institu-

Table 1. Some theoretical opinions as to the value of the lower critical dimension, d_{lc}, for Ising-like ferromagnets in random fields. Apart from Imry and Ma, only the institutions* of the various authors are recorded.

	$d_{lc} = 2$		$d_{lc} = 3$
1975	(Imry and Ma) UCSD	1976	Tel-Aviv and UCSD
1982	UCSD and IBM	1979	Rome and Paris
1982	ILL, Grenoble	1981	IBM and Weizmann Inst.
1983	Tel-Aviv and IBM	1982	KFA Jülich, Tel-Aviv and IBM
1983	IBM	1981	Edinburgh
		1982	MIT and Helsinki
		1983	UCSB
1983	Geilo vote — 4		17

*Some of the abbreviations are: UCSD and UCSB: University of California at San Diego and at Santa Barbara; IBM: T. J. Watson Laboratory, New York.

tions around the world as time progressed. Evidently there was no consensus. Indeed, matters came to such a pass that at the 1983 Spring School in Geilo, Norway, the issue was put to a vote! As I recall, not everybody cared to express an opinion; nevertheless, 17 voted for $d_{lc} = 3$ while only 4 (or 5) voted for $d_{lc} = 2$. So $d_{lc} = 3$ won!

Well, let us look into the issues involved more closely and learn how this story turned out. How can one determine d_{lc}? The first conclusion, $d_{lc} = 2$, actually follows by extending the Peierls' argument outlined above for a pure system. The appropriate picture resembles that in Figure 3(b) except that now the interface of the the counter-domain may, owing to the quenched or frozen-in random fields, want to *wander* in order to incorporate more sites on which the local field already points 'down'. This effect can be represented by adding a new term to the estimate (13) for the total free energy increment, ΔF_L, of the counter-domain. Imry and Ma argued[17] that the typical net total random field in a domain of volume V, should, by the central limit theorem, scale like $\sqrt{V}$. This suggests that the term needed should be

$$\delta F_{\text{random}} \sim -M_0 (\overline{h_i^2} L^d)^{1/2}. \tag{28}$$

Minimizing ΔF_L with this term included leads immediately to the conclusion that the order is stable if, for large L, $L^{d-1} > L^{d/2}$; in other words $d_{lc} = 2$. However, the argument, and its later developments, was clearly not absolutely convincing — recall the bubbles-within-bubbles problems with the original Peierls argument.

Now, as mentioned, there are real materials in which one can examine both two- and three-dimensional random magnets. It is not possible to produce directly a local magnetic field that points randomly up or down; but Aharony and Fishman[18] showed that one could, instead, study impure *anti*ferromagnets in *uniform* magnetic fields. Theoretically, that becomes essentially the same problem. So what had seemed a purely abstract model — a theorists' plaything – turned out to apply to some real materials, namely, diluted antiferromagnetic crystals in uniform fields. Some of the best real examples of such systems are[19] $Rb_2Mn_{1-c}Mg_cF_4$ for $d = 2$ dimensions and $Fe_{1-c}Zn_cF_2$ for $d = 3$ dimensions.

Now the experimentalists were having trouble isolating a clear transition to antiferromagnetism in the three-dimensional systems; that fact seemed to favor $d_{lc} = 3$ so the Imry-Ma conclusion was further doubted. And then modern field theory came to our aid! But, as we will see, in this case it was not truly of assistance. There is, now, in field theory the wonderful principle of supersymmetry wherein fermions and bosons are seen to be aspects of the same more basic field. It was further discovered that one could do perturbation theory for a random-field system, starting at $d = 6$ dimensions, writing $d = 6 - \varepsilon$, and treating $\varepsilon = 6 - d$ as an expansion parameter. The formal perturbation expansion for the random systems about six dimensions turned out to be exactly the same as around four dimensions for the corresponding pure systems — and field theory explained why by revealing a supersymmetry connection between the models. Thus the random system in $d = 6 - \varepsilon$ dimensions looked like the pure system in $d = 4 - \varepsilon$ dimensions. Well, we knew what happens in the pure case: for an Ising system one has $d_{lc} = 1$ — that is just Peierls' argument again! The change $\Delta d = 2$ from $d = 4 - \varepsilon$ to $d = 6 - \varepsilon$ thus gives $d_{lc} = 1 + 2 = 3$ for the random system. In other words, Ising-like order should survive random fields only for $d > 3$. Many people examined this argument, and it seemed to hold up well: see Table 1.

Here was a serious impasse! It was one of those situations where experiment really needed guidance from theory. Sometimes experiment serves as a true guide to theory; but this was not such a case. In fact, it finally took some really hard and careful mathematics — truly "mathematical physics," with all its epsilons and deltas — to settle the matter.[20] Sad to say, physics is not a democratic subject: the Geilo vote was wrong; democracy was dismissed! John Z. Imbrie,[20] at Harvard, succeeded in showing that d_{lc} was strictly less than 3 (in fact, less than $2 + \delta$ for any positive δ). So we can be convinced that $d_{lc} = 2$ after all![g]

Now I dislike criticising colleagues — but what tends to happen when somebody really proves something that was not previously obvious, is that all those good theorists who guessed right before-hand, say, "Well, why do we need a rigorous proof? It was really physically obvious!" And all those who did not guess correctly keep discreetly quiet; or, perhaps, they refer grudgingly to Imbrie's work saying, "He has a reasonable argument." What Imbrie actually proved, with full rigor, was that in more than two dimensions there is long-range ferromagnetic order at $T = 0$ even in the presence of small random fields. So we must conclude that Ising-like order certainly survives in $d > 2$ dimensions. This is really a striking result, but I fear that Imbrie's name will be forgotten except, I trust, by the historians of science, because those who were proved right will march on with little reflection while the others will turn to fresh problems. Maybe I am too pessimistic but I do feel that this episode has something to teach us about Science.

Modulated Matter

My next topic concerns an unusual type of phase, or set of phases of matter that are observed in certain alloys and other systems as well. Consider the alloy Ag_3Mg: other like examples are Au_3Zn, Cu_3Pd, Au_3Cu, Al_3Ti, etc. As illustrated in Figure 11(a), this forms a face-centered lattice structure with the silver ions (open circles) on the cube faces and the magnesium ions (closed circles) on the corner sites. However, this same alloy or ones with closeby composition (say $Ag_{3-x}Mg_{1+x}$, with x small) can form a quite different phase, the so-called $\langle 2 \rangle$ phase, wherein alternate planes of Mg atoms normal to one crystal axis are displaced slightly — actually by half a basic cube diagonal — with respect to one another. As shown in Figure 11(b), two planes shifted "up" are followed by two planes shifted "down", and so on: the two-fold repeat justifies the terminology $\langle 2 \rangle$. So here is a crystal structure that is seemingly more complicated than it had any need to be; but, as the micrograph in Figure 12 demonstrates, it is actually found in the

---

[g]It is still not, at this point in time, clear precisely *how*, the supersymmetry field-theoretic arguments fail. However, in another simpler but analogous case it is seen that there is no true renormalization group fixed point on which to base a super-symmetric perturbative expansion. [D. S. Fisher, *Phys. Rev. Lett.* **56**, 1984 (1986).]

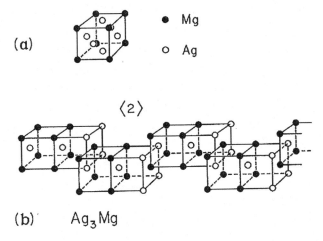

Figure 11. (a) A cubic unit cell illustrating the fcc structure of Ag_3Mg and similar alloys. (b) The "two-up/two-down" or $\langle 2 \rangle$ phase of Ag_3Mg. As in (a) solid circles denote Mg ions, open circles Ag ions; not all the face-centered ions are shown.

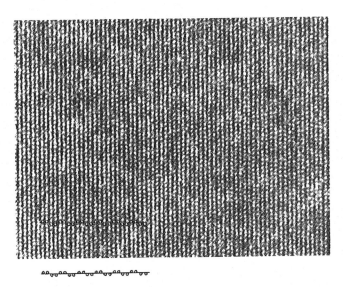

Figure 12. A photomicrograph of the $\langle 2 \rangle$ phase in Au_3Zn. [After Guymont, Portier and Gratias, Ref. 21.]

laboratory![21] The ordered state of this alloy could have been a perfect fcc structure, and that probably is the ultimate, low temperature ground state; but at moderately high temperatures and appropriate composition range one finds this unexpected $\langle 2 \rangle$ phase — our first, simple example of what we may call *spatially modulated matter* in which there appear extra periodicities, significantly larger than those more-or-less imposed by the atomic or ionic sizes and the gross composition. In this case a wavelength $\Lambda \approx 16$ Å appears compared to the basic cubic cell edge of approximately 4 Å.

Now where do these larger periodicities come from? And is this simple $\langle 2 \rangle$ phase typical of the possibilities? The answer to this latter question is a resounding "No!". Indeed, in this same silver-magnesium alloy, with very slightly differing composition, one discovers much more complicated patterns: see Figure 13.[22] The general nature of the structures visible in these photomicrographs is illustrated schematically in Figure 14(a). The patterns consist of two layers of ions displaced "up", then two layers "down", and so on k times but, finally, only *one* layer "up" (or "down"); this whole pattern of $(2k+1)$ layers then repeats. Figure 14(a) represents the case $k = 4$, while the real examples in Figure 13 correspond to $k = 3$ and $k = 2$: in short one sees a $\langle 2^3 1 \rangle$ phase with a total repeat distance of $\Lambda \simeq 28$Å and a $\langle 2^2 1 \rangle$ phase with $\Lambda \simeq 40$Å. But this is not all: in Ag$_3$Mg one also finds the equilibrium phases $\langle 2^4 1 \rangle$, $\langle 2^5 1 \rangle$ and $\langle 2^6 1 \rangle$ while in Au$_3$Zn one observes, as illustrated schematically in Figure 14(b), the phase $\langle 2^3 3 \rangle$. There are further variants: for example, one discovers 'mixed' or 'branched' phases[21,22] such as $\langle 2^3 1 2^4 1 \rangle$ in which *three* pairs of layers, "two-up/two-down" are followed by a single layer which, in turn, is followed by *four* pairs of layers before the next single layer. The alternate spacing, three-pairs/four-pairs, continues generating an overall period exceeding 120 Å.

How can one understand these large equilibrium periodicities? It is possible that ground-state quantum-mechanical effects associated, for example, with the location of the Fermi surface in the alloys play a role. However, thinking of the ground state alone and attempting to explain only that is *not* adequate; indeed, as mentioned, the ground state might well be just the ordinary fcc lattice. This approach, indeed, illustrates a sinful attitude apt to be taken by 'quantum physicists': it was epitomized earlier in this Symposium when one participant said, no doubt mainly in jest, that he wished he had been the first to say "To understand the hydrogen atom is to understand the whole of physics!" What a revealing and distressing overstatement! Understanding the hydrogen atom, even with its most highly excited states will, alas, not carry one far beyond the ground state of any condensed matter system — and, as alluded to, even leaves one many problems in understanding the quantal states of a simple molecule such as methane,

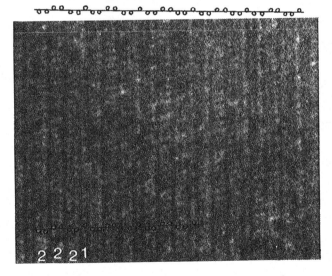

Figure 13. Photomicrographs of Ag$_3$Mg crystals revealing $\langle 2^3 1 \rangle$ and $\langle 2^2 1 \rangle$ phases (lower and upper parts, respectively). The total repeat distances are $\Lambda \simeq 28$Å and 40Å. [After Kulik, Takeda and de Fontaine, Ref. 22. Reprinted with permission from *Acta Metallurgica* **35**, 1137 (1987), Pergamon press plc.]

containing only four hydrogen atoms! No, the long-wavelength spatially modulated patterns found at high temperatures, and seen to be sensitive to the temperature, depend on entropy and fluctuations — the effects may be subtle but they would have been appreciated by Gibbs, Boltzmann and Maxwell.

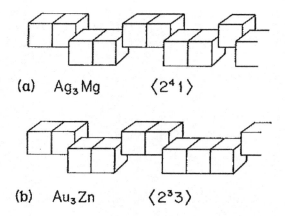

Figure 14. Schematic depiction of (a) the $\langle 2^4 1 \rangle$ phase, as found in Ag$_3$Mg, and (b) the $\langle 2^3 3 \rangle$ phase which is observed in Au$_3$Zn.

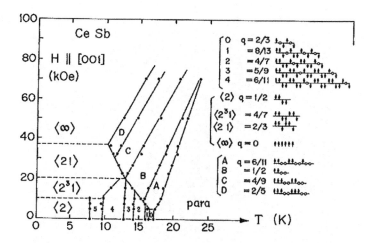

Figure 15. The magnetic phase diagram of cerium antimonide for magnetic field parallel to a [001] axis. The orientation of the spins in successive layers in the various phases is indicated on the right; ○ denotes a layer of zero average magnetization. (The symbol q denotes the reduced wave number characterising the basic periodicity of a phase.) [After J. Rossat-Mignod et al., Ref. 23].

In thinking about spatially modulated phases and their interrelations it must be realized that their occurrence is not confined to metallic alloys or represented only by purely structural transformations. Figure 15 shows the phase diagram of the magnetic crystal CeSb, cerium antimonide, in the temperature/magnetic field plane as determined by specific heat measurements made by J. Rossat-Mignod and coworkers.[23] One observes an amazing multiplicity of phases. The nature of the individual phases is revealed by neutron scattering.[23] As indicated in the figure, the majority prove to be long-period, *spatially modulated magnetic phases*. The spins, which are Ising-like in this material, are aligned parallel in ferromagnetic layers but the orientation of successive layers alternate to produce patterns like those found in the alloys. Note, at low T and as a function of H, the sequence of phases $\langle 2^k 1 \rangle$ with $k = 1, 3$ and ∞: it is not unlikely that some higher order phases with $k = 5, 7, \ldots$ have been missed. Similarly, for $H \simeq 0$ when the temperature is reduced below the disordered paramagnetic phase one finds a sequence which may be written $\langle 2^k \tilde{3} \rangle$ where $\tilde{3}$ denotes a band of three layers, one "up", one "zero" or *unmagnetized* (indicated by ∘ in the figure), and one "down", or vice-versa. As T falls, equilibrium phases corresponding to $k = 0, 2, 3, 4$, and ∞ appear.

Now what is going on? Clearly, whatever the underlying quantum mechanical basis, it is different in the alloys than in the spin systems. But the overall similarities of the phenomena suggest that some common principles may be operating. One approach to the issue is to return to the Ising model, more particularly, to a simple three-dimensional Ising model with just one new feature. Take a spin system on a three-dimensional simple cubic lattice with nearest-neighbor couplings of strength J_0 in two of the directions, say, x and z, see Figure 16, and, J_1 in the third direction, y. If J_0 and J_1 are positive this is just a three-dimensional Ising model, which can have a standard ferromagnetic transition but nothing more. But now, in the third direction put in an *extra* coupling of strength J_2 between second neighbors: see Figure 16. This yields a three-dimensional model with second-neighbor interactions parallel to one axis which has been dubbed the "ANNNI" model, for "axial next-nearest-neighbor Ising model."

To explore some of the physics underlying long-period phases one should take the second-neighbor coupling J_2 to be negative or *anti*ferromagnetic so that it *competes* with the nearest-neighbor couplings J_1. Then spins along the y direction get confused: J_1 tends to align them but J_2 tries to make them point antiparallel! If one writes $J_2 = -\kappa J_1$, then the parameter κ measures the degree of competition. It is easy to see that the ground state remains totally ferromagnetic for $\kappa < \frac{1}{2}$; but for $\kappa > \frac{1}{2}$ a $\langle 2 \rangle$-phase configuration provides the lowest energy. Nothing more appears at $T = 0$. However, at nonzero temperatures, where fluctuations and entropy come into play, matters change dramatically. In the first place, at

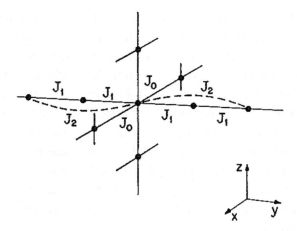

Figure 16. Part of a simple cubic lattice of Ising spins showing the couplings of strength J_0, J_1 and J_2 defining the ANNNI or axial next-nearest-neighbor Ising model.

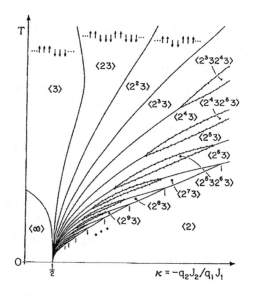

Figure 17. Schematic phase diagram of the ANNNI model in terms of temperature and competition parameter, κ, showing infinitely many modulated phases near the multiphase point $(\kappa, T) = (\tfrac{1}{2}, 0)$. (The axial coordination numbers, q_1 and q_2, extend the model beyond the simple cubic lattice version described in the text, for which $q_1 = q_2 = 2$.) [After Szpilka and Fisher, Ref. 24.]

$\kappa = \tfrac{1}{2}$ the simple modulated phase $\langle 3 \rangle$ consisting of three-up/three-down layers appears; but then as κ increases the competition shifts and yields further modulated phases $\langle 23 \rangle$, $\langle 2^2 3 \rangle$, etc. Indeed, an *infinite* number of new phases arise and all are spatially modulated![24] Figure 17 shows the low-temperature region of the phase diagram in the (κ, T) plane near the so-called *multiphase point* at $\kappa = \frac{1}{2}$, $T = 0$. In addition to the simple periodic phases $\langle 2^k 3 \rangle$ for $k = 0, 1, 2,\ldots$, all the mixed phases $\langle 2^k 3 2^{k+1} 3 \rangle$ also appear.[24] This is not the place to enter further into the details, but it is worth mentioning that the behavior of these, and many other modulated phases can be understood and systematized in terms of effective interactions, mediated largely by statistical fluctuations, between 'domain walls,' 'discommensurations,' 'phase slips,' or 'solitons.'[24,25] In the ANNNI model at low T, the three-layer bands, separated by $2k$ layers of the underlying $\langle 2 \rangle$ phase, constitute the domain walls; pairwise, triplet and higher order effective forces depending on T and κ can be elucidated.[24]

Quasicrystals

To conclude our brief tour of condensed matter physics two very recent and striking discoveries demand attention. Figure 18 records the experimental evidence produced in 1984 by D. Shechtman, I. Blech, D. Gratias and J. W. Cahn:[26] it shows an x-ray scattering pattern of sharp spots which they obtained from a special, quench-cooled aluminum-manganese alloy, roughly of composition Al_6Mn. The crucial feature of this pattern is that it

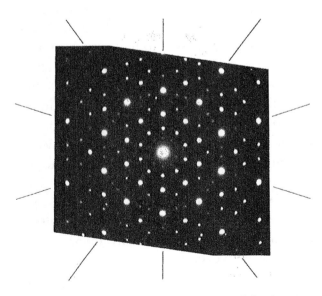

Figure 18. X-ray diffraction pattern from the iscosahedral phase of aluminum manganese. Note tenfold symmetry axis. [After Schechtman *et al.*, Ref. 26.]

exhibits a ten-fold axis of symmetry. Anyone familiar with x-ray crystallography should become uncomfortable on contemplating this figure! Why? Because a ten-fold or five-fold axis of symmetry is quite incompatible with the translational invariance that is usually taken as defining a crystal. That is a classic result of group theory. This state of matter — whatever it may be — cannot belong to one of the classical crystallographic groups which, it has long been known, exhaust all the possibilities. Now one may and, indeed, should worry lest the ten-fold effect is due merely to the phenomenon of twinning. However, after careful investigation, the answer is unequivocal: the crystal concerned are *not* twins. This new Al_6Mn material was first produced by quenching the liquid alloy by the technique of melt-spinning.[26] The grains of the new material are some microns in diameter. If one examines microscopically different parts of a grain the same scattering pattern is always seen. The full x-ray scattering displays essentially perfect *icosahedral symmetry*. Figure 19(a) portrays an icosahedron and indicates some of its symmetries: note that it has many "forbidden" five-fold axes. The Al_6Mn icosahedral crystals violate some of the basic, well-established facts of crystallography. Furthermore, the phenomenon is not isolated: precisely similar icosahedral *quasicrystals* have been grown in the alloys $Al_{6+x}M_{1-x}$ with M denoting Cr, Fe, Pd, Pt and Ru.

Now is this a quantum mechanical effect? At one level of description the atoms involved are surely quantum mechanical. But the way they fit together does not, at first sight, seem to have much to do with that; rather the atomic sizes must matter. Perhaps the fact that aluminum is a light atom and combines with transition-metals atoms in which the electrons act in complex ways will eventually be seen as crucial. But, in the meanwhile, let us at least try to understand the "new" crystallography: after all, one thought one had all of these issues straight, from "day one" of solid state physics! As one can imagine, there has been a lot of fun in understanding the nature of this novel type of material.

Consider, first, two-dimensions and ask about five-fold symmetry. Indeed, pentagonal symmetry, as depicted in Figure 19(b), poses the same problems: one cannot have a translationally invariant crystal in a plane with such a symmetry. However, Roger Penrose showed sometime back,[27] that one can construct two *tiles*, parallelograms of special shapes as illustrated on the left of Figure 20, that can be fitted together as shown on the left, to build up a pattern that goes on forever, i.e., that fills the plane. The

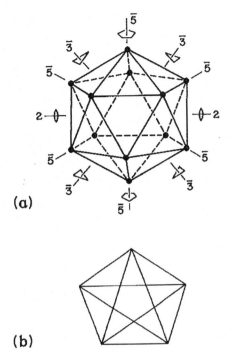

Figure 19. (a) An icosahedron and some of its symmetry axes. (b) A pentagram: all line segments are related to one another via powers of the golden ratio, τ.

proportions of the tiles entail the so-called *golden ratio*, $\tau = \frac{1}{2}(1 + \sqrt{5})$, known to the ancients for its appealing properties not least of which is its intimate connection with the geometry of a pentagram: indeed, all the line segments in Figure 19(b) are divided in the ratio $\tau:1$. Clearly this tiling pattern has *long-range bond-orientational order*: all the 'bonds' linking 'sites' at the corners of the tiles are parallel to one of the five pentagonal directions. Now the diffraction pattern of this two-dimensional pentagonal quasicrystal cannot have Bragg spots, because it is not a classical, periodic crystal. Nevertheless, as suggested by the shading in Figure 20, there are scattering rows in the plane which exhibit a *quasi-periodic* translational order. Consequently the diffraction pattern contains spots distributed with full five-fold symmetry as indicated in Figure 21. These spots are generated from a set of *five incommensurate* wave-vectors, $q_1, \ldots, q_5$, and are thus actually dense in the plane. However, their intensities fall off with increasing order.

Happily, there are three-dimensional generalizations of the Penrose tilings. One can find two distinct rhombohedra, again with proportions

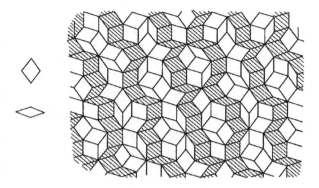

Figure 20. A Penrose tiling of the plane. The acute angles of the two types of tile, shown on the left, are $\pi/5$ and $2\pi/5$, respectively. The shaded tiles indicate a set of quasi-periodic scattering rows. [Adapted from Nelson and Halperin, Ref. 27. ©1985 by the AAAS.]

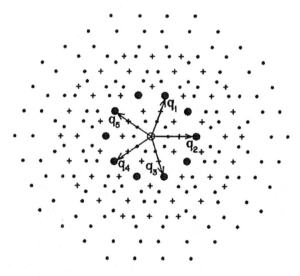

Figure 21. Diffraction pattern arising from the Penrose tiling in Figure 20. The ten bold spots are generated in first order from the five incommensurate wave-vectors. $q_1, \ldots q_5$, the crosses in second order, the light dots in third order. [After Nelson and Halperin, Ref. 27.]

determined by the golden ratio. They fit together to fill space and produce an icosahedral quasicrystal. The diffraction pattern is now generated by six incommensurate wave-vectors, $q_1, \ldots, q_6$, in direct analogy to Figure 21. The rhombohedra may be decorated with two sorts of atoms, representing the Al and Mn atoms, so that a given overall packing corresponds to an array of atoms spaced in such a way as to respect the steric restrictions. The icosahedral symmetry of the resulting diffraction pattern matches that seen in experiments on the real systems — essentially by construction. More significantly, the relative intensities of the diffraction spots correspond reasonably well. The details of the spot intensities should tell us about the locations of the atoms but that information is harder to extract and check unambiguously.

At this point the subject of quasicrystals is well launched and actively "en route". Many research workers, both experimental and theoretical, have been attracted to the area. It is clear that quasicrystals constitute a new phase of matter — one originally thought to have been ruled out! Basic questions still remain: "Can quasicrystals be genuine ground states?" One does not know. However, it is plausible that if one takes two sorts of atoms of different sizes and treats them quantnm-mechanically one will find quasicrystalline ground states if one adjusts the sizes to be just right. It will be most interesting to see. Yet, in the recent past, some of my theoretical colleagues have been trying to prove that all ground states are truly periodic.

Quantum Hall Effect

Now the final topic is a most exciting one, and certainly a case where quantum mechanics matters, namely, the quantum Hall effect, for which Klaus von Klitzing recently received the Nobel prize. This represents two-dimensional physics again, but now two-dimensional *electron* physics. Quantum mechanics always matters for electrons at low temperatures, and in this case temperatures of a few degrees or fractions of a degree are entailed. The inset in Figure 22 shows some dimensions of the small device, through which one drives a current, I, of some tens of microamperes. The essential feature is that electrons are confined to a two-dimensional layer (of thickness a 100 Å or so). One then measures the Hall voltage, V_H, across the planar strip as a function of the external magnetic field, H, imposed normal to the strip. Nothing unusual happens at small magnetic fields ($H \lesssim 0.5$ T $\equiv 5$ kG), but as H increases one observes well-marked plateaus in V_H: see Figure 22.[h] These plateaus get broader and flatter with increasing H and decreasing T. Moreover, they satisfy the following remarkable

---

[h]One also sees from the righthand scale and lower plot in Figure 22 that the standard, longitudinal voltage drop, V_x, falls to zero when V_H takes a plateau value. Thus in the quantized regions the ordinary longitudinal resistance ρ_{xx} vanishes! Indeed, a driveless, pseudosupercurrent will persist for some minutes in this regime.

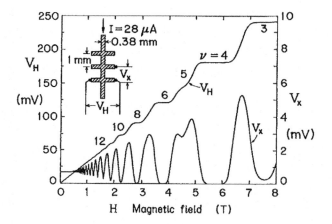

Figure 22. Data illustrating the integral quantum Hall effect. The inset indicates the dimensions of the two-dimensional disordered conducting sample and the voltages measured: the Hall voltage, V_H, on the left-hand scale, exhibits plateaus as a function of the magnetic field, H, at levels accurately given by $V_H/I = h/e^2\nu$ with ν integral; the longitudinal voltage, V_x, on the right hand scale, exhibits sharp dips at the Hall voltage plateaus. [After M. E. Cage et al., Ref. 29. ©1985 IEEE]

rule, namely, the Hall resistance $\rho_{xy} \equiv V_H/I$, is given by[28,29]

$$\rho_{xy} = \frac{R_Q}{\nu} \quad \text{with} \quad R_Q = \frac{h}{e^2} = 2.5812.80 \text{ ohms}, \tag{29}$$

where ν is an integer. This holds with a totally unexpected precision of one part in 10^8, and an accuracy of one part in 10^6 or better! The question is, of course, "Why do we see such a precise, universal quantized Hall resistance?" and, from a practical viewpoint "Is this phenomenon going to prove a better way of measuring h and e and the fine structure constant $\alpha = 2\pi e^2/hc$?"

Now the physics of this effect is subtle: but what is even more complicated and surprising is that at higher magnetic fields Tsui, Störmer and Gossard[30] discovered plateaus in which ν is a rational fraction — typically the most strongly-marked one is $\nu = \frac{1}{3}$ but $\frac{1}{5}$ and $\frac{1}{7}$ are frequently observed — and other fractions are also seen, not just any fraction but only certain special rational fractions!

Robert B. Laughlin has been awarded the 1986 Buckley Prize for his contributions to our understanding of both the integral and fractional quantized Hall effect[31] — naturally, quantum mechanics comes in with a vengence![32] To round off the discussion I will sketch some of the main features of the theory.[31,32] (Readers not inclined to follow the somewhat technical details may wish to skip to the 'Concluding Remarks'.)

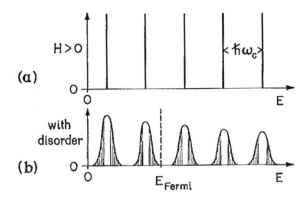

Figure 23. (a) Schematic illustration of the Landau levels for a charged quantal particle moving in two dimensions in a magnetic field. (b) Single-particle density of states derived from the ideal Landau levels in a disordered system: the shading denotes levels with localized wave functions; the states near the centers of the bands are extended. [After R. B. Laughlin, Ref. 31.]

The first point to make is that, in the presence of a magnetic field, the momentum operator, $\mathbf{p}$, for a quantum-mechanical particle carrying a charge e must be modified by including the vector potential A to get $\mathbf{p} - (e/c)\mathbf{A}$ — a subtle and beautiful result. On applying a uniform magnetic field to such a two-dimensional quantal particle one discovers that the usual continuum of energy levels condenses into a set of discrete, highly degenerate 'Landau levels': see Figure 23(a). The uniform level spacing is $\Delta E = \hbar \omega_c$ where $\omega_c = eH/cm^*$ is the cyclotron frequency, while m^* is the effective mass. Now if one ignores the electron interactions one can just fill these Landau states up to the Fermi level, E_F. Real electrons, of course, interact both with the 'substrate' or ionic medium, which usually will be somewhat disordered, and with one another. These interactions are not easy to deal with theoretically. The first thing one believes is that the Landau levels spread out into bands: see Figure 23(b). More subtley, one expects the band 'tails', shaded in Figure 23(b), to consist of localized, nonconducting states. Indeed, some degree of disorder, leading to localized states, seems essential to the full quantum Hall effect.

Following Laughlin,[31] let us suppose that the Fermi level lies in a region of localized states between the bands of extended states, as indicated in Figure 23(b). As the field, H, is changed the cyclotron frequency, ω_c, changes which moves the energy bands through E_F; equivalently, one can think of scanning the Fermi level through a set of fixed levels. Now we will invoke guage invariance: the way that enters quantum mechanics is one of

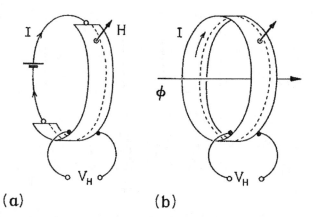

Figure 24. Illustration of Laughlin's argument explaining the quantum Hall effect. The external battery in (a) which drives a current through the planar sample in a magnetic field H is replaced in (b) by a closed sample threaded by a time-varying auxiliary flux, ϕ, which induces a current I. [After Ref. 31.]

the crucial ingredients. In reality one has some sort of external battery driving current through the planar strip, as shown schematically in Figure 24(a); but one can, instead, remove the battery and complete the circuit to form a closed strip, as in Figure 24(b). One may then induce a current in the strip by passing an auxiliary time-varying magnetic flux, $\phi(t)$, through the loop: see Figure 24(b). Increasing ϕ uniformly drives a steady current around the circuit. If this flux is changed by an integral multiple of the flux quantum

$$\phi_0 = hc/e, \tag{30}$$

the electronic system will, by guage invariance, look just the same since the Hamiltonian has not changed except for a trivial factor. On the other hand, if ϕ is increased slowly other considerations come into play. Whenever E_F lies in a region of localized states between the extended bands, there is an energy gap and one cannot excite low-energy quasiparticles. Thus a slow process will be adiabatic. On following what happens to the individual two-dimensional electron states one finds that the states that change are those whose wave-functions run completely around the loop: roughly speaking their quantization condition is modulated by ϕ. As ϕ increases adiabatically these extended states are driven sideways across the strip. Thus an electron or a fraction of an electron is effectively moved across the strip. By the time ϕ has changed by one flux quantum, ϕ_0, all the states have been restored to their original form and an integral number of electronic charges have been transported across the width of the strip. Basically one expects one electronic charge transferred for each of the Landau levels lying below E_F: thus if ν labels the highest filled level, $n = \nu$ electrons move across in total.

To complete the argument we appeal to a statistical/thermodynamic relation. If U is the total electronic energy, the current is given by

$$I = c(\partial U/\partial \phi). \tag{31}$$

But for $\Delta \phi = \phi_0 \equiv hc/e$ the change in energy resulting from the transport of n electronic charges is just $\Delta U = neV_H$ — by definition of the Hall voltage V_H. Thus one obtains

$$I = \frac{necV_H}{(hc/e)} = \frac{ne^2}{h}V_H, \tag{32}$$

which implies the quantization rule (29) for ρ_{xy}.

One might well object that the argument seems to rest very heavily on a noninteracting or one-electron, picture. That is certainly true and, although great success has been had with one-electron models in the past, one should not, these days, be content with such simplifications; rather one should attempt to understand the electronic interactions in greater depth. Indeed, fascinating progress has been made: on the one hand, it is found that the integer quantum Hall effect reflects profound topological properties of the phase of wave functions — full many-body wave functions, Ψ_N — in the presence of varying magnetic fluxes.[32] On the other hand, it transpires, wonderfully, that replusive interactions between electrons are essential to the *fractional* quantum Hall effect at higher magnetic fields. One now believes that at zero temperature a system of electrons in a fractional Hall regime should be regarded as an *incompressible quantum fluid* — a quite new state of matter.

To attack the problem of describing such states while allowing for the strong Coulombic repulsions between electrons, Laughlin proposed a new type of many-body wave function. It turns out to be a remarkably good guess which embodies the essential ordering features of these new states of matter. To represent Laughlin's wave function it is convenient to take advantage of the two-dimensionality by expressing electronic coordinates as complex numbers via

$$z - x + iy. \tag{33}$$

Then the proposed N-electron ground state wave function is

$$\Psi_0(z_1, z_2, \ldots, z_N) = \prod_{i<j}^{N}(z_i - z_j)^{m_l} \prod_{k=1}^{N} e^{-|z_k|^2/4l^2}. \tag{34}$$

Here $l = (hc/eH)^{1/2} = (\phi_0/H)^{1/2}$ is the cyclotron radius. (Alternatively, one can say that l^2 is the area per state in a given Landau level.) This wave function is not quite as mysterious as appears on first sight. One can, indeed, show that the exponential factor shown *times* a polynomial in the z_i is the form to be expected. More concretely, the prefactors $(z_i - z_j)^{m_l}$ serve to keep different electrons apart so reducing the total repulsive energy. In order to enforce proper Fermi statistics, m_l must be an odd integer. The first case, $m_l = 1$ corresponds simply to a Slater determinant; but the new states with $m_l = 3, 5, \ldots$ turn out to describe the fractional quantized Hall states with index $\nu = \frac{1}{3}, \frac{1}{5}, \frac{1}{7}$. One also discovers that the low lying excitations carry fractional electronic charges, $e^* = e/m_l$! To understand the other fractional states with $\nu = \frac{2}{3}, \frac{2}{5}$, etc. one has to think further: a hierarchy of similar states is required; but the ingenious wave function (34) provides the basis for our understanding.

Amusingly and instructively, one route to gaining an insight into the properties of Laughlin's wave function turns back to classical statistical mechanics. If one squares the wave function to obtain the corresponding many-particle distribution function one sees that it can be written as a Boltzmann weight, i.e., the exponential of a classical Hamiltonian divided by $k_B T$. The Hamiltonian turns out to describe a one-component plasma, in two dimensions. If e_0 is the charge of the classical particles, the plasma parameter $\Gamma \equiv 2e_0^2/k_B T$ is given just by $2m_l$. This classical system had, indeed, been studied previously for its independent interest. So once again we see that quantum-mechanics, if one takes the many-body aspects and the interactions seriously, is so intertwined with statistical mechanics that one has to recognize that classical ideas of fluctuations, ordering and entropy play a significant, if somewhat hidden, role even in this most quantum-mechanical of problems!

Concluding Comments

I was asked by Herman Feshbach to mention major unsolved problems. Certainly, a full understanding of the magnetic properties of solid helium-three[33] and of the microscopic basis of the so-called 'heavy fermion' systems[34] represent outstanding questions. Basically, we have not yet gained a very good grasp of all that can happen when fermions interact strongly. We are lucky that the outcome is frequently that the system generates simpler, more independent entities like atoms, molecules, quasiparticles, Cooper pairs, etc. whose internal structures may be ignored for many purposes. But there are important cases in which the overall behavior is not so readily characterized. Simple models, like the Hubbard model — a lattice description of electrons with short-range repulsions — still defy detailed analysis in many regimes despite manifold attacks including numerical aid from powerful computers. Thus the task of obtaining a reliable and thorough understanding of interacting Fermi particles remains a basic challenge. Indeed, these words, spoken in 1985, have been strengthened by the discovery of high-T_c superconductors[1] fifteen months later!

In conclusion, I hope to have given a picture of the multifaceted character of modern condensed matter physics and the varying degrees to which quantum mechanics is of direct relevance *or* almost total irrelevance. I feel that if Niels Bohr were here with us, in the role in which I would best like to recall him, namely, as a young, active theorist grappling with problems at the blackboard, he would rise to the challenges of condensed matter physics regardless of whether his own earlier work in quantum mechanics had anything to do with the problem or not. Niels Bohr certainly cared for the phenomena seen in the natural world and that, we should agree, is the crucial feature in doing physics.

Acknowledgments

I am grateful to many colleagues and scientific friends, most of whose names do not appear above or in the references, who have educated me through the literature, through correspondence and through patient discussions and arguments. All those who allowed reproduction of data and figures are especially thanked. I am indebted to Professor Peter H. Kleban for preparing, from the taped and transcribed version of the original lecture, a first draft of the manuscript. Professor B. I. Halperin, Professor R. L. Jaffe, Dr. Matthew P. A. Fisher and Professor Daniel S. Fisher kindly commented on and criticized various specific points. Last, the hospitality of the Aspen Center for Physics, where the manuscript was finally written, and the support of the National Science Foundation, primarily through the Condensed Matter Theory program,[35] are gratefully acknowledged.

References

1. J. G. Bednorz and K. A. Müller, *Z. Phys. B* **64**, 189 (1986); M. K. Wu, J. R. Ashburn, C. J. Torng, P. H. Hor, R. L. Meng, L. Gao, Z. J. Huang, Y. Q. Wang and C. W. Chu, *Phys. Rev. Lett.* **58**, 908 (1987).
2. F. J. Dyson and A. Lenard, *J. Math. Phys.* **8**, 423 (1967); F. J. Dyson, *J. Math. Phys.* **8**, 1538 (1967); A. Lenard and F. J. Dyson, *J. Math. Phys.* **9**, 698 (1968).
3. See e.g., M. E. Fisher, *Arch. Ratl. Mech. Anal.* **17**, 377 (1964).
4. E. H. Lieb and J. L. Lebowitz, *Advan. Math.* **9**, 316 (1972).
5. See the book P.-G. de Gennes, *"Scaling Concepts in Polymer Physics"*, (Cornell Univ. Press, Ithaca, New York, 1979).
6. See the review Y. Oono, *Adv. Chem. Phys.* **61**, 301 (1985).
7. See e.g., (a) M. E. Fisher in *"Contemporary Physics: Trieste Symposium 1968"*, Vol. I (Internat. Atomic Energy Agency, Vienna, 1969), pp. 19–46; (b) in *"Essays in Physics"*, Vol. 4 (Academic Press, London,1972), pp. 43–89; (c) *J. Appl. Phys.* **38**, 981 (1967).
8. R. B. Griffiths, *Phys. Rev.* **136 A**, 437 (1964).
9. L. Onsager, *Phys. Rev.* **65**, 117 (1944).
10. P. Nordblad, D. P. Belanger, A. R. King, V. Jaccarino and H. Ikeda, *Phys. Rev. B* **28**, 278 (1983).
11. H. K. Kim and M. H. W. Chan, *Phys. Rev. Lett.* **53**, 170 (1984).
12. A. A. Belavin, A. M. Polyakov and A. B. Zamolodchikov, *J. Stat. Phys.* **34**, 763 (1984); *Nucl. Phys. B* **241**, 333 (1984).
13. Vl. S. Dotsenko and V. A. Fateev, *Nucl. Phys. B* **240**, 312 (1984); D. Friedan, Z. Qiu and S. H. Shenker, *Phys. Rev. Lett.* **52**, 1575 (1984).
14. J. L. Cardy in *"Phase Transitions and Critical Phenomena,"* Eds. C. Domb and J. L. Lebowitz, Vol. 11 (Academic Press, New York, 1987), p. 55; M. E. Fisher, *J. Appl. Phys.* **57**, 3265 (1985); *J. Mag. Magn. Matls.* **54–57**, 646 (1986).
15. B. C. Crooker, E. Hebral, E. N. Smith, Y. Takano and J. D. Reppy, *Phys. Rev. Lett.* **51**, 666 (1983).
16. (a) M. Rasolt, M. J. Stephen, M. E. Fisher and P. B. Weichman, *Phys. Rev. Lett.* **53**, 798 (1984); (b) P. B. Weichman, M. Rasolt, M. E. Fisher and M. J. Stephen, *Phys. Rev. B* **33**, 4632 (1986).
17. Y. Imry and S.-K. Ma, *Phys. Rev. Lett.* **35**, 1399 (1975).
18. S. Fishman and A. Aharony, *J. Phys. C* **12**, L729 (1979).
19. D. P. Belanger, A. R. King and V. Jaccarino, *Phys. Rev. B* **31**, 4538 (1985); R. J. Birgeneau, Y. Shapira, G. Shirane, R. A. Cowley and H. Yoshizawa, *Physica* **137B**, 83 (1986).
20. J. Z. Imbrie, *Phys. Rev. Lett.* **53**, 1747 (1984); see also *Commun. Math. Phys.* **98**, 145 (1985) and D. S. Fisher, J. Fröhlich and T. Spencer, *J. Stat. Phys.* **34**, 863 (1984); J. Chalker, *J. Phys.* **16**, 6615 (1983).
21. (a) M. Guymont, R. Portier and D. Gratias, *Acta Cryst. A* **36**, 792 (1980); (b) R. Portier, D. Gratias, M. Guymont and W. M. Stobbs, *Acta Cryst. A* **36**, 190 (1980).
22. J. Kulik, S. Takeda and D. de Fontaine, *Acta Metall.* **35**, 1137 (1987).
23. J. Rossat-Mignod, P. Burlet, H. Bartholin, O. Vogt and R. Lagnier, *J. Phys. C* **13**, 6381 (1980).
24. A. M. Szpilka and M. E. Fisher, *Phys. Rev. Lett.* **57**, 1044 (1986).

25. See P. Bak, *Repts. Prog. Phys.* **45**, 587 (1982).
26. D. Schechtman, I. Blech, D. Gratias and J. W. Cahn, *Phys. Rev. Lett.* **53**, 1951 (1984).
27. See the review by D. R. Nelson and B. I. Halperin, *Science* **229**, 233 (1985) and C. L. Henley, *Comm. Cond. Mat. Phys.* **13**, 59 (1987).
28. K. von Klitzing, G. Dorda and M. Pepper, *Phys. Rev. Lett.* **45**, 494 (1980).
29. M. E. Cage, R. F. Dziuba and B. F. Field, *IEEE Trans. Instr. Meas.* **IM-34**, 301 (1985).
30. D. C. Tsui, H. L. Störmer and A. C. Gossard, *Phys. Rev. Lett.* **48**, 1559 (1982).
31. R. B. Laughlin, *Phys. Rev. B* **23**, 5632 (1981); *Phys. Rev. Lett.* **50**, 1395 (1983).
32. See also R. E. Prange and S. M. Girvin, Eds. "The Quantum Hall Effect" (Springer Verlag, Berlin, 1987).
33. M. Cross and D. S. Fisher, *Rev. Mod. Phys.* **57**, 881 (1985).
34. P. A. Lee, T. M. Rice, J. W. Serene, L. J. Sham and J. W. Wilkins, *Comm. Cond. Mat. Phys.* **12**, 99 (1986).
35. Under Grants No. DMR 81-17011 and 87-01223.

6

Phases and Phase Diagrams: Gibbs's Legacy Today*

Michael E. Fisher

Institute for Physical Science and Technology, University of Maryland, College Park, MD 20742, USA

GIBBS introduced the fundamental thermodynamic potentials, $U(S, V)$ and $G(p, T)$, and their representation as convex surfaces. The developments and implications of these ideas, in particular as they arise in phase diagrams, are examined. These include: the nonexistence of "metastable extensions" of free energies, violations of the phase rule on critical lines, current concepts for characterizing phases going beyond broken symmetries, especially the response to "twisted" boundary conditions and the relevance to spin glasses, and the importance of dimensionality. Unanticipated singularities in phase diagrams at critical endpoints and multicritical points are remarked and the theory of anomalous first-order transitions at which, e.g., the pressure jumps discontinuously when the volume is changed isothermally, is reviewed. Issues in the thermodynamics and phase transitions of random systems are mentioned.

It is an honor to be invited to this occasion which commemorates the birth, 150 years ago, of Josiah Willard Gibbs, one of the outstanding, if not *the* outstanding theoretical physicist born and bred in the United States. For me it is also personally rewarding to lecture at Yale. Twenty-five years ago I was invited to present four lectures here on the nature of critical points. I

---

*Reprinted with permission of the American Mathematical Society from *Proceedings of the Gibbs Symposium, Yale University, May 15–17, 1989*, edited by D. G. Caldi and G. D. Mostow, 39–72. Providence: American Mathematical Society, 1990. © 1990 American Mathematical Society.

would like to re-express my thanks now for that invitation which provided the stimulus and encouragement to think broadly about a subject that was just entering a renaissance. It was a pleasure then, as it is now, to address a distinguished audience in New Haven.

1. Introduction

Thermodynamics and statistical mechanics provide the basic theoretical tools which we use to understand the everyday, observable macroscopic physical world—on its own terms and in terms of its microscopic constituents, atoms, ions and electrons. Gibbs was, perhaps, the leading founding father of these two disciplines; certainly he was the one who set out the fundamental principles in greatest depth, breadth and logical clarity.

Gibbs, in treating thermodynamics, stressed the value of geometrical and graphical representations. In 1873 he introduced his famous fundamental surface relating U, the energy of a system, to S, its entropy, and V, its volume; he showed how the nature of the tangent planes to this surface related to the *coexistence of phases* and to *critical points*. This work greatly impressed James Clerk Maxwell who was moved to construct his well-known model of the surface. As an aside, I might mention that on looking through Gibbs's collected works, one finds remarkably few diagrams or figures for one advocating graphical methods. Nevertheless, it is obvious from his writings that Gibbs had an excellent geometrical sense and frequently employed geometrical images in his thinking. Here I will freely use schematic depictions of the energy surface, and of the various related surfaces and phase diagrams, in order to emphasize the significance of Gibbs's basic ideas. Indeed, in this review, I will start with the $U(S, V)$ surface and explain some of the general consequences which follow from its *convexity*—the crucial property that was enunciated so clearly by Gibbs. Many of these consequences are overlooked today. One can go through almost any journal publishing phase diagrams and see that the authors are unaware that their proposed diagrams, in the light of Gibbs's work, must violate the Second Law of Thermodynamics!

Then, I will take up some aspects of phase equilibria and phase diagrams that were left unexplained by Gibbs and others where we might have had arguments with Gibbs if he were here today: points that he did not, perhaps, get right. I will address other matters, like multisingularity and violations of the phase rules associated with symmetries, of which he did not know and so could not think about. If one reads Gibbs, however, one soon learns that he was a cagey thinker who phrases what he says with great care. Indeed, he obviously thought a good bit beyond what he actually wrote down. Thus, I will point out the existence of certain surprising singularities in phase diagrams with which, nonetheless, Gibbs would probably have been perfectly happy and, I think, interested to contemplate.

I will also touch on aspects of thermodynamics and statistical mechanics entailed in "anomalous" first-order phase transitions that are never seen in the real world, but are allowed by general Gibbsian principles: the absence of such behavior thus tells us of certain special aspects of our own world. Finally, I will raise one last question to illustrate that the basic issue Gibbs addressed, namely, what are the right foundations for thermodynamics and statistical mechanics, is still with us today when one comes to consider systems with intrinsic randomness.

2. Gibbs's Fundamental Surface

Figure 1 depicts, schematically, Gibbs's fundamental energy surface, $U = U(S, V)$, associated with the basic differential relation

$$dU = TdS - pdV, \tag{1}$$

where T is the absolute temperature and p is the pressure. Recall, as Gibbs demonstrated, that $U(S, V)$ is a master function; full knowledge of $U(S, V)$ determines *all* the thermodynamic properties of a system. (By contrast, for example knowledge of the equation of state, $p = P(T, V)$, is *inadequate* in this respect.)

Especially for experts, I put in a word of warning here: since, for normal thermodynamic systems, T and p are necessarily positive some of the slopes depicted in Figure 1 are unrealistic. However, the figure has been drawn to emphasize Gibbs's central thesis, namely, that the energy surface must be *convex* ("upwards" in my own idiosyncratic terminology) like a cup or bowl. More precisely, if a plane like that illustrated on the right of the figure, is brought into tangential contact with any point of the surface then any other point of the surface lies either *on* this tangent plane or else lies *above* it. This convexity is the central mathematical fact underlying thermodynamics and is crucial for much that follows.

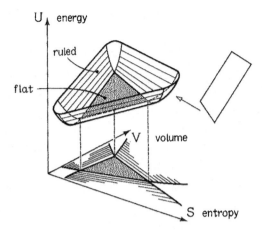

Figure 1. Schematic depiction of the energy surface, $U(S, V)$, introduced by Gibbs, illustrating its convexity and the occurrence of ruled and flat regions, which are also shown projected down onto the (S, V) plane. (Note that the surface is *not* realistic as shown since it neglects the positivity of p and T.)

Gibbs goes on to stress that multiple points of tangency are to be expected and have an important physical meaning. Regions where the tangent plane makes contact along a line are *ruled*; these describe the *coexistence* of two distinct physical phases each characterized by the values of U, V and S corresponding to the end points of the contact line. See Figure 1. But flat faces, leading to tangential contact over an *area* are also possible: see the stippled region in Figure 1. As Gibbs illustrated, such regions typically correspond to the coexistence of three phases with thermodynamic properties specified by the external points of the contact plane.

The coexistence of phases will be one of my main themes. Gibbs showed that a good grasp of it can be obtained by projecting the surface features down onto the (S, V) plane as shown in Figure 1. An example of such an (S, V) diagram (which more or less follows a rather more elaborate one drawn by Gibbs himself) is shown in Figure 2. If we suppose that it describes a simple substance, then the original flat face, shown here as a stippled triangle, describes the coexistence of solid, liquid and vapor phases—as indicated by the peripheral cartoon. The original ruled areas correspond, respectively, to regions of vapor-solid, vapor-liquid, and liquid-solid coexistence. Note that, following Gibbs, I have put in the straight "tie-lines" connecting points representing the pure phases and running through all possible associated mixed phases. Incidentally, if more authors of text books in physical chemistry, physics, metallurgy, etc. would be meticulous in putting in tie-lines

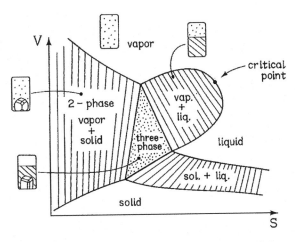

Figure 2. An (S, V) diagram for a simple substance with vapor, liquid, and solid phases illustrating a three-phase region (stippled), two-phase regions (ruled with tie-lines), and single phase regions (clear).

(especially in (p, V), (x, T) diagrams, etc.) there would be fewer generations of students left profoundly confused by such diagrams! Single-phase regions, of solid or of fluid (liquid or vapor), are, of course, left unshaded; they correspond to (doubly) curved regions of the $U(S, V)$ surface.

It is interesting to note that Andrews had discovered the critical point of carbon dioxide in 1869 only a few years before Gibbs penned his magnum opus *On the Equilibrium of Heterogeneous Substances*; Gibbs made a special point of explaining how such critical points fit naturally into his geometrical picture as illustrated in Figure 2.

Let us look in a little more detail at the quadruple points in Figure 2 where two tie-lines bounding a three-phase region meet two loci separating the one-phase region from the adjacent two-phase regions: see Figure 3, which illustrates four ways in which one might imagine these lines meeting. Of course, any such arrangement can be specified by fixing three of the four angles θ_a, θ_{ab}, θ_{ac}, and θ_{abc} shown in part (b). Now, surprisingly, it transpires that the *convexity* of the $U(S, V)$ surface imposes strong restrictions on the four angles: as stated in the figure, all must be *less* than 180° (two right angles). Note that the limiting case of equality is *not* allowed. In addition, the tangential extensions of the two boundary lines of the single-phase region, shown dashed in Figure 3, must *both* lie *either* in the two-phase regions, as in part (a), *or* in the three-phase regions, as in part (b). These are Schreinemaker's rules.[1] Evidently, then, possibilities (a) and (b) in Figure 3

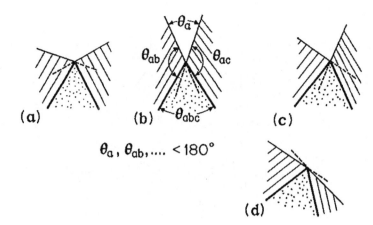

Figure 3. Putative ways of arranging the four boundary loci and tie-lines meeting at a quadruple point in an (S, V) diagram: see Figure 2. Note the tangential extensions shown dashed. Thermodynamic convexity allows (a) and (b) but (c) and (d) are not permitted.

are thermodynamically allowed (and are observed in practice). However, both configurations (c) and (d) are impossible: in essence they violate the Second Law of Thermodynamics! If reported on the basis of experiment or theoretical calculations they are simply due to error (or, for angles near 180°, lack of adequate numerical precision).

Evidently convexity has nontrivial implications for the (S, V) diagrams and the inter-relations of different phases.

3. The Phase Diagram

As we have observed, the (S, V) diagram can be rather complicated with many different sorts of regions corresponding to different phases and their possible pairwise or multiple coexistence. I would, therefore, like to focus attention instead on the conjugate diagram which Gibbs introduced and explained although he did not emphasize it to the degree which I feel its relatively much greater simplicity warrants.

To this end, following Gibbs and naming it, as we now do, in his honor, consider the conjugate thermodynamic function to the energy, namely, the Gibbs free energy

$$G = G(p, T) = U - TS + pV, \qquad (2)$$

for which we have

$$dG = Vdp - SdT. \qquad (3)$$

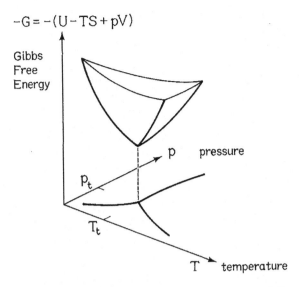

Figure 4. Schematic representation of the Gibbs free energy surface, $G(p, T)$ illustrating its convexity. Note that the vertical axis corresponds to $-G$. The coordinates (p_t, T_t) label the triple point and peak in G.

This relation, of course, confirms that p and T are the appropriate variables for G. Again, $G(p, T)$ is a master function and, again, it is a totally *convex function* of its two arguments. This is illustrated in Figure 4 where $-G$ has been sketched in order to match the sense of convexity to that in Figure 1.

However, as Gibbs explains, what were originally ruled regions of the energy surface, describing two-phase coexistence, now become lines of sharp bends or "creases" on the Gibbs surface: see Figure 4. Projected down on to the (p, T) plane, therefore, one has phase boundary lines *on which* the two adjoining phases can coexist in any proportions. Alternatively, but quite equivalently, we can say the creases mark points of first-order phase transitions; on crossing them first derivatives of the thermodynamic potential, namely, by (3), V and/or S, exhibit discontinuities. The flat, three-phase region of the energy surface in Figure 1 becomes a sharp three-sided peak on the Gibbs surface which projects onto a *triple point* in the (p, T) plane at which three first-order phase boundaries meet.

Thus as illustrated in Figure 5 (see p. 262), the *phase diagram*, as, following current custom we shall term the (p, T) plane, is simple in that different regions correspond to different *pure phases*.[a] Only on the boundaries

---
[a] The corresponding (H, T) plane for a magnetic material, etc., will also be termed a phase diagram.

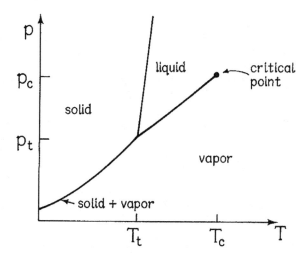

Figure 5. The (p, T) plane or phase diagram for a simple substance. Note the various one-phase regions and the triple and critical points, (p_t, T_t) and (p_c, T_c), respectively.

separating such regions may two phases coexist; only at isolated triple points on the boundaries can three phases coexist. A *critical point* arises at the isolated terminus of a first-order locus. As demonstrated by Andrews for the liquid and vapor phases of CO_2, one has continuity of phase around a critical point despite the first-order discontinuity that appears below the critical point (or more generally, on one side of it).

Let us examine a few geometrical properties of the phase diagram focusing, first, on the triple-point region. Figure 6 depicts a triple point in the (p, T) plane with three angles labelled corresponding to the three adjoining phases. Once more, convexity restricts the angles as stated in the figure; each must be less than 180°. Hence Figure 6(a) is allowed, and corresponds to what is invariably observed; but Figure 6(b) is forbidden since θ_a greatly exceeds 180°. Thus one should never see a phase diagram resembling Figure 6(b) in the real world. But this actually overstates what Gibbs would have said since, in analogous situations, he was careful to use wording that implied that exceptions should not be expected, would be rare, etc., but not that they were necessarily impossible.

Indeed, have we gone a little too rapidly? Recall that every theorem has conditions: do they matter? May we assume that a diagram like Figure 6(b) is thermodynamically *impossible*? No, that is certainly not the case! Nevertheless, Gibbs's approach stressing the convexity is still powerful. To

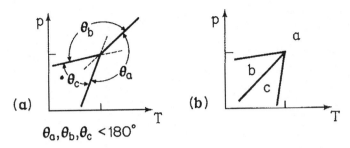

Figure 6. Triple points in the phase diagram: (a) a normal triple point satisfying the 180° rule; (b) an anomalous triple point violating the rule.

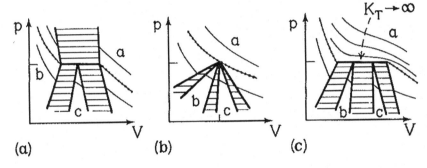

Figure 7. Isotherms in the (p, V) plane: (a) in the vicinity of a normal triple point, as in Figure 6(a); (b) and (c), in the neighborhood of anomalous triple points, like that in Figure 6(b), which entail elements of criticality such as vanishing volume jumps or a divergent isothermal compressibility, K_T.

illustrate this, consider the traditional (p, V) diagrams shown in Figure 7. The first, (a), corresponds to a typical, realistic triple point as in Figure 6(a). Note, in particular, that *at* the triple point the three phases, a, b and c, have three distinct volumes, V_a, V_b, and V_c (or, equivalently, three distinct densities).

Now, what if we were to observe a phase diagram like Figure 6(b)? Convexity tells us[2] that one *cannot* have *three, normal, distinct phases* meeting at such an anomalous triple point; in fact, there must be peculiarities present which resemble those that characterize a critical point. Two of the possibilities[2] are sketched in Figures 7(b) and (c). In (b) all three phase volumes (or densities) coincide at p_t and across each phase boundary the volume (or density) jump goes to zero at p_t, just as it does at a typical critical point. In (c) the volumes (or densities) remain distinct but phase a displays *two* different extremal densities at p_t! Consequently, the corresponding isotherm is horizontal at p_t which implies an infinite compressibility, K_T. Thus, the compressibility of phase *a diverges* when the triple point is approached from above, just as at a normal critical point. Evidently, then we cannot regard a "triple point" with the configuration shown in Figure 6(b) as normal in any way! Rather it must represent, so convexity tells us, some exceptional type of critical behavior.

4. Metastability and the Singularities at Condensation

Now there is one point which Gibbs seems not to discuss explicitly but which I regard as extremely important conceptually and about which it is not unfair to say that he was wrong. Nevertheless, he has enjoyed very good company over the years, including van der Waals and Maxwell. This is the question of the status of metastable phases and the continuation of the free energy beyond coexistence.

Figure 8(a) shows plots of the (negative) Gibbs free energy versus the pressure at fixed temperature. The solid branch labelled a corresponds to one phase, say vapor; the solid branch b would then describe liquid. At $p = p_\sigma$ a first-order phase transition occurs. In our example this is gas-liquid condensation. Now Gibbs writes and even draws diagrams as though one can take the free energy branch a and continue it unambiguously through the condensation point, with no problem, to obtain the "metastable" extension a'. Likewise it is supposed that b can, without question, be extended to b'. And, of course, this viewpoint has become

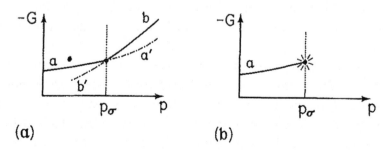

Figure 8. Isotherms of the Gibbs free energy in the vicinity of a first-order transition at $p = p_\sigma$ illustrating, (a) the traditional but unjustified extension of the stable branches, a and b, into "metastable" branches, a' and b', by contrast with the realistic situation (b) where the isotherm of a single phase stops at the condensation point on encountering an essential singularity. (Of course, the isotherm for phase b, which is not illustrated in part (b), also stops at an essential singularity as p approaches p_σ from *above*.)

embodied in nearly all texts. Now there are very good reasons for believing that this diagram, Figure 8(a), is just wrong in the terms that Gibbs would have desired: such unambiguous or, in mathematical terms, analytic continuations do not actually exist in realistic systems. But what, one may ask, happens when, as in Figure 8(b), one starts in the vapor phase a and proceeds towards the condensation point at $p = p_\sigma$? What goes wrong with the mathematical expression for G as a function of pressure? In fact, there are excellent arguments,[3,4] based on the droplet picture of condensation of a vapor (or, for any similar first-order transition), demonstrating that the $G(p)$ isotherm simply comes to an *end* at $p = p_\sigma$, as suggested in the figure; there is *no* natural extension of the isotherm which can be regarded as describing the metastable state.

To forestall objections that the droplet arguments are merely heuristic, I must stress that this result has been proved in full mathematical rigor by the Russian theorist Isakov[5] for a very fundamental statistical mechanical model, namely, the Ising ferromagnet or simple lattice gas. Isakov's achievement, published in 1985, deserves wider recognition than it has gained so far. To be more technical for a moment, what the droplet picture demonstrates and Isakov proves is that the nth derivatives of the Gibbs free energy at the condensation point vary as

$$G_{(n)} \equiv \left. \frac{\partial^n}{\partial p^n} G \right|_{p=p_\sigma} \approx A b^n (n!)^\delta, \tag{4}$$

for large n, with A and b temperature-dependent constants and $\delta > 1$. (Actually one has, at least at low temperatures, $\delta = d/(d-1)$, where d is the spatial dimensionality.) It follows from this that one *cannot* use the coefficients $G_{(n)}$ to construct a convergent Taylor expansion about $p = p_\sigma$ with which to continue the isotherm beyond $p = p_\sigma$: the $G_{(n)}$ simply increase too rapidly as $n \to \infty$. In other language, the condensation point marks an essential singularity on the isotherm.

It is sometimes objected that such an essential singularity is nonphysical since in the case, say, of a near-ideal gas condensing into a liquid it is almost inconceivable that an experimentalist could measure sufficiently many derivatives of the pressure isotherm at condensation with sufficiently high precision to see the $(n!)^\delta$ behavior with $\delta > 1$, as predicted by the droplet arguments and proved by Isakov. First, however, one must recognize that a variety of profound physical facts are not actually accessible to *direct* experimental verification; perhaps the existence of quantum-mechanical wavefunctions is one of the most fundamental examples. Secondly, as the theory reveals, the physical fact underlying the essential singularity is the existence, in a vapor in true equilibrium near condensation, of indefinitely large molecular clusters or droplets which resemble the bulk fluid phase. Droplets of significant size may be exceedingly rare statistically and will be transitory in existence. Nevertheless, by the well-verified principles of statistical mechanics, they must be present with nonzero probability. In this way one sees that the essential singularity is very "physical". It can be avoided only in models with artificial interactions of infinite-range and vanishing strength.

Furthermore, what our conclusion implies theoretically is that one cannot solve the problem of representing metastable states properly in thermodynamics simply by analytically continuing the stable isotherms, as is traditionally assumed and usually conveyed in the text books. Instead, this issue must be thought through and reformulated. Considerations of nonequilibrium time-dependence must enter; a crucial question, both theoretically and experimentally, is "What is the lifetime of a metastable state and what physical processes determine it?"

These conclusions would, I feel sure, have engaged Gibbs's attention. Indeed, he might well have argued strongly against them! However, as the mathematician Mark Kac was wont to say, the beauty of a mathematical proof is that it convinces even a stubborn proponent. Gibbs might have been firm in his convictions but he would surely have recognized the power of a rigorous mathematical proof, always so valuable in physics when one faces subtle or controversial basic questions.

5. The Phase Rule and its Violations

One of Gibbs's most renowned contributions (although he, himself, did not especially stress it) is the *phase rule*. It has proved of particular value to chemists, to metallurgists, and to others dealing with complicated, many-component systems with the possibility of various numbers of coexisting phases. The issue then is: "Can one still have the same number of phases coexisting when one changes the pressure or varies the temperature or chemical potentials?" Gibbs answers this by showing that if r distinct phases coexist in a system of n chemical components then the number of degrees

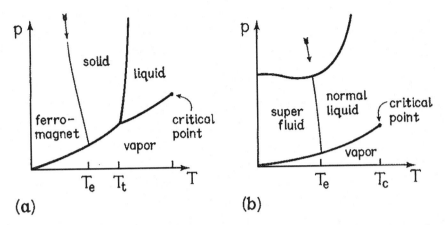

Figure 9. A phase diagram for a metallic element, like nickel, exhibiting an ordinary gas-liquid critical point and a critical line, marked by an arrow, separating paramagnetic and ferromagnetic solid phases; (b) sketch of the phase diagram of helium-four showing the ordinary critical point and the superfluid lambda line.

of freedom of the state is

$$f = n + 2 - r. \tag{5}$$

One can rephrase this by noting: (1) the total number of variables required to specify the thermodynamic state is $D = n + 1$ (say, n number densities or n chemical potentials plus, e.g., the temperature), and (2), the *codimension* of a manifold of r coexisting states is $\bar{D}_r = r - 1$ (representing one restriction for each additional phase). Then the net variability, or dimension of the manifold is just $f \equiv D_r = D - \bar{D}_r$. The phase diagram for a single-component system shown in Figure 9(a) illustrates the rule that on melting, liquid and solid can coexist and so, with $n = 1$ and $r = 2$, we have $f = 1$; thus melting should occur along a linear locus as observed. Likewise, vapor and liquid coexist along the vapor pressure line. The coexistence of *three* phases, liquid, solid and vapor, yields $f = 1 + 2 - 3 = 0$ and so can occur only at a fixed or invariant point, namely, the triple point, (p_t, T_t).

Now, as mentioned, Gibbs was aware of the gas-liquid critical point. Accordingly, he presented a modification of his rule adapted for critical manifolds, namely,

$$f_{crit} = n - 1. \tag{6}$$

(One might say that two phases coexist ($r = 2$) but have identical properties which yield one extra condition.) This can also be checked in Figure 9(a), which we may envisage as describing a ferromagnetic chemical element

($n = 1$), such as nickel. One has $f_{crit} = 1 - 1 = 0$ and so criticality should occur only at an invariant point. This is indeed true of the gas-liquid critical point (as marked in the figure). The same applies in Figure 9(b) which corresponds, schematically to helium-four.

However, in both Figures 9(a) and 9(b) violations of Gibbs's criticality rule (6) occur; these are the *lines* of critical points ($f_{crit} = 1$) marked by the arrows. In Figure 9(a) one has a line of ferromagnetic critical points below which the atomic spins spontaneously align to produce a net bulk magnetic moment. This critical line terminates at a *critical endpoint*, at $T = T_e$, on the solid-vapor phase boundary. In Figure 9(b) the critical line, or *lambda line*, runs between upper and lower critical endpoints and separates *superfluid* helium from normal liquid helium. Of course, we must first agree that Gibbs would have considered these lines as true critical lines. I do not think it would be hard to persuade him by drawing attention to the various *singularities* in the second derivatives of the $G(p, T)$ surface that actually characterize critical points in the laboratory. In particular, I would first present data demonstrating the divergence of the specific heat at constant volume at a normal gas-liquid critical point according to

$$C(T) \approx A^{\pm}|t|^{-\alpha} + B, \qquad (7)$$

as $t = (T - T_c)/T_c \to 0\pm$ with $\alpha \simeq 0.11$. This behavior was unknown to Gibbs but today one has excellent observations.[6] Then I would produce evidence for the fully analogous specific heat singularity occurring at the transition to ferromagnetism in nickel, iron, etc., and point out that the same behavior occurs under pressure. (For isotropic ferromagnets one usually has $\alpha \simeq -0.1$.) In the case of helium-four the corresponding behavior, with $\alpha \simeq -0.02$, has been demonstrated with remarkable precision[7] along the full length of the lambda line.

What has gone wrong with Gibbs's rule? Or what has been overlooked which allows these violations for critical points? Had the issue been put to Gibbs for ferromagnetism he would, I believe, have quickly recognized the crucial issue, which is one of *symmetry*. For magnetic systems we normally draw the phase diagram for zero magnetic field $H = 0$; but the spontaneous magnetization is then, in the absence of an externally selected direction, equally happy to point "up" or "down". More generally, one has the exact symmetry

$$H \Rightarrow -H, \quad M \Rightarrow -M.$$

It is this special symmetry which undermines the "generic" argument Gibbs gave for his phase rule. Nevertheless, if we insist that whenever one has such a symmetry one must imagine a suitable "ordering field" that breaks the symmetry—which, in the case of ferromagnetism, is just the magnetic field, H—then one may enlarge the phase diagram by including this field (in addition to p and T). Once this is done, all Gibbs's phase rules are fully restored in the extended phase space.

Antiferromagnetic systems, which exhibit similar phase-rule violating critical lines, equally admit an exact magnetic symmetry. In that case, however, the required "staggered" or "up-down" magnetic ordering field cannot (normally) be generated in the laboratory although its conceptual basis is firm. Liquid helium would, also, have been more puzzling to Gibbs. The symmetry now is that of the wave function, Ψ, which, as a complex number, admits multiplication by any complex factor of modulus one, say $e^{i\phi}$; the phase, φ, of the wave function can be changed arbitrarily with no change in the physics. This rotational, $O(2)$ symmetry allows critical lines to appear in the (p, T) diagram. However the corresponding bulk "off-diagonal ordering field" cannot, as far as we know, be realized physically in any way.

6. Phases and their Characterization

Our consideration of the phase rule raises a basic question, namely, "What is a phase?" Gibbs more or less takes for granted that one can see a phase with one's eyes, as in phase separation within a test tube. (Recall, again, the cartoons in Figure 2.) But that is inadequate today: one of the current issues in modern condensed matter theory is to understand what properties properly distinguish one phase from another when density and composition fail to do so. An important tool is the concept of "order". In a disordered phase like a gas or liquid, the correlations between local properties, density, magnetization, etc., at spatially nearby points decay rapidly to zero as the separation, r, increases to ∞; typically the decay is exponentially fast, i.e., as $\exp(-r/\xi)$. By contrast, when *long-range order* is present, signaling a new phase, the correlations of one or more properties do *not* decay to zero; rather, as in a ferromagnet, they may approach a nonzero value (related to the spontaneous magnetization), or they may eventually oscillate with constant amplitude, as does the density in a crystal. In addition, the importance, especially in $d = 2$ dimensions, of phases characterized by *slow* or *algebraic* decay of correlations, say, as a power law like $1/r^n$, has been recognized: superfluid films are a case in point. The correlations in phase and amplitude of the quantized wave function, $\Psi(r) = e^{i\phi(r)}|\Psi(r)|$, are characterized by power law decays.

More recently another and in many ways more profound concept for detecting phases has gained significance: this is via the response of a system to changes in the boundary conditions at the walls or enclosing surfaces. This matter I am confident would have interested Gibbs because of his concerns for surfaces and interfaces. To explain this in a little more detail, consider first Figure 10 which shows two copies of the system in question in the form of d-dimensional cubes of sides L. Suppose we suspect that there may be a phase connected with the ordering of some microscopic parameter which, conceptually, we can control at a surface even if, perhaps, it is not amenable to actual experimental manipulation. For simplicity, we may further suppose that the microscopic variable is like a local magnetization which has a definite sense, "up" or "down", or like the wave function Ψ which has a phase angle which may be pinned by some appropriate local field. In Figure 10(a) the sense of the order parameter is imposed in "like" fashion on one pair of opposite faces of the cube, as suggested by the arrows (the remaining sides being left "free" or "neutral" or subject to periodic boundary conditions); in Figure 10(b) opposite or "unlike" conditions are imposed on the pair of faces. We then ask for the difference, ΔG, in Gibbs free energy between the two situations. (With realistic conventions the total free energy will be larger for unlike conditions.) Clearly, ΔG must depend on L. Indeed as L becomes large it is reasonable, as we will see, to postulate a power law,

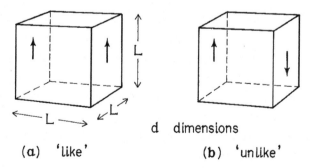

Figure 10. Illustrating the detection of phases by the change in free energy resulting on changing boundary conditions on opposite faces of a cube from (a) "like" to (b) "unlike": see (8).

namely,
$$\Delta G_L \sim L^\theta \quad \text{as} \quad L \to \infty. \tag{8}$$

Since changes at a surface cannot alter the bulk free energy density the characteristic exponent θ must satisfy $\theta < d$.

Now, as to more specific values of θ suppose, first, that θ is *negative*. Then in a large system there is a vanishing sensitivity to the boundary conditions. That is really the hallmark of a *disordered system* such as a fluid or a paramagnet: asymptotically there is *no* coupling between far boundaries.

Next refer to Figure 11(a) which illustrates, in essence, the situation of coexisting phases envisaged by Gibbs. As the illustration suggests, the opposing boundary conditions induce distinct bulk phases ("up" and "down") on opposite sides of the cube. Between these coexisting phases there should appear a definite physical region of inhomogeneity or changeover which we normally characterize as an *interface* or *domain wall*, etc. If the phases differ in density or chemical composition, one can usually hope to see such an interface by eye; but that certainly need not be the case. In any event, the total free energy will be least for minimal interfacial area; the cross-sectional area, in turn, cannot be of smaller magnitude than L^{d-1}. Thus this case is characterized by $\theta = d - 1$, as stated in the figure. A little reflection shows that the coefficient of (asymptotic) proportionality in (8) must simply be the interfacial tension, say, $\Sigma(T)$. Note that if the cube is a "black box" but one then observes the value $\theta = d - 1$, it is fair to conclude that some sort of interface between domains must be present.

Figure 11(b) illustrates a contrasting situation appropriate to an order parameter, like $\Psi(r)$, that can be smoothly "twisted" or "bent" from one side of the cube to the other. When the order parameter is subject to a

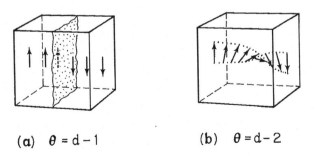

(a) $\theta = d - 1$ (b) $\theta = d - 2$

Figure 11. Depiction of an interface induced by opposing boundary conditions, leading to $\theta = d - 1$; (b) the twisting of an order parameter embodying a continuous symmetry, yielding $\theta = d - 2$.

continuous, rotational type of symmetry, such a twisted or helical configuration characterizes the lowest free energy: there is no longer a localized interface. It is then not hard to see that the exponent takes the new, smaller value $\theta = d - 2$.[8] Clearly, the concept of coexisting phases needs generalization for this situation; in some senses one can say there is a continuous infinity of distinct bulk phases. Superfluids and various other systems are, in fact, characterized by $\theta = d - 2$; the coefficient of proportionality in (8) now determines the superfluid density, $\rho_s(T)$, or, more generally, an appropriate *helicity modulus* (or stiffness or rigidity) $\Upsilon(T)$.[8] It is worth noting that a *two-dimensional* superfluid or XY-like ferromagnet *cannot* display a broken symmetry or corresponding spontaneous order, off-diagonal or magnetic, etc., as arises in three dimensions. Nevertheless, such two-dimensional systems *can* have a $\rho_s(T)$ or helicity modulus detectable through the influence of boundary conditions and defined via (8).

Systems with quenched impurities or other forms of frozen-in randomness present an especial challenge currently. It seems likely that the exponent θ plays a crucially important role in understanding such systems, particularly random magnets including the so-called *spin-glasses*.[9–11] Because of the intrinsic randomness, the criterion (8), as stated, is no longer fully adequate. Indeed, one expects the *average* over a random ensemble to yield $\langle \Delta G_L \rangle = 0$ for reversal of one boundary condition. However, replacing ΔG_L in (8) by $\Delta G_L^{\text{rms}} \equiv \langle (\Delta G)^2 \rangle^{1/2}$ should suffice to define θ for random systems. The inequality $\theta < \frac{1}{2}(d-1)$ ought then hold rather generally (for short range interactions).[11] Numerical studies of model d-dimensional spin glasses confirm this and yield[9–11]

$$\theta = -1, \quad \text{for } d = 1,$$
$$\theta \simeq -0.2, \quad \text{for } d = 2, \quad (9)$$
$$\theta \simeq +0.2, \quad \text{for } d = 3.$$

Note that for one- and two-dimensional systems θ is negative so that no true, ordered spin-glass phase should exist. By contrast, θ is positive in three dimensions so that a distinct, ordered spin-glass phase, and corresponding sharp thermal transition *should* then exist.[11] I believe that the discovery of this way of characterizing spin glasses and other random systems represents an important theoretical development.

Incidentally, these results highlight the importance of the spatial dimensionality for the existence and nature of various phases, something that

Gibbs might, perhaps, have suspected although he does not really address the issue. For simple spin-glass models one can, in fact, obtain an expansion for the critical point, $T_c(d)$, in inverse powers of d. If J is the appropriate spin-spin coupling energy, one obtains[12]

$$\tanh(J/k_B T_c) = \sigma^{-1/2}\left(1 - \frac{3\frac{1}{2}}{\sigma^2} - \frac{10\frac{1}{2}}{\sigma^3} - \frac{91\frac{1}{8}}{\sigma^4} - \frac{699\frac{5}{12}}{\sigma^5} - \cdots \right), \quad (10)$$

where $\sigma = 2d - 1$. On extrapolation to low values of d, one finds[12] that $T_c(d)$ goes to zero as $d \to d_c \simeq 2.5$. Since a vanishing T_c implies the nonexistence of the corresponding ordered phase, this agrees with the conclusions based on the θ-exponent values in (9).

7. Singularities in the Phase Diagram

Let us return to the phase diagram itself and ask if the actual phase boundaries in the (p, T) plane may have any mathematical peculiarities. In particular, consider Figure 12 which displays the vicinity of a critical endpoint at (p_{end}, T_{end}). One might think of the phase lying below the first-order phase boundary, $p_\sigma(T)$ (shown bold), as vapor while the phases lying above represent solids. A critical line, labelled $T_c(p)$, separates a low-temperature ordered phase from a high-temperature disordered one. On crossing the critical line the specific heat at constant pressure, $C_p(T)$, diverges with an exponent α as indicated in the figure. In many cases one will have $\alpha \simeq 0.11$.

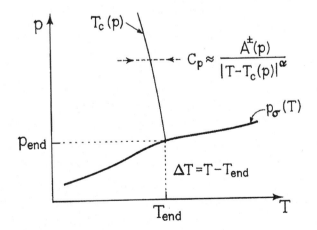

Figure 12. The vicinity of a critical endpoint in the (p, T) plane. The first order phase boundary, $p_\sigma(T)$, displays a singularity at T_{end}.

Now, what form will the vapor-pressure line, $p_\sigma(T)$, take as T passes through T_{end}? At the critical endpoint, p has a definite value and so does the slope, $(dp_\sigma/dT)_{\text{end}} \equiv p_1$. Thus we can write

$$p_\sigma(T) = p_{\text{end}} + p_1 \Delta T + \cdots, \qquad (11)$$

with $\Delta T = T - T_{\text{end}}$. But what comes next? It seems likely that Gibbs and most theorists of his generation would answer: "clearly $p_2(\Delta T)^2$ and further analytic terms in the Taylor series expansion of $p_\sigma(T)$." However, that answer is almost certainly wrong! Indeed, one can show, using arguments that I think Gibbs would have accepted, that the divergence of $C_p(T)$ on the critical line induces a *singularity* in the vapor-pressure line itself. More concretely, for $\alpha > 0$ the curvature, $(d^2 p_\sigma/dT^2)$, of the vapor pressure line *diverges*, like $-|\Delta T|^{-\alpha}$, at the critical endpoint. Thus (11) becomes

$$p_\sigma(T) = p_{\text{end}} + p_1 \Delta T - B^\pm |\Delta T|^{2-\alpha} + p_2 (\Delta T)^2 + \cdots, \qquad (12)$$

as $\Delta T \to 0\pm$, where B^+ and B^- are positive, and typically $2-\alpha \simeq 1.89$. One concludes, furthermore, that the ratio B^+/B^- will be universal over a wide range of systems with, when $\alpha = 0.11$ (for $d = 3$), $B^+/B^- = A^+/A^- \simeq 0.52$.[13] (In $d = 2$ dimensions $|\Delta T|^{2-\alpha}$ should be replaced by $-(\Delta T)^2 \log |\Delta T|^{-1}$ and one has $B^+/B^- = 1$.)

To my knowledge the conclusion (12) has not been checked experimentally although it would certainly be interesting to do so since the theoretical arguments are not rigorous. However, I am rather confident of their correctness and one can, in particular, demonstrate that the droplet effects[3] that lead to the essential singularities at condensation discussed above do not affect the validity of (12).

8. Bicriticality

While Gibbs would not, I feel, have been greatly surprised by the analysis leading to the critical endpoint singularity in the vapor-pressure line, he might well have reacted differently to the rather more dramatic singularities that appear in the phase diagrams for certain antiferromagnetic crystals. In this case the appropriate diagram is formed by replacing the pressure by the magnetic field, H. Figure 13 (see p. 275) illustrates the phases that are observed in materials such as MnF_2 and $GdAlO_3$. At high temperatures and small fields one finds the usual disordered, paramagnetic phase. On cooling, the system reaches a critical line, labelled $T_c^\parallel(H)$, below

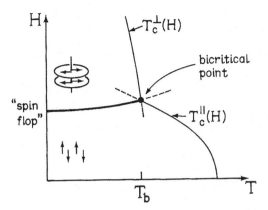

Figure 13. Schematic phase diagram of a weakly anisotropic antiferromagnet displaying a spin-flop transition and a bicritical point at (H_b, T_b) where two critical lines, $T_c^{\parallel}(H)$ and $T_c^{\perp}(H)$, meet. The correct behavior close to the bicritical point is *not* as shown here: see the experimental results in Figure 14.

which the spins order antiferromagnetically and parallel to the *c*-axis of the crystal (which is also the direction of the applied magnetic field). On increasing the magnetic field at low temperatures, the system undergoes an abrupt, first-order transition: this is the so-called "spin flop" transition marked by the bold line in the figure. In the new, high-field phase the spins remains antiferromagnetically aligned but the direction of alignment is now in the plane *perpendicular* to the *c*-axis as suggested by the arrowheads in the figure. Finally, on raising the temperature, the spin-flop phase attains criticality on the locus labelled $T_c^{\perp}(H)$; further heating destroys the order and the system returns to the paramagnetic phase. In total, then, there are two distinct critical lines and a first-order, spin-flop line. As shown in the figure, these three lines meet at a unique point—the *bicritical point* at (H_b, T_b).[14] In practice the bicritical field, H_b, is large, of the order of 100 kOe.

Now it is perfectly natural to expect the two critical lines and the spin-flop line to approach the bicritical point from three distinct directions, as indicated in the figure. Indeed, the standard phenomenological and mean-field theories yield this result with, as at a standard triple point, a 180° rule applying. Furthermore, the initial exploratory experiments appeared to confirm this picture.[14] However, renormalization-group and scaling arguments lead to very different conclusions! Rather generally, they demonstrate that two critical lines in a phase plane do not meet like two simple curves of the sorts treated by Euclid's geometry; instead the bicritical point is special and markedly singular.

To approach the issue theoretically it is reasonable to study the simplest models that display the essential phenomenon. By considering the interplay of the two distinct phase symmetries characterizing the antiferromagnetic, or $\parallel$ phase, and the spin-flop, or $\perp$ phase, one concludes that the spin-flop boundary represents, asymptotically, a locus of higher-order symmetry. Specifically, it embodies an axis of effective *isotropy* in spin space. One is thus led to consider an effective magnetic Hamiltonian of the form

$$\mathcal{H}_{\text{mag.}} = \mathcal{H}_{\text{iso.}} + g\mathcal{H}_{\text{aniso.}}, \tag{13}$$

in which the last term represents symmetry-breaking interactions which for $g > 0$ favor $\perp$ (or spin-flop) ordering while for $g < 0$ they favor $\parallel$ ordering. In rather more technical terms,[14] one may take $\mathcal{H}_{\text{iso.}}$ as a fully symmetric or $O(3)$ Heisenberg Hamiltonian with nearest-neighbor spin-spin couplings of the form $\vec{S}_i \cdot \vec{S}_j$ while $\mathcal{H}_{\text{aniso.}}$ embodies so-called Ising or $O(1)$, and XY or $O(2)$ couplings in the form

$$S_i^z S_j^z - \tfrac{1}{2}(S_i^x S_j^x + S_i^y S_j^y).$$

For $g = 0$ this system will order into a fully isotropic, $O(3)$-symmetric state at a critical temperature, say T_b, which corresponds to the bicritical point. For nonzero g we can envisage a correspondence $g \sim (H - H_b)$.[b] Then for $g > 0$ there will be a critical locus, say, $g_c^+(T)$, corresponding to $T_c^\perp(H)$; likewise for $g < 0$ a locus $g_c^-(T)$ corresponds to $T_c^\parallel(H)$.

Now general scaling theory principles and renormalization-group analysis indicate that the effects of any perturbation on a fully isotropic near-critical system should scale with a characteristic, nontrivial power of the temperature deviation, which we may take as

$$t = (T - T_b)/T_b. \tag{14}$$

Asymptotically, as $t \to 0$ all properties should, thus, depend only on the scaled combination $g/|t|^\phi$. It follows,[14] in particular, that the two critical loci should obey the law

$$g_c^\pm(T) \approx W^\pm |t|^\phi, \tag{15}$$

where W^+ and W^- are constants (linked by a universal ratio).[14] Now for a ($d = 3$)-dimensional system the value of the crucial exponent ϕ can be

---

[b]Strictly speaking one has a correspondence $g \sim (H - H_b) + c(T - T_b)$ where c is a system-dependent coefficient; however, this does not affect our principal conclusions.

found only by explicit calculations. Renormalization group $\epsilon = 4 - d$ expansions and high-temperature series extrapolations[14] yield

$$\phi = 1.25 \pm 0.02. \tag{16}$$

The significant feature is that ϕ *exceeds* unity. This means that the two critical loci, $g_c^+(T)$ and $g_c^-(T)$, approach the bicritical point with a *common tangent* which, for the model (13), coincides with the symmetry axis.

If these results are translated into the (H, T) phase diagram of Figure 13 one concludes that contrary to appearances the two critical lines $T_c^\perp(H)$ and $T_c^\parallel(H)$ should curve around and approach the bicritical point *tangentially to the spin-flop line*. (If ϕ were equal to unity the form of Figure 13 would be correct.) This unexpected tangency prediction prompted experiments by Rohrer and collaborators.[15] Figure 14 depicts some of their first data for manganese fluoride. As is evident from the range of $T - T_b$ and $H - H_b$ displayed, the true asymptotic bicritical behavior is found only within 1 or 2% of the bicritical point: the initially almost vertical $T_c^\perp(H)$ locus hooks sharply back towards lower T at less than 0.1°K above the bicritical point! Both critical lines become tangent to the spin-flop transition line. By

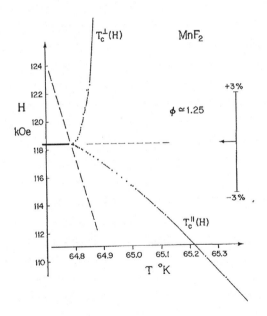

Figure 14. Data near the bicritical point of manganese fluoride revealing the asymptotic tangency of the critical lines to the spin-flop line. The degree of tangency is described by the crossover exponent $\phi > 1$: see text and (16). (After H. Rohrer: see also Ref. 15.)

measuring the degree of tangency at T_b the exponent ϕ can be estimated; the experimental data[15] agree fully with (16).

Precise observations of the bicritical point in gadolinium aluminate and other materials leads to the same conclusion, namely, boundaries in a phase diagram can display strongly singular behavior. The origin of these singularities lies in the *fluctuations* in the equilibrium state which are enormously enhanced near criticality. It is appropriate to remind ourselves that Gibbs's systematic formulation of statistical mechanics provides the foundations on which modern calculations of critical fluctuations are based.

9. Phase Diagrams in More Variables

I mentioned, but did not stress, that the magnetic field referred to in Figures 13 and 14 must be aligned parallel to the *c*-axis of the crystal. In fact, one of the severe practical difficulties of the experiments is that the alignment must be very precise indeed. The real magnetic field, $\vec{H}$, is, of course, a vector with three components H_x, H_y, and $H_z \equiv H_\parallel$. If $H_x \equiv H_\perp$ and H_y are not accurately zero one cannot find the bicritical point. Since a real crystal does not come with its axes labelled for the convenience of the experimenter, finding the bicritical point is like looking for the proverbial needle in a haystack! Indeed it is worse since the "haystack" here is really four-dimensional, the coordinates being H_x, H_y, H_z and T.

Figure 15 (see p. 279) is presented in order to give a feel for these circumstances (and also because I cannot resist displaying a phase diagram in a space of more than two thermodynamic fields). The figure represents a schematic bicritical phase diagram, appropriate to MnF_2, in the three variables $H_\parallel$, $H_\perp$, and T (with $H_y \equiv 0$). I think it would have appealed to Gibbs and I invite you to appreciate its form! First, in the shape of a balloon, one notices a critical surface—for the experts, this is of Ising-like or $O(1)$ character—which separates the magnetically disordered, paramagnetic, high-temperature phase from a region of antiferromagnetic order (of varying spin axes). Where this surface approaches the upper part of the $H_\perp = 0$ plane there appears a shallow furrow. On the intersection with this plane lies the $T_c^\perp(H_\parallel)$ locus of Figures 13 and 14; it forms a singular "seam" of different critical character (XY-like or $O(2)$ in nature) which joins the two sides of the furrow. The bicritical point itself appears at the end of this seam and represents an umbilicus on the surface. (As mentioned before, the

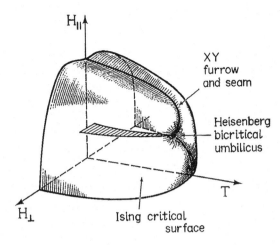

Figure 15. Phase diagram in the space of three variables, T, $H_\parallel$, and $H_\perp$, of the vicinity of a bicritical point in an antiferromagnet like MnF$_2$.

criticality here has Heisenberg or $O(3)$ character.) Finally, running back from the bicritical point, inside the ordered regime, is a narrow pointed "shelf" which represents a surface of first-order transitions; its intersection with the $H_\perp = 0$ plane is the spin-flop line. The edges of this shelf constitute two critical lines (of Ising-like character).

This figure serves to indicate the topological and geometric complexities that can appear in phase diagrams in three or more variables. In fact, to reveal the full nature of the thermodynamic behavior in the vicinity of a bicritical point requires a thermodynamic space of nine or ten dimensions! A *tricritical* point at which, on one thermodynamic plane, a critical line becomes a first-order phase boundary requires "only" four dimensions. We will not, however, enter further into such details here. Nevertheless, it is worth noting that more complicated phase diagrams occur even with just two variables. An example is provided by the occurrence of *multiphase points* at which infinitely many distinct first-order phase boundaries converge and meet.[16] These have been found specifically in statistical mechanical models, such as the so-called ANNNI model (or axial next-nearest neighbor Ising model), which represent physical systems with competing interactions leading to many spatially modulated phases of increasing periods. Various binary metallic alloys exhibit sequences of long-period modulated phases as the temperature and composition are varied;[16] it is thus possible that true multiphase points will appear in some real materials.

10. Anomalous First Order Transitions

Let us now venture out from the world as it is into the world as it might have been. Consider again the Gibbs energy surface, $U(S, V)$, as depicted schematically in Figure 1. Recall that is a convex surface; but, beyond that it has ruled regions corresponding to two-phase coexistence and flat faces which describe the coexistence of three distinct phases. The conjugate surface, $-G(p, T)$, is shown in Figure 4 (again only schematically). It is convex in its variables in precisely the same sense as is $U(S, V)$. On the other hand, the two-phase regions are now represented merely by sharp bends or curvilinear creases on the surface; likewise, the flat face describing three-phase coexistence is transformed simply into a pointed peak on the G surface—the triple point, where three linear phase boundaries meet together.

Now let us ask: "Why did Gibbs describe the two potential surfaces this way round?" Could not the appearances have been reversed so that the surface in Figure 1, with ruled regions and a flat face, would represent $-G$ as a function of p and T? Is there something wrong with that? After all, convexity is fully respected in both cases. Would it violate thermodynamics or, perhaps, statistical mechanics in some way if the $U(S, V)$ surface had bends and a peak as in Figure 4? And, if there is nothing wrong with these alternatives, why did Gibbs not discuss them or, at least mention them?

One can ask an almost equivalent but visually rather simpler question by contemplating Figure 16. Part (a) illustrates the behavior of a (p, V) isotherm associated, in the standard way, with gas-liquid condensation. On reducing the volume at fixed T, the pressure rises smoothly until one

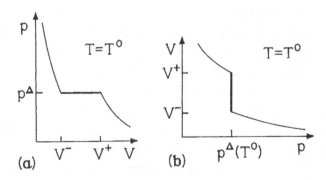

Figure 16. An isotherm illustrating a normal first-order phase transition which can be regarded, in (b), as a discontinuity of V as a function of p.

reaches the condensation point V^+. Then the isotherm becomes level at an unvarying pressure p^Δ over an interval down to V^- as gas is compressed entirely into liquid. Thereafter the pressure once again rises continuously. Evidently, the figure represents a perfectly ordinary first-order phase transition as seen innumerable times in experiments. The situation looks more dramatic in Figure 16(b): the same data are plotted but now V is regarded as a function of p at fixed T. In this case, when the pressure is reduced from high values the volume rises smoothly and continuously until the pressure p^Δ is reached. At that point, however, the isotherm jumps *discontinuously* from V^- to V^+. Beneath p^Δ a continuous increase is resumed. From this viewpoint, a normal first-order transition is most conveniently characterized as a *discontinuity* of a V versus p isotherm.

Now consider the theoretical or experimental question: "Why could not matters be reversed as in Figure 17?" The pressure may again be regarded as a function of volume at fixed T but on reducing V, a point of transition, V^Δ, is reached. On further compression the pressure jumps discontinuously from p^- to p^+. Thereafter p rises smoothly and continuously as V decreases. Is anything wrong with such an *anomalous first-order transition* in which the pressure is a discontinuous function of V (or, as in Figure 17(b), V remains constant while p varies)?

Well, the main thing wrong with pressure discontinuities like that in Figure 17 is that no good experimentalist has ever seen one! When something is never observed one is strongly inclined to believe that some basic law of nature is involved. The question thus becomes: "What is the law that stops us from discovering anomalous first-order transitions?" Or: "Is there an

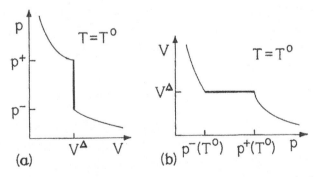

Figure 17. An *anomalous* first-order transition in which the pressure isotherm is a discontinuous function of volume or equivalently, as in (b), V is constant over an interval of p.

extra law of thermodynamics which Gibbs appreciated (but did not explain) which accounts for his describing only normal, first-order transitions like those implied by Figures 1 and 4 and illustrated in Figure 16?"

The answer is that there is *no* law of thermodynamics that forbids anomalous first-order transitions of the sort we have been contemplating.[17,18] Perhaps, then, anomalous transitions are excluded by the principles of statistical mechanics which, as a microscopic theory, underlies thermodynamics? Again the response must be "No".[17,18] But, if that is so, why does one never see a discontinuity in a p versus V isotherm? The answer now is that, in fact, one *does* find anomalous first-order transitions in appropriate models![18,19] Indeed, one can write down perfectly good Hamiltonians describing the interactions of particles, even in a single spatial dimension, which, when the statistical mechanics is worked out exactly, lead to pressure discontinuities and other anomalous transitions.[17-19]

If such models exist why are anomalous transitions not seen in the real world? To understand this one must appreciate that the particular models in question entail *many-body interactions* in a crucial way.[18,19] It is a commonplace that the forces between two particles at positions r_1 and r_2 can be described by a pair potential, $\varphi_2(r_1, r_2)$. If a third particle is introduced at r_3, the total *pairwize* energy of interaction is

$$\varphi_2(r_1, r_2) + \varphi_2(r_2, r_3) + \varphi_2(r_3, r_1).$$

However, to obtain the full interaction energy one must in general also allow for intrinsic three-body forces with a potential, say, $\varphi_3(r_1, r_2, r_3)$. Similarly when a fourth particle is introduced one must include, in addition to further pair interactions, $\varphi_2(r_4, r_1), \ldots$ and three-body or triplet terms, $\varphi_3(r_2, r_3, r_4), \ldots$, a four-body interaction $\varphi_4(r_1, r_2, r_3, r_4)$, and so on. In order to have sensible statistical mechanics and thermodynamics, the set of many-body potentials $\Phi = \{\varphi_\ell(r_1, \ldots, r_\ell)\}$ must satisfy some appropriate conditions. It turns out[18,20,21] that there is a natural condition which may be expressed mathematically as the bound

$$\|\Phi\| \equiv \sum_\ell \frac{1}{\ell} \|\varphi_\ell\| < \infty. \tag{17}$$

This is a sufficient condition for the existence of the so-called *thermodynamic limit* (and it is also close to a necessary condition); it basically expresses the fact that in a large system the total energy per particle should be finite. (Note that ℓ particles share in the ℓ-body potential $\varphi_\ell(r_1, \ldots, r_\ell)$; this accounts for the factor $1/\ell$ in (17).)

While satisfaction of condition (17) (together with a few further requirements[18,20]) ensures behavior satisfying the basic laws of Gibbsian thermodynamics, it does allow *both* normal *and* anomalous first-order transitions. In other words, a jump in the pressure at constant volume (or density) is permitted and, as stated, this will occur for suitable interaction potentials. On the other hand, Griffiths and Ruelle[21] invoked the *stronger* bound

$$\|\Phi\|^+ \equiv \sum_\ell \|\varphi_\ell\| < \infty, \tag{18}$$

which differs from (17) only by omission of the factor $1/\ell$. This condition is certainly not necessary for proper thermodynamic behavior. However, Griffiths and Ruelle proved[21] that if a system of particles[c] satisfies (18) it *cannot* exhibit any type of anomalous first-order transition since $G(p,T)$ must be *strictly* convex with *no* ruled sections or flat faces. In particular, the pressure must be a continuous function of the volume at fixed T. The behavior shown in Figure 17 is then impossible.

We may conclude that the absence of the crazy but thermodynamically allowed anomalous transitions in the real world teaches us something beyond the general validity of thermodynamics and statistical mechanics. Specifically, we learn that although many-body forces of indefinitely high order, ℓ, are doubtless present in real systems they fall off more rapidly as ℓ increases than might have been so in another universe which, nonetheless, would still obey bulk thermodynamics. Thus not everything that is mathematically natural—fully respecting the basic convexity—need occur in reality!

11. Classifying First-Order Transitions

It is interesting to pursue anomalous and normal first-order transitions a little further since it transpires that not every imaginable type is actually possible; indeed, the dictates of convexity have something further to contribute. Consider Figure 18: part (a) extends Figure 16(a) by embedding a particular isotherm in a set displaying condensation and leading up to a critical point; part (b) exhibits the same situation in a way that makes it clear that the embedded discontinuity in V versus p at fixed T, which characterizes this normal transition, is *movable*. In other words, when T varies

---
[c]The work of Ref. 21 is actually confined to lattice gas systems, quantal or classical, but this is not important for the issue here.

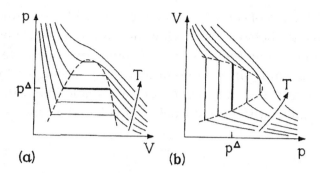

Figure 18. A set of isotherms illustrating an ordinary first-order phase transition which, in (b), can be regarded as embodying a *movable* discontinuity.

so does the point of discontinuity, $p^\Delta(T)$. Of course, such movement is not mandatory. For example, the (M, H) or magnetization-field isotherms of an ordinary ferromagnet, which are the analogues of Figure 18(b), display a discontinuity below the Curie point, T_c, which occurs only at $H = 0$ and so does not move. However, discontinuities in M versus H isotherms *are* movable and do, in fact, move in certain magnetic materials with transitions at $H \neq 0$.

In Figure 19 sets of isotherms illustrating anomalous transitions are exhibited. As drawn, the discontinuity point, $V^\Delta(T)$, does not move when T varies (although the discontinuity vanishes at certain critical temperatures). But could these discontinuities equally be movable? Said in other words, may one merely interchange the labels p and V in Figure 18 and still describe an acceptable thermodynamic system? It surprised me, personally, that the answer is "No". The convexity of the underlying thermodynamic

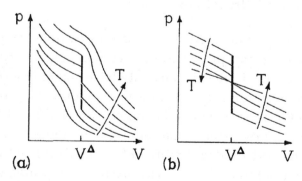

Figure 19. Sets of isotherms showing anomalous transitions which do not move. In fact, transitions of this class must be immovable: see text and Table. (After Ref. 17.)

potentials stressed by Gibbs dictates[17] that any discontinuity in a *p versus V* isotherm must be *fixed* and immovable. That is the only possibility (unless the discontinuity is *isolated* and disappears as soon as T changes). No proper statistical mechanical model can escape this result, try as one might!

More generally, one may try to classify[17] all possible types of first-order transition allowed by thermodynamics in systems described by two thermodynamic fields, h_a and h_b, like p or T, and their two conjugate densities, Q_a and Q_b, like V and S. For x, y, and z chosen from h_a, h_b, Q_a and Q_b, one may suppose that $y(x)$ at fixed z is discontinuous or, equivalently, that $x(y)$ at fixed z is constant over an interval. One finds[17] that the resulting twenty-four possible thermodynamic discontinuities or first-order transitions fall into six classes as listed in Table 1. Each discontinuity, **D**, might be *movable*, **M**, *fixed*, **F**, or *isolated*, **I**, as mentioned. In addition, a discontinuity may be *expanded*, **E**, describing a locus of constant z that actually fills a domain of the (x, y) plane (see Ref. 17). The table indicates that not all classes of discontinuity may actually exhibit fixed, movable, or expanded discontinuities.

Table 1. Classification of first-order transitions. A total of twenty-four possible first-order transitions may be represented as discontinuities, $D(x,y;z)$, in a plot of y versus x at fixed z, where x, y, and z are selected from two thermodynamic fields, h_a and h_b and their conjugate densities Q_a and Q_b. The discontinuities representative of the various classes are listed in lexicographical order in the second column; the third column is obtained by interchanging Q and h; interchanging the subscripts a and b also leaves the class unchanged and generates the twelve discontinuities not listed. The fourth column gives the particular embeddings allowed: see text. These various embedded discontinuities are realized in the specific transition types listed in the last column, each of which represents a special type of geometric feature on the Gibbs free energy surface: no others may occur. (After Ref. 17.)

Class	x	y	z	x	y	z	D	Transition types
1	h_a	h_b	Q_a	Q_a	Q_b	h_a	I, F, M, E	V, VIII; II, VII; II, IX; II, III
2	h_a	h_b	Q_b	Q_a	Q_b	h_b	I, E	V, VII; II, III
3	h_a	Q_a	h_b	Q_a	h_a	Q_b	I, F, M	III–V; II, VII; I, VI
4	h_a	Q_a	Q_b	Q_a	h_a	h_b	I, F	V, VII; II, III
5	h_a	Q_b	h_b	Q_a	h_b	Q_b	I, M, E	III–V; I, VI; II, III
6	h_a	Q_b	Q_a	Q_a	h_b	h_a	I, F, M, E	V, VIII; II, III; II, IX; II, VII

As an illustration of the use of the table consider, as previously, a discontinuity in a pressure *versus* volume isotherm. For this one has $(x, y; z) = (V, p; T) \equiv (Q_a, h_a; h_b)$ since V is conjugate to p. This combination appears in the table only under class 4. The only possibilities respecting the convexity are then, from the fourth column, *fixed* or *isolated*. The possible behavior of other first-order transitions may be elucidated similarly.

The roman numerals in the table identify nine different *transition types* corresponding to specific, characteristic, geometrical features on the thermodynamic potential surfaces. These transition types generate, in turn, twenty *groups of associated discontinuities* to which any one particular discontinuity must belong. Details are given in Ref. 17, which presents the full classification of all first-order transitions, normal or anomalous, that may occur in a space of two thermodynamic variables. It is surprising that so many distinct possibilities arise in this conceptually simple situation.

12. Random Thermodynamic Systems

We have examined, rather rapidly, various consequences that follow from Gibbs's majestic formulation of thermodynamics. I have stressed the geometrical facts underlying thermodynamics and the various images, particularly in the form of phase diagrams, which reflect them. But Gibbs's program to set the foundations of thermodynamics on a rational basis and to find the viewpoint of greatest simplicity is not yet completed. I alluded before to random systems. That topic is currently a focus of interest for many researchers and, thus, a few remarks about it may provide a fitting close to this discourse. Specifically, one should ask what, if any, are the extensions or modifications of the principles of thermodynamics applicable to physical systems in which some of the dynamical degrees of freedom have been "frozen" or "quenched" in irregular, random fashion? This is a challenge which Gibbs himself would surely have accepted.

Figure 20 (see p. 287) provides a topical example. It represents a schematic phase diagram, more or less as proposed by Aharony and coworkers,[22] for the renowned high-temperature superconducting material lanthanum barium copper oxide, $La_{2-x}Ba_xCuO_4$. The chemical components may be prepared at high temperatures, the crystal being grown with a relatively small concentration, measured by the mole fraction x, of barium, which acts as a controlled substitutional impurity or dopant. On quenching to lower temperatures, the barium ions become essentially immobile occupying random positions in the crystalline lattice. Barium has fewer valence electrons than lanthanum which leads to a density of electronic holes in the crystal and, thence, to dramatic changes in the properties of the phases.

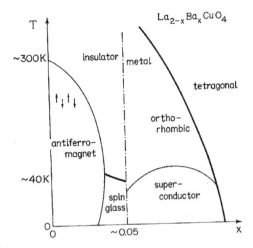

Figure 20. Suggested phase diagram for the high-temperature superconducting crystal lanthanum barium copper oxide showing various magnetic and electronic phases. (After Aharony et al., Ref. 22.)

Now all the thermodynamic variables which Gibbs considered come, within experimental time scales, to full equilibrium. It is this crucial feature which leads to the universality of thermodynamics and to a clear separation from nonequilibrium behavior. The concentration variable x in the lanthanum barium copper oxide crystals, however, does *not* describe equilibrium configurations of the barium. These would vary in response to changes in temperature, pressure, etc.; rather, to an extremely good approximation, the dopant ions may be regarded as permanently frozen in a configuration which, at best, might be typical of some sort of high-temperature equilibrium state but which does *not* change with temperature or other thermodynamic variables. What, then, can be said about the thermodynamic potentials and the phase diagrams as functions of the random impurity concentration x?

For small x one sees in Figure 20 a critical curve (thin) separating an insulating paramagnetic phase from an ordered antiferromagnetic phase. As x increases the critical line drops sharply from around 300°K and is, so it seems, met at about 40°K by a slowly falling first-order transition line (thick) bounding a spin-glass phase. This line is apparently cut off sharply when $x \simeq 0.05$ by a transition from insulator to metal, indicated by a vertical dot-dash line; probably this does not represent a sharp transition when $T > 0$. At high temperatures and larger values of x a first-order orthorhombic-to-tetragonal crystalline transition occurs. Lastly, the famous superconducting state appears at lower temperatures in the orthorhombic metallic phase. Are there any general principles governing phase diagrams such as this? Does Figure 20 correctly reflect them? Or, due to current experimental shortcomings, does it embody violations of the appropriate thermodynamic rules? Should certain free energies be convex in the randomness variable x? Or, if not, do some weaker constraints apply?

To my knowledge these various questions have not been answered but two recently proved general theorems for random systems can be reported and would surely have pleased Gibbs. In both the spatial dimensionality, d, enters in a crucial way. Consider first a critical locus, such as those shown in Figure 20, which can be crossed by varying a disorder variable like x. One expects very generally that there will be a correlation length, $\xi(x)$, which diverges as $(x - x_c)^{-\nu}$ when the critical line is crossed at x_c. For the critical exponent ν, Chayes, Chayes, D. S. Fisher, and Spencer[23] have proven, the inequality

$$\nu \geq 2/d \tag{19}$$

under quite weak assumptions. Since Ising-like criticality in $d = 3$ dimensions is characterized by $\nu = 0.632 \pm 0.002$, the condition (19), which implies $\nu \geq \frac{2}{3} \simeq 0.667$, has real experimental significance.

Finally, a striking result has been proved by Aizenman and Wehr[24,25] for $(d = 2)$-dimensional systems. It amounts to the destruction of first-order transitions by randomness, however weak. To be more explicit, suppose that $Q(r)$ represents a bounded local variable, like a spin or energy density, which is conjugate to a thermodynamic field h and that the system Hamiltonian includes a *random component* of the field, $\delta h_R(r)$, with mean $\langle \delta h_R \rangle = 0$. Then the theorem states[24] that the conjugate order parameter averaged over the randomness, namely,

$$\langle Q \rangle \equiv \frac{\partial \langle F \rangle}{\partial h}, \tag{20}$$

where $\langle F \rangle$ is the average free energy, *cannot show any discontinuity* as a function of h. Thus, for example, in the Ising model in a random field (where $Q(r) = s_r$ denotes a spin on lattice site r) there can be no spontaneous magnetization in two dimensions. When the local variables $Q(r)$ have a suitable continuous symmetry, as in a Heisenberg or XY Hamiltonian with $O(n \geq 2)$ spin symmetry, analogous results extend to dimensionalities $d \leq 4$. Evidently, a small amount of quenched randomness may have profound affects on a phase diagram.[25]

These remarks on the general properties of random thermodynamic systems bring my discussion to an end. In conclusion, let us continue to enjoy the magnificent legacy that Gibbs has left us in his writings on thermodynamics and statistical mechanics and be grateful for the opportunity to pay homage to his genius and to follow in his footsteps.

Acknowledgements

I am most grateful to the organizers of the Gibbs Symposium for the invitation to speak at Yale. Many authors and colleague have contributed to my thoughts and understanding of the matters discussed above but particular thanks are due to my longstanding friend and colleague Benjamin Widom, to my eldest son Daniel S. Fisher, and to my talented and enthusiastic former student Graeme W. Milton. The support of the National Science Foundation over more than two decades has been much appreciated.

References

1. See, e.g., J. C. Wheeler, *J. Chem. Phys.* **61**, 4474 (1974); *Phys. Rev. A* **12**, 267 (1975).
2. L. E. Wold, M. Sc. thesis, Cornell University (1974).
3. M. E. Fisher, IUPAP Conf. Stat. Mech. (Brown University, 1962); see S. Katsura, *Adv. Phys.* **12**, p. 416 (1963); *Physics* **3**, 255 (1967).
4. A. F. Andreev, *Sov. Phys. JETP* **18**, 1415 (1964).
5. S. N. Isakov, *Commun. Math. Phys.* **95**, 427 (1985).
6. See, e.g., J. A. Lipa, C. Edwards and M. J. Buckingham, *Phys. Rev. Lett.* **25**, 1086 (1970).
7. See, e.g., G. Ahlers, *Phys. Rev. A* **8**, 530 (1973).
8. M. E. Fisher, M. N. Barber and D. M. Jasnow, *Phys. Rev. A* **8**, 1111 (1973).
9. J. R. Banavar and M. Cieplak, *Phys. Rev. B* **26**, 2662 (1982); *J. Phys. C* **16**, L755 (1983).
10. W. L. McMillan, *Phys. Rev. B* **31**, 340 (1985); A. J. Bray and M. A. Moore, *Phys. Rev. B* **31**, 631 (1985).
11. D. S. Fisher and D. A. Huse, *Phys. Rev. Lett.* **56**, 1601 (1986); *J. Phys. A* **20**, L1005 (1987); D. A. Huse and D. S. Fisher, *J. Phys. A* **20**, L997 (1987).
12. R. R. P. Singh and M. E. Fisher, *J. Appl. Phys.* **63**, 3994 (1988); M. E. Fisher and R. R. P. Singh in *Disorder in physical systems* (G. Grimmett and D. J. A. Welsh, eds.), Oxford Univ. Press, 1990, pp. 87–111.
13. See A. J. Liu and M. E. Fisher, *Physica A* **156**, 35 (1989).
14. See, e.g., M. E. Fisher in *Magnetism and magnetic materials-1974*, AIP Conf. Proc. **24**, p. 273 (AIP, New York, 1975).
15. H. Rohrer and Ch. Gerber, *Phys. Rev. Lett.* **38**, 909 (1977); *J. Appl. Phys.* **49**, 1341 (1978).
16. See, e.g., A. M. Szpilka and M. E. Fisher, *Phys. Rev. Lett.* **57**, 1044 (1986).
17. M. E. Fisher and G. W. Milton, *Physica* **138A**, 22 (1986).
18. M. E. Fisher, *Commun. Math. Phys.* **26**, 6 (1972).
19. G. W. Milton and M. E. Fisher, *J. Stat. Phys.* **32**, 413 (1983); M. E. Fisher and B. U. Felderhof, *Ann. Phys.* **58**, 176, 217, 268 (1970).

20. M. E. Fisher, *Arch. Ratl. Mech. Anal.* **17**, 377 (1964).
21. R. B. Griffiths and D. Ruelle, *Commun. Math. Phys.* **23**, 169 (1971).
22. A. Aharony, R. J. Birgeneau, A. Coniglio, M. A. Kastner and H. E. Stanley, *Phys. Rev. Lett.* **60**, 1330 (1988).
23. J. T. Chayes, L. Chayes, D. S. Fisher and T. Spencer, *Phys. Rev. Lett.* **57**, 2999 (1986). (See also R. R. P. Singh and M. E. Fisher, *Phys. Rev. Lett.* **60**, 548 (1988) [C].)
24. M. Aizenman and J. Wehr, *Phys. Rev. Lett.* **62**, 2503 (1989).
25. See also K. Hui and A. N. Berker, *Phys. Rev. Lett.* **62**, 2507 (1989).

7

How to Simulate Fluid Criticality: The Simplest Ionic Model has Ising Behavior but the Proof is Not so Obvious!*

Michael E. Fisher[†]

*Institute for Physical Science and Technology,
University of Maryland, College Park, MD 20742, USA*

THE essence of simulating criticality in *asymmetric* fluid models is to discover effective, *unbiased* finite-size scaling methods that (i) recognize that *both* the critical temperature, T_c, and the critical density, ρ_c, are unknown and (ii) that can resolve 'nearby' critical universality classes. To this end *precise*, focused simulations for a *range* of 'box' sizes, L^d (in d dimensions), are imperative.

Indeed, since a true nonclassical (i.e., non-van-der-Waalsian) critical point cannot be realized in a finite system, some form of systematic extrapolation on the linear size, L, of a system is crucial to drawing reliable conclusions not only as regards the location of the critical point but – of greater interest theoretically – as to the nature of the criticality. A long standing problem[2] in this regard concerns the critical character of the simplest ionic model, the so-called Restricted Primitive Model (RPM), namely, a 1:1 hard sphere electrolyte (half the spheres of diameter a carrying charge $+q$, half $-q$).[2,4,5] As the title above indicates, recent work[5] has rather convincingly solved this problem: but new methods of analyzing simulation data had to be developed.

---

*Reprinted from *Ann. Henri Poincaré* 4, Suppl. 1, 437–440 (2003), © Birkhäuser Verlag, Basel, 2003, with permission of Springer.
[†]Active coworkers: Y. C. Kim, E. Luijten, G. Orkoulas and A. Z. Panagiotopoulos; support from the National Science Foundation under Grant No. CHE 99-81772.

In order to navigate the temperature-density, (T,ρ), plane various special loci were devised that, in the thermodynamic limit, $L \to \infty$, all spring from the critical point (T_c, ρ_c). Specifically, the k-loci are defined by the points of inflection of ρ^{2-k} vs. pressure isotherms.[3,5–8] Studying the L-dependence of these loci and using appropriate *unbiased* finite-size extrapolation techniques led to estimates of unprecedented precision and, it is believed, comparable accuracy for the critical parameters of the hard-core (attractive) square-well fluid (HCSqW).[3] In addition, the data (which did *not* require a massive computational effort) confirmed the expected Ising ($n = 1$) behavior, with critical exponents $\gamma = 1.23_9$, $\nu = 0.630_3$, and $\beta = 0.326_6$ (for the susceptibility, correlation length, and coexistence curve, respectively). Furthermore, and importantly, Ising character was clearly resolved from the 'nearby' XY ($n = 2$) or self-avoiding walk (SAW, $n = 0$) universality classes with $(\gamma, \nu, \beta) \simeq (1.31_6, 0.670, 0.34_7)$, and $(1.15_9, 0.58_8, 0.30_2)$.[3]

However, the critical behavior of the RPM turns out to be *highly asymmetric* – much more so than the HCSqW fluids. Indeed, the observed asymmetry seems likely to entail significant *mixing of the pressure*, p, into the linear scaling fields for the chemical potential μ, and the temperature, T, as is needed to describe a *Yang-Yang anomaly*[1] in which the second derivative $(d^2\mu/dT^2)$ on the critical isochore, $\rho = \rho_c$, diverges like the specific heat when $T \to T_c-$ [in contrast to traditional expectations that $(d^2\mu/dT^2)$ remains finite]. As a result, the previous approach using the k-loci proved inadequate. Instead, extending to noncritical densities an idea originally exploited by K. Binder, the primary analysis for locating T_c precisely[5] was based on an examination of the dimensionless moment ratios

$$Q_L(T; \langle \rho \rangle) \equiv \langle m^2 \rangle^2 / \langle m^4 \rangle \quad \text{with} \quad m = \rho - \langle \rho \rangle, \tag{1}$$

computed within a finite ($L < \infty$) grand canonical ensemble. At fixed T the ratio $Q_L(\langle \rho \rangle)$ exhibits a (unique) maximum at a density $\rho_Q(T;L)$. As $L \to \infty$ the resulting "Q-loci" tend to a limiting locus which, in turn, approaches the critical point when $T \to T_c+$.[5,6,8] More to the point, however, if one evaluates Q_L on the *corresponding* Q-locus, i.e., $Q_L(T; \rho_Q(T;L))$ one discovers[5,6,8] that the plots vs. T display tightly spaced successive intersections as L increases (say, in steps of size a) that rapidly converge to T_c *and* to a limiting value Q_c. Knowing a precise estimate for T_c one can obtain unbiased and reasonably precise estimates of the critical density. This is known to be very low for the RPM: indeed, the simulations yield $\rho_c^* \equiv \rho_c a^3 \simeq 0.079(3)$ (see Ref. 5) where the figure in parentheses represents the confidence limits.

On the other hand, extensive previous work[9] has shown that the critical value, Q_c, for grand canonical simulations in a cubic $L \times L \times \cdots \times L$ box is *universal* with values $Q_c = 0$, 0.6236(2), and 0.8045(1) for short-range, ($d = 3$)-dimensional $n = 0$, 1, and 2 (i.e., SAW, Ising, and XY) systems. For long-range $n = 1$ systems with attractive potentials decaying as $1/r^{3+\sigma}$ (for $d = 3$), one finds that $Q_c(\sigma)$ and also $\gamma(\sigma)$ increase almost linearly from their classical (van der Waals) values, namely, $Q_c = 0.4569\cdots$ and $\gamma = 1$, in the interval $\frac{3}{2} < \sigma \leq (\gamma/\nu)_{n=1} \leq 1.96_6$ with $Q_c(\sigma = 1.9) \leq 0.600$ and $\gamma(\sigma = 1.9) \leq 1.20_5$.[9] Consequently, the value of Q_c is a rather clear and robust indicator of universality class!

To complete the story, we note that the simulation data for the RPM[5] yield $Q_c = 0.624(2)$. This is remarkably close to the Ising value quoted above; it is also far from the classical, XY and SAW values and, as an extra bonus, appears to exclude also long-range effective power-law interactions decaying more slowly than $1/r^{4.9}$ (i.e., with $\sigma \leq 1.9$). Granted the reliable values for T_c and ρ_c, studying the effective susceptibility exponents, $\gamma_{\text{eff}}(T; L)$, leads to the estimate $\gamma = 1.24(3)$: this upholds Ising behavior while SAW and XY values are clearly less plausible.[5] Estimating the correlation exponent calls for further care in order to identify various estimators, $T_j(L)$ – extrema of various derivatives and moments – *above* T_c, which can, therefore, be reliably and *precisely* evaluated by the simulations and then suitably extrapolated to estimate ν.[5,6,8] The value $\nu = 0.63(3)$ is obtained[5] which, once again, points clearly to Ising criticality.

Finally, it should be noted that, in order to provide some easing of the computational burden the precise simulations so far reported[5] were performed using a ($\zeta = 5$)-level of fine discretization.[5] The estimated critical density and critical temperature, $T_c(\zeta) = 0.05069(2)$, must clearly depend on ζ; however, with reasonable confidence, one may believe that the universality class of critical behavior does *not* change with $\zeta > 3$. Subsequent calculations for larger values of ζ (in progress with Y. C. Kim) confirm this surmise, yielding essentially identical estimates for Q_c with no dependence on ζ. Furthermore, one finds that convergence to the continuum limit as ζ increases is quite rapid.

In another direction, by studying the behavior of $Q_L(T, \langle \rho \rangle)$ systematically *below* T_c, greatly improved methods have also been developed for estimating the coexistence curve even very close to T_c; thence, independent estimates of the exponent β can be obtained and pressure mixing in the scaling fields for the RPM is unequivocally revealed (indicating a significant Yang-Yang ratio[1] with $R_\mu = 0.26$).[15] In parallel work,[11,12] the nature of the *ionic screening* near gas-liquid criticality can be estimated by studying the Lebowitz length, $\xi_L(T, \rho)$.[10–12] In the RPM it appears that $\xi_L(T, \rho_c)$ displays a critical behavior matching that of the entropy[12] as may be anticipated on general grounds. As yet, the Ising (or other?) type of criticality in *asymmetric* ionic models (see e.g., Refs. 13 and 14) remains an open question as does the issue of the possible *divergence* of some of the ionic screening lengths.[10] More recent developments addressing these and related problems are listed below.[16–29]

References

1. M. E. Fisher and G. Orkoulas, The Yang-Yang anomaly in fluid criticality: Experiment and scaling theory, *Phys. Rev. Lett.* **85**, 696–699 (2000).
2. See, e.g., H. Weingärtner and W. Schröer, Criticality in ionic fluids, *Adv. Chem. Phys.* **116**, 1 (2001).
3. G. Orkoulas, M. E. Fisher and A. Z. Panagiotopoulos, Precise simulation of criticality in asymmetric fluids, *Phys. Rev. E* **63**, 051507:1–14 (2001).
4. E. Luijten, M. E. Fisher and A. Z. Panagiotopoulos, The heat capacity of the restricted primitive model electrolyte, *J. Chem. Phys.* **114**, 5468–5471 (2001).
5. E. Luijten, M. E. Fisher and A. Z. Panagiotopoulos, Universality class of criticality in the restricted primitive model, *Phys. Rev. Lett.* **88**, 185701:1–4 (2002).
6. Y. C. Kim, Fluid Criticality: Experiment, Scaling and Simulations, Ph.D. Thesis (University of Maryland, 2002).
7. Y. C. Kim, M. E. Fisher and G. Orkoulas, Asymmetric fluid criticality I. Scaling with pressure mixing (arXiv:cond-mat/0212145, 6 Dec 2002), *Phys. Rev. E* **67**, 061506:1–21 (2003).
8. Y. C. Kim and M. E. Fisher, Asymmetric fluid criticality II. Finite-size scaling and applications, *Phys. Rev. E* **68**, 041506:1–23 (2003).
9. See references cited in [5].
10. S. Bekiranov and M. E. Fisher, Fluctuations in electrolytes: the Lebowitz and other correlation lengths, *Phys. Rev. Lett.*, **81**, 5836–39 (1998); Diverging correlation lengths in electrolytes: exact results at low densities, *Phys. Rev. E* **59**, 492–511 (1999).
11. E. Luijten, M. E. Fisher and A. Z. Panagiotopoulos, Criticality and charge fluctuations in the restricted primitive model, *Bull. Amer. Phys. Soc.* **46**(1), 71 (2001) A11 4.
12. Y. C. Kim, E. Luijten and M. E. Fisher, Screening in ionic systems: Simulations for the Lebowitz length, *Phys. Rev. Lett.* **95**, 145701:1–4 (2005).
13. J. M. Romero-Enrique, G. Orkoulas, A. Z. Panagiotopoulos and M. E. Fisher, Coexistence and criticality in size-asymmetric electrolytes, *Phys. Rev. Lett.* **85**, 4558–61 (2000).
14. A. Z. Panagiotopoulos and M. E. Fisher, Phase transitions in 2:1 and 3:1 hard-core model electrolytes, *Phys. Rev. Lett.* **88**, 045701:1–4 (2002).

15. Y. C. Kim, M. E. Fisher and E. Luijten, Precise simulation of near-critical fluid coexistence (arXiv: cond-mat/0304032, 1 Apr 2003) *Phys. Rev. Lett.* **91**, 065701:1–4 (2003).
16. Y. C. Kim, M. E. Fisher and A. Z. Panagiotopoulos, Universality of ionic criticality: Size- and charge-asymmetric electrolytes, *Phys. Rev. Lett.* **95**, 195703:1–4 (2005).
17. V. Kobolev, A. B. Kolomeisky and M. E. Fisher, Lattice models of ionic systems, *J. Chem. Phys.* **116**, 7589–7598 (2002).
18. J.-N. Aqua, S. Banerjee and M. E. Fisher, Criticality in charge asymmetric ionic fluids (arXiv:cond-mat/0410692), *Phys. Rev. E* **72**, 041501:1–25 (2005).
19. Y. C. Kim and M. E. Fisher, Discretization dependence of criticality in model fluids: a hard-core electrolyte (arXiv:cond-mat/0402275), *Phys. Rev. Lett.* **92**, 185703:1–4 (2004).
20. J.-N. Aqua and M. E. Fisher, Ionic criticality: an exactly soluble model (arXiv:cond-mat/0311491), *Phys. Rev. Lett.* **92**, 135702:1–4 (2004).
21. J.-N. Aqua and M. E. Fisher, Criticality in multicomponent spherical models: Results and cautions, *Phys. Rev. E* **79**, 011118:1–13 (2009).
22. J.-N. Aqua and M. E. Fisher, Critical charge and density coupling in ionic spherical models [in preparation].
23. J.-N. Aqua and M. E. Fisher, Charge and density fluctuations lock horns: ionic criticality with power-law forces, *J. Phys. A: Math. Gen.* **37**, L241–L248 (2004).
24. Y.C. Kim and M. E. Fisher, Fluid coexistence close to criticality: Scaling algorithms for precise simulation (arXiv: cond-mat/0411736), *Comp. Phys. Commun.* **169**, 295–300 (2005).
25. M. E. Fisher, J. N. Aqua and S. Banerjee, How multivalency controls ionic criticality (arXiv:cond-mat/0507077), *Phys. Rev. Lett.* **95**, 135701:1–4 (2005).
26. Y. C. Kim and M. E. Fisher, Charge fluctuations and correlation lengths in finite electrolytes, *Phys. Rev. E* **77**, 051502:1–7 (2008).
27. S. K. Das, Y. C. Kim and M. E. Fisher, When is a conductor not perfect? Sum rules fail under critical fluctuations, *Phys. Rev. Lett.* **107**, 215701:1–4 (2011).
28. S. K. Das, Y. C. Kim and M. E. Fisher, Near critical electrolytes: Are the charge-charge sum rules obeyed?, *J. Chem. Phys.* **137**, 074902:1–12 (2012).
29. See also subsequent work by Angel Alastuey.

8

Molecular Motors: A Theorist's Perspective*

Anatoly B. Kolomeisky[1] and Michael E. Fisher[2]

[1]*Department of Chemistry and Chemical and Biomolecular Engineering, Rice University, Houston, Texas 77005*
[2]*Institute for Physical Science and Technology, University of Maryland, College Park, Maryland 20742*

INDIVIDUAL molecular motors, or motor proteins, are enzymatic molecules that convert chemical energy, typically obtained from the hydrolysis of ATP (adenosine triphosphate), into mechanical work and motion. Processive motor proteins, such as kinesin, dynein, and certain myosins, step unidirectionally along linear tracks, specifically microtubules and actin filaments, and play a crucial role in cellular transport processes, organization, and function. In this review some theoretical aspects of motor-protein dynamics are presented in the light of current experimental methods that enable the measurement of the biochemical and biomechanical properties on a single-molecule basis. After a brief discussion of continuum ratchet concepts, we focus on discrete kinetic and stochastic models that yield predictions for the mean velocity, $V(F, [ATP], \ldots)$, and other observables as a function of an imposed load force F, the ATP concentration, and other variables. The combination of appropriate theory with single-molecule observations should help uncover the mechanisms underlying motor-protein function.

Keywords: Motor proteins; kinesin; myosin; single-molecule experiments; discrete stochastic models.

*Reprinted from *Annual Review of Physical Chemistry*, **58**, 675–695 (2007).

INTRODUCTION

Biological cells are complex heterogeneous systems that undergo many dynamic biochemical processes, such as gene replication, transcription and translation, transport of vesicles and organelles between different locations, and segregation of chromosomes during mitosis (i.e., cell division).[1-3] A cell's ability to sustain these processes in a fast and effective way relies heavily on a class of protein molecules generally called motor proteins or molecular motors.[1-6]

Although many types of motor proteins are currently known (such as myosins, kinesins, dyneins, DNA and RNA polymerases, and helicases), and new motor species are constantly discovered, it is widely believed that all function by converting chemical energy into mechanical motion. The most common source of chemical energy for motor proteins is, first, the hydrolysis of ATP (adenosine triphosphate) or related compounds, and, second, the polymerization of nucleic acids and proteins such as tubulins. These transformations of chemical energy into mechanical work typically involve a complex network of biochemical reactions and physical processes. They often take place on millisecond or shorter timescales with a high thermodynamic efficiency.[4] However, the microscopic details of the mechanochemical couplings in motor proteins remain largely unknown.[1-6] Understanding these mechanisms is one of the more challenging problems that require concerted efforts by chemists, physicists, and biologists.

From the mechanical point of view, motor proteins can be considered as submicroscopic nanometer-size motors[4] that consume fuel (via chemical processes) to produce mechanical work. However, in contrast to macroscopic engines, molecular motors operate mainly at the single-molecule level in nonequilibrium but isothermal conditions. The state of the local molecular environment and thermal fluctuations are critically important. A successful theoretical description of motor-protein mechanisms should recognize their multiple conformational transitions, account for the complex mechanochemical processes involved, and explain their efficiency.

The past decade has seen great progress in experimental studies of motor proteins [see the monograph by Howard[4] and Refs. 7–47]. It is now possible to monitor and control the motion of a single motor-protein molecule under a variety of external conditions and measured loads with high spatial and time resolution. These investigations have revealed many previously unknown microscopic details, and their quantitative results have stimulated various theoretical discussions of the mechanisms underlying the dynamics of molecular motors.[48-83]

In this brief review we summarize recent experimental advances and selected theoretical developments in the field of motor proteins. Because the information gained from experiments is growing rapidly, we focus only on the principal biochemical and biophysical features. There are important classes of molecular motors that rotate,[4,7,14,18,36,42,43,51,58] in particular, bacterial flagella motors and F_0F_1-ATPase, which generates ATP in mitochondria. However, we consider here only motor proteins that transform chemical energy into linear translational motion: they might be called translocases. This is not inappropriate because many of the experimental and theoretical approaches to rotary and linear motor proteins are essentially the same.[4]

Similarly, we focus theoretically on discrete stochastic or biochemical-kinetic models because at this stage in the subject, they seem the most appropriate for concrete quantitative understanding.

Motor proteins

The variety of biological functions that molecular motors must perform in cells determines their complex multidomain structure (see Figure 1, which depicts three important motor proteins).[1,3,4,11,22,84] For these motors, on which we focus attention, the most crucial parts are the motor domains, often called "heads," where the enzymatic activity takes place and which bind strongly to specific molecular tracks, such as microtubules and actin filaments (or, in other cases, to DNA and RNA molecules). The catalytic activity of a motor domain is strongly diminished when it unbinds from its linear filament. For most motor proteins, there is only one active site for enzymatic transformation per motor domain, as in kinesins and myosins.[3,4] However, the motor domains of cytoplasmic dyneins have at least four binding sites for ATP;[39,46] the existence of additional potentially active sites might be related to the regulation of motor activity.

As seen in Figure 1, the motor domains are connected by tethers or stalks (often of coiled-coil structure) to tail domains that also play a role in the activity of the motor.[1,3,4,84] These connect to cellular cargo, such as vesicles and organelles, and in the absence of a suitable load, the tail domains may bind to the motor domains and thereby cut off the enzymatic activity.[4]

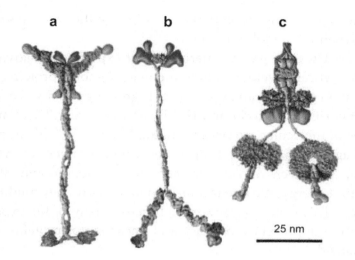

Figure 1. Domain structures of (a) conventional kinesin, (b) myosin V, and (c) cytoplasmic dynein. The tail domains are at the top, and the motor domains or heads are at the bottom. The (approximate) scale bar indicates 25 nm. Figure extracted from a review by Vale[6] with permission from Elsevier.

Several classes of motor proteins function as single independent entities as does a locomotive: they move on their tracks by repeatedly hydrolyzing ATP molecules (at rates of order one per 10 ms), taking hundreds of discrete, close-to-equisized, nanoscale steps before finally dissociating. Among such processive motor proteins are conventional kinesin, cytoplasmic dynein, and myosins V and VI. The first pair walk on microtubules, kinesin towards the plus (or fast-growing) end, whereas dynein is minus-end directed. Myosins, however, bind to actin filaments, myosin V moving towards the barbed or plus end, whereas myosin VI moves oppositely towards the pointed/minus end.

Most single-molecule experiments have been performed on these enzymes. Many motor proteins, however, most notably the muscle myosins,[4] function biologically only in large groups, although the details of the cooperative mechanism are largely unresolved.[3,4] Such nonprocessive motors normally complete only one or a few steps, or strokes, before completely detaching from their filaments. It is widely believed that the specific processivity of a motor protein is closely related to its particular structural features.[85,86] Nonprocessive motors are often monomers, whereas processive motor proteins exist in dimeric or even oligomeric forms.[3,4] This latter observation explains why processive motor proteins can stay attached to their filaments for long times: thus while one motor domain or head moves forward (presumably in an unbound or weakly bound state), the other head (or heads) can remain bound and carry the load imposed by the cargo.[4,48]

EXPERIMENTS

Structural information about motor proteins (such as seen in Figure 1) results principally from diffraction-based techniques and cryomicroscopy.[4] Although such data are vitally important and can also lead to insights concerning intermediate structural states in a motor, our present understanding of the dynamics of molecular motors has largely rested on two classes of in vitro investigations. On the one hand, bulk solution observations of ensembles of motor molecules principally determine the chemical-kinetic properties of the various biochemical processes they undergo. On the other hand, single-molecule experiments uncover the fluctuations and mechanochemical responses of individual molecules. The approaches are complementary, and both are important for elucidating the mechanisms of motility.[4,22,87] Furthermore, both can be enhanced by the study of mutated versions of the motor, which can reveal the roles of specific structural domains and their interactions.

Studies of motor proteins in bulk solutions constitute a convenient approach because the well-developed chemical-kinetic methods (such as stopped flow, isotope exchange, fluorescent labeling, and temperature quenching) can be applied to determine equilibrium and nonequilibrium properties of motor enzymes.[15,20,88-92] The results of such experiments demonstrate that the functioning of a motor protein may include multiple states and conformations coupled in a complex biochemical network. For many motor proteins, however, one or a few biochemical pathways prove dominant and control the overall dynamics. Thus for conventional kinesins and for myosins V and VI,[15,20,88-90,92] the dominating biochemical pathway always includes a sequence of at least four states of ATP hydrolysis.

Single-molecule observations

The most informative data concerning the dynamics of motor proteins have recently come from single-molecule experiments, which include optical-trap spectrometry, magnetic tweezers, Förster resonance energy transfer

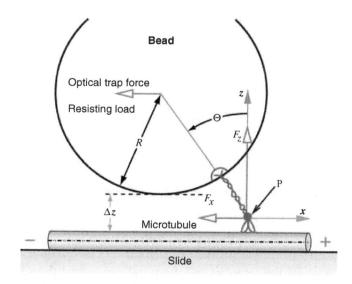

Figure 2. Schematic diagram of a kinesin/microtubule/bead complex in an optical-trap experiment. The microtubule is fixed to a glass slide that can be moved relative to the (fixed) optical trap. The force F_x exerted on the bead by the optical trap is transmitted by the tether to the point of attachment P on the motor at which the two heads are joined: see Figure 1(a). The mean offset $\Delta z \simeq 5$ nm is a result of thermal fluctuations; to scale, Δz should be much smaller, whereas the bead diameter, $R \simeq 250$ nm, should be twice as large. Figure adapted from Ref. 69. Copyright (2005) National Academy of Science, U.S.A.

(FRET), dynamic force microscopy, fluorescent imaging, and many other techniques.[7,8,10-14,16-19,21-37,40-44,46] The ability to passively monitor and actively influence the dynamics of individual single molecules (in particular by imposing forces and torques) provides a powerful tool for uncovering motor mechanisms.

One of the most successful and widely used methods is optical-trap spectrometry.[11,12,17,21,23-27,32,34,35,39,41,46] In this approach a single motor protein is chemically attached to a micron-sized or smaller bead that is captured by an external laser beam. The bead follows the motion of the motor molecule as it binds to its track and proceeds to move (Figure 2). Because the external electromagnetic field is nonuniform, the bead is trapped close to the focal point at which the light is most intense. Any nanometer-scale displacement of the bead from the focal point produces a restoring force, of the order of pico-Newtons, that is almost proportional to the displacement that in turn can be measured by differential outputs from a quadrant photo-diode.[10,12,17] Thus optical tweezers generate a harmonic potential well that can be calibrated with high precision.

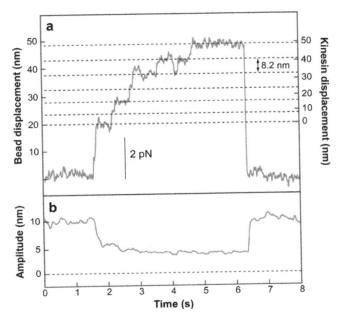

Figure 3. Optical-trap measurements of the motion of a single kinesin molecule at 20 μM ATP.[10] (a) Bead displacement as a function of time. Note this is proportional to the load force (see the 2-pN scale bar). (b) Measurements of time-dependent kinesin-to-bead stiffness. Figure taken from Yanagida and coworkers.[10] Copyright (1997) National Academy of Science, U.S.A.

Figure 3 shows the striking results of such a force-displacement-time experiment for kinesin moving on a microtubule. The binding of the motor to its track is detected by the sharp drop at time $t \simeq 1.5$ s in motor-to-bead stiffness (as monitored dynamically) (Figure 3(b)). Immediately, in the presence of ample ATP (typically at millimolar levels), the motor starts to drag the bead out of the trap not continuously but rather by taking a series of plus-end-directed discrete steps. The step length, d, proves close to 8.2 nm which is the periodicity of a microtubule protofilament, in other words, the (α, β) tubulin one-dimensional lattice spacing.[3,6] For myosin V moving on an actin filament, the mean step length $d \simeq 36 \pm 3$ nm is, similarly, close to the half-period of the actin double helix.[3,6,22] Furthermore, other experiments demonstrate[93–95] that each forward step (at low loads) corresponds to the hydrolysis of a single ATP molecule; this basic observation is referred to as tight coupling. When the bead is drawn further out of the trap, the resisting force (F_x in Figure 2) increases, and the motor slows down, reaching stall conditions (i.e., a zero mean velocity, V) at loads of 7 to 8 pN. As evident in Figure 3, reverse or back steps may then occur until, after some time, the motor detaches from the track (at $t \simeq 6.3$ s). Repeating such an experiment many times with the same identical molecule reveals the intrinsic stochastic fluctuations and yields, for example, the mean velocity, $V(F_x, [\text{ATP}])$, as a function of load and fuel supply.

By incorporating feedback controls, a force clamp can be imposed, enabling processive runs of tens to hundreds of steps to be observed under steady, controlled loads. Then one can measure further statistical parameters such as

$$r(F_x, [\text{ATP}]) = 2D/Vd \approx \langle [\Delta x(t)]^2 \rangle / d\langle x(t)\rangle, \qquad (1)$$

which has been called the randomness.[8,17] Here $x(t)$ is the displacement of the motor along the track as a function of time; the angular brackets, $\langle \cdot \rangle$, denote averages over many runs; and $\Delta x(t) = x(t) - \langle x(t)\rangle$ so that $D \approx \frac{1}{2}\langle [\Delta x(t)]^2 \rangle / t$ measures the diffusivity or dispersion.[8,96]

More recently the force-clamp set-up has been extended to allow the observation of single-protein dynamics under controlled vectorial forces $\mathbf{F} = (F_x, F_y, F_z)$, assisting as well as resisting, and sideways at an arbitrary angle.[25,26] As yet, however, simultaneous control of F_z (see Figure 2) has not been implemented although it is desirable.[9,69,70] In addition to high spatial resolution (of order 1 nm), time resolutions of order 10 μs or better can be achieved.[23,41]

Closely related to the optical-trapping technique is magnetic tweezers spectroscopy.[19,29,36] One end of a motor protein is again fastened chemically to a magnetic bead while the other end is fixed to a surface. The motor is maintained under tension by an imposed magnetic field gradient normal to the surface. The distance, z, of the bead from the surface and the observed magnitude of the transverse fluctuations of the bead, $\langle \delta x^2\rangle$, yield (via the equipartition theorem) the force exerted as $F_z = k_B T/\langle \delta x^2\rangle$. Controlled torques can also be exerted. Magnetic tweezer experiments are especially suitable for studying motors such as topoisomerases and helicases[19,29] that serve to unwind, untangle, and remove supercoiling in double-stranded DNA. Although magnetic tweezers are simpler to construct and use than optical traps, they are currently less sensitive and of lower resolution.

Selvin and coworkers[28,37,38] have developed another experimental approach of particular value, which they dubbed FIONA, standing for fluorescent imaging with one-nanometer accuracy. This method enables one to track the position of a single dye molecule attached to a specific location on a motor-protein molecule with nanometer accuracy at subsecond resolution. Although the fluorescent image has a diffraction-limited spot-size of several hundred nanometers, the brightest point, which corresponds to the desired position of the dye molecule, can be determined with a precision down to 1 nm, provided sufficiently many photons can be collected. With the aid of this technique it has been proved unambiguously that individual double-headed motors, such as kinesins and myosins V and VI, step in a so-called hand-over-hand fashion, meaning that the two heads exchange leading and trailing positions as the motor walks along its track.[28,37,38] Recent evidence[46,47] suggests that cytoplasmic dynein moves in a similar fashion by alternately shuffling its relatively large motor domains past one another.

THEORETICAL ASPECTS

The significant quantitative data resulting from single-molecule experiments have stimulated notable efforts in theory.[48-73,81] A framework for describing motor-protein dynamics should respect the basic laws of physics and chemistry and recognize the symmetries of the system such as periodicity, polarity, and chirality. A fully successful theory should not only provide a minimal consistent and reasonably quantitative description but should also yield mechanistic insights and experimentally testable predictions.

The central task of theoretical models for molecular motors is to connect biochemical processes to directed mechanical motion. It is fundamental that all biochemical transitions are reversible, even when available data may not provide direct evidence. The reverse transitions might be slow, but they should not be neglected in a comprehensive analysis as that may lead to unphysical conclusions.[41,54,69] This observation implies that under some conditions motor proteins that hydrolyze ATP when they step forward can resynthesize ATP when they step backward. This fact is well established for the rotary motor F_0F_1-ATP synthase.[36,43] For typical processive motor proteins, the situation is open experimentally, but recent observations[92,97] suggest that ATP can similarly be synthesized by such motors.

We can divide current theoretical approaches into two main groups: continuum ratchet models[49-52,57,63-65,72,73,82] and discrete stochastic (or chemical-kinetic) descriptions.[48,53-55,59-62,66-71,81-83]

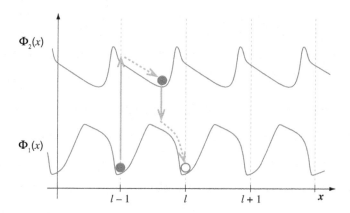

Figure 4. Illustration of the dynamics of a motor protein in the simplest two-potential periodic continuum ratchet model in the absence of a load. The vertical arrow represents the input of chemical energy, for example, via the hydrolysis of an ATP molecule; this is followed by diffusion, a drop to the lower potential surface, and further diffusion.

Continuum ratchets

In this physics-oriented approach, a motor protein at point $r = (x,y,z)$ close to its track is viewed as diffusing on two or more spatially parallel, periodic but in general asymmetric coarse-grained free-energy surfaces (see Figure 4).[49-52,57,63-65,72,73,82] The corresponding potentials, say $\Phi_j(\mathbf{r})$, describe distinct biochemical states of the motor. The ratchet-like character of the potentials depicted in Figure 4 cannot itself induce directed motion in an isothermal environment. Under the input of chemical energy, however, the motor switches stochastically between different potentials. Then, as illustrated in Figure 4, the system evolves according to a set of coupled Fokker-Planck equations[4] so that, in general, a biased diffusion ensues. This might well be called processive Brownian motion; but sustaining such directed, albeit thermally fluctuating, movement demands the continued overall positive supply of chemical energy.

Such chemically driven ratchets,[50,57,65] which might more descriptively be termed Markov-Fokker-Planck models,[72,74] provide a physically appealing, rather concrete picture of motor-protein dynamics that can be handled by established mathematical tools. But there are several troublesome aspects. With the exception of a few oversimplified and mostly unrealistic potentials, general analytical results cannot be found. One can, of course, resort it to a full numerical approach, but because the necessary computations are relatively demanding and many functional parameters are entailed, determining the range and uniqueness of fits to real data is a nontrivial task. Furthermore, in light of currently available knowledge of the relevant protein structures and their motions, deriving appropriate realistic potential functions, $\Phi_j(\mathbf{r})$, that are meaningfully detailed presents significant challenges. [Indeed, corresponding transition-rate functions[72,74] $k_{ij}(\mathbf{r})$ are also required.] As a consequence, although successful fits to experimental dynamical data have been obtained [notably for the F_0F_1-ATPase system and the bacterial flagella motor[51,72,74]], it is hard to judge the reliability and instructiveness of the resulting implications for real motor proteins. Thus, for the present, we believe such continuum models can most profitably be utilized to describe various qualitative rather than quantitative features of motor dynamics. Nonetheless, as experiments reveal further structural and dynamical information at a molecular level, the continuum aspects of motor-protein motion seem likely to demand modeling at the more intrinsically mechanical levels that ratchet models enable.

Discrete stochastic models

A rather different approach adapts the discrete stochastic models of traditional chemical kinetics.[48,53-55,59-62,66-71,78-82] The simplest model supposes that during each enzymatic cycle [processing a single fuel molecule, typically[93-95] ATP] the motor steps from a binding site l to the next one, $l+1$, at distance d along the track and passes through a sequence of N intermediate biochemical states (see Figure 5). These states should, in general, be associated with distinct spatial locations, as suggested in Figure 5. A motor in the mechanochemical state j_l ($j = 0, 1, 2, \ldots, N-1$) moves forward to state $(j+1)_l$ at a rate u_j or backward to state $(j-1)_l$ at rate w_j. (Detachment from the track at rate δ_j may also be recognized.) We take 0_l to specify the long-lived state in which the motor is strongly bound to its track, awaiting the arrival and binding of a fuel molecule. Note that reverse transitions are taken into account in accordance with the observations of back steps (see Figure 3 and Refs. 23, 26, and 41).

The dynamics of discrete kinetic models are governed by linear master equations that specify the net gain/loss, $dP_j(l,t)/dt$, where $P_j(l,t)$ is the probability that the motor is in state j_l at time t. Mathematically, this simple sequential model (with $\delta_j = 0$, all j) describes a particle that hops randomly on a one-dimensional periodic lattice of sites (of period N). One may thus

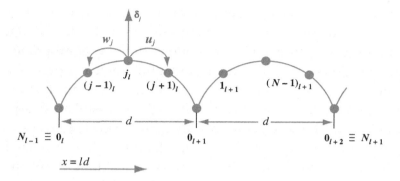

Figure 5. Kinetic scheme for the simplest N-state periodic stochastic model. A motor in state j_l can move forward at a rate u_j, backward at a rate w_j, or can dissociate irreversibly from the track at a rate δ_j.

use (and extend as needed) the powerful theoretical formalism constructed by Derrida in 1983.[54,55,59-61,81,98] This yields exact and explicit expressions for the asymptotic mean velocity and the dispersion for all N in terms of the rates u_j and w_j.[53-55] For example, the velocity and dispersion for $N = 2$ are given by

$$V = d\frac{u_0 u_1 - w_0 w_1}{u_0 + w_0 + u_1 + w_1}, \quad D = \frac{d^2}{2}\frac{(u_0 u_1 + w_0 w_1) - 2(V/d)^2}{u_0 + w_0 + u_1 + w_1}. \quad (2)$$

Randomness and number of states

It is valuable to know that the randomness, $r = 2D/Vd$, an observable measure of the dynamical fluctuations of a motor (see Eq. (1)), obeys the inequality $r \geq 1/N$.[96] Indeed, if reverse rates are neglected (setting $w_j \equiv 0$), one finds $r = 1/N$ when all the forward rates are equal (i.e., $u_j = u$). As an aside, if one also accepts $N = 1$ as a first-level motor model, the step size can be estimated experimentally, via $d = 2D/V$, by measuring D and V.[99] More informatively, $r \simeq 0.39 \leq \frac{1}{2}$ has been observed for kinesin[12] at so-called saturating ATP levels (beyond which V does not increase). This means that $N \geq 3$ [ATP]-independent intermediate transitions contribute to the motor dynamics, with, indeed, comparable weight. This conclusion is in accord with the basic ($N = 4$)-state biochemical view of ATP hydrolysis, namely for, say, kinesin, K, bound to a microtubule, M:

$$\mathrm{M \cdot K + ATP} \underset{w_1}{\overset{u_0}{\rightleftharpoons}} \mathrm{M \cdot K \cdot ATP} \underset{w_2}{\overset{u_1}{\rightleftharpoons}} \mathrm{M \cdot K \cdot ADP \cdot P_i} \underset{w_3}{\overset{u_2}{\rightleftharpoons}} \mathrm{M \cdot K \cdot ADP} \underset{w_0}{\overset{u_3}{\rightleftharpoons}} \mathrm{M \cdot K}, \quad (3)$$

where ADP denotes adenosine diphosphate, and P_i represents inorganic phosphate. Note that neither the water of hydrolysis nor the released P_i and ADP are shown in Eq. (3). Most of the rates exhibited here can be determined in bulk biochemical experiments.[87-90] Furthermore, the first forward rate may be written $u_0 = k_0$ [ATP] because it must depend on the ATP concentration.

Detachment and processivity

A major advantage of the discrete stochastic models is that they can readily handle more complex biochemical reactions than the linear sequence in Eq. (4) or Figure 5. As a basic example, note that an active motor cannot stay forever bound to its track: as seen in Figure 3 it will eventually dissociate — say, with state-dependent rates δ_j as in Figure 5. In single-molecule experiments such detachments may be regarded as irreversible (because the motor rarely reattaches rapidly). The analysis for V and D under stationary conditions can then be extended by mapping onto a renormalized model with no detachments.[59,62] The effects on velocity and dispersion are relatively small, but the processivity, as measured by mean run lengths or in other ways,[4,17] is strongly affected.[17,62] Amusingly, allowance for detachments can, in principle, yield enhanced velocities if, at slower speeds, the motor spends more time in states with higher rates of detachment.

Parallel pathways

Beyond simple detachments, biochemical observations indicate that motor proteins need not follow one simple, sequential reaction path in stepping from a state 0_l to 0_{l+1}. Specifically, experiments on single-headed kinesins[100] and RNA polymerases[101] demonstrate the possibility of parallel biochemical pathways, whereas studies[32] of myosin V in which many but not all 36-nm steps exhibit distinctive substeps seem to demand some branched parallelism. The single-pathway analysis[54,55,62] can again be generalized to provide exact results for models with general branches[59] and for parallel paths.[61] Simple, direct site-to-site diffusive or weakly bound parallel transitions correspond to what might be regarded as slipping: the increase in randomness seen for kinesin at low loads ($|F_x| \lesssim 1$ pN) may demand such a mechanism.[4,62]

Waiting-time distributions

A central tenet of traditional chemical kinetics implies that once a motor reaches a state j, its subsequent stochastic motion is independent of how it arrived. Accordingly, the time a motor spends in a state j represents a Poisson process. More concretely, suppose $\psi_j^+(t)dt$ is the probability of jumping forward to state $j+1$ in the time interval t to $t+dt$ after arriving in state j at $t=0$, and $\psi_j^-(t)dt$ is the corresponding probability of a reverse transition to $j-1$. Then these two waiting-time distributions are pure exponents with a common decay rate; i.e., $\psi_j^\pm(t) \propto e^{-(u_j+w_j)t}$. In principle such distributions are directly observable in single-molecule studies, but in practice, identifying the arrival and subsequent departure of a motor in a particular mechanochemical state is seldom possible. Then if various intervening states are missed, the distributions $\psi_j^\pm(t)$ will be nonexponential. Similarly, in a continuum ratchet model, the diffusive motions entailed in a transition between two relatively long-lived substates, such as the potential wells in Figure 4, again mean that the associated waiting-time distributions will be nonexponential. It is thus of interest to extend the exact analysis to arbitrary waiting-time distributions $\psi_j^+(t)$ and $\psi_j^-(t)$, etc. This can be achieved[60] by appealing to the theory of generalized master equations in which the relaxation kernels or memory functions are directly related to the waiting-time distributions.[102] It transpires that nonexponential distributions still yield the same rate-dependent relations for mean velocities; the dispersion expressions are, however, significantly changed. The departures from simple chemical-type processes are conveniently quantified by defining mechanicities, M_j^+, etc., which can provide a more economic description (with fewer parameters) of randomness observations.[60,62,69]

Substeps and load dependence

A unique feature of single-molecule experiments is the ability to impose a measured force, $F \equiv (F_x, F_y, F_z)$, directly on a single motor protein and to observe changes in velocity, V, and other properties. But what might this teach us? The point of application of the load, say, $r \equiv (x, y, z)$ (see P in Figure 2), may be identified as the dynamic position, $r(t)$, of the motor as it moves. However, because the motor shifts from, for example, r_0 in state 0 to $r_0 + d\hat{x}$ (where $\hat{x}$ is a unit vector parallel to the track) as one enzymatic cycle is completed, $r(t)$ may also be regarded as a mechanochemical reaction coordinate. As the reaction progresses, r passes through positions, r_j, that physically locate the intermediate mechanochemical states $j = 1, 2, \cdots$. The successive differences $d_0 = r_1 - r_0, d_1 = r_2 - r_1, \cdots$ then represent substeps characterizing the motor mechanism. En route from state j to $j+1$, the path should pass through a transition state, say, at $r_j^+ \equiv r_{j+1}^-$. If substeps are sufficiently large (greater than 1 nm or so), they should be detectable experimentally.

Now, as customary, one may also visualize the state point r as exploring a free-energy landscape with a potential $\Phi(r)$ and a valley (or potential well) at each r_j and a col (or saddle point) for a transition state at r_j^+.[4,57,69,70] One may then view the imposition of a load F most simply as adding a term $-F \cdot r$ to $\Phi(r)$ which tilts the landscape in the direction of the force.[a] This, in turn, changes the relative heights of the transition-state barriers that separate successive mechanochemical states; thereby the corresponding rates u_j and w_{j+1} become force dependent. The standard reaction-rate theories (see, e.g., Ref. 4) then lead to[70]

$$u_j(F) = u_j(0) \exp(\theta_j^+ \cdot Fd/k_BT), \qquad (4)$$

$$w_j(F) = w_j(0) \exp(\theta_j^- \cdot Fd/k_BT), \qquad (5)$$

in which the dimensionless load-distribution vectors, θ_j^+ and θ_j^-, describe how the work $F \cdot d$ performed by the external force is apportioned between the various forward and reverse transitions. Furthermore, the vectors $\theta_j^\pm$ relate simply to the substeps via $d_j = (\theta_j^+ + \theta_{j+1}^-)d$, while the transition-state displacements satisfy $d_j^\pm = \theta_j^\pm d$. In Eqs. (4) and (5), terms of order F^2 have been neglected in the exponent (see also Refs. 1, 54, 69, and 70). More elaborate treatments yield F-dependent prefactors as well. However, the linear exponential factors normally dominate strongly, so that, unless $\theta_j^\pm \cdot F$ vanishes for some j,[69] there is little merit in employing more elaborate but inevitably approximate expressions.

The strategy is now clear. Using Eqs. (4) and (5) in the expressions for V and other parameters: see Eq. (2), one may attempt to fit force-velocity-[ATP] and other dynamical data.[12,17,25,26,32,40,41] Even for the simplest $N = 2$ models in which $F_y = 0$ and one neglects F_z and possible z excursions of $r(t)$,[53-55,62,67] there are seven parameters: specifically, the rates at zero-load $u_0^0 = k_0^0[\text{ATP}]$, u_1^0, w_1^0, and w_0^0 [which requires special consid-

---

[a]Following Refs. 69 and 70, we take $F_x > 0$ to specify an assisting load on a motor, whereas $F_x < 0$ represents a normal resisting load (see Figure 2).

eration[62,66,70]] and the (now-scalar) load-distribution factors θ_0^{x+}, θ_1^{x+}, and θ_1^{x-} [since $\sum_j (\theta_j^{x+} + \theta_j^{x-}) = 1$ is required]. However, to a large degree the separate regimes of high and low loads ($|F_x| \simeq 1$ pN to 10 pN) and limiting and saturating [ATP] ($= \mu$M to mM) are dominated by distinct parameter sets. Thus the stall force, F_S, at which V vanishes is given by

$$F_S = (k_B T/d) \ln \prod_{j=0}^{N-1} (u_j^0 / w_j^0). \qquad (6)$$

Furthermore, some rates can be checked against bulk biochemical observations.

Some lessons and predictions

A first conclusion that may be rather generally valid for processive motors is that the observed decrease of V under resisting loads ($F_x < 0$) is governed primarily by the force dependence of the reverse rates, w_j, rather than the relatively insensitive forward rates. This conflicts with traditional views and ad hoc models that typically neglect backward rates and assume all the force dependence resides in a single, forward power stroke.

The original analysis of the extensive kinesin data of Block and coworkers[12,17] neglected the F_z dependence and predicted an initial substep, on binding ATP, of $d_0 = 1.8$–2.1 nm,[62] seemingly supported by structural studies.[13] Subsequent observations at 1-nm resolution ruled this out;[23,41] however, the discrepancy was resolved by allowing for F_z dependence and incorporating later data[26] that also encompassed assisting (and sideways) loads.[69] The new treatment predicts that on binding ATP, the motor might be said to crouch so that the initial movement, d_0 is directed downward towards the track with $d_0^z \simeq -(0.5$–$0.7)$ nm (while $|d_0^x| \leq 0.2$ nm).[69] Also predicted and as yet unverified is a strong sensitivity to bead size under assisting loads. At the same time, the analysis uncovered a mechanism that, by increasing F_z by approximately 2 pN, opposes assisting and leftward loads. On the other hand, extending the fitted (V, F_x) plots to superstall loads ($F_x \lesssim -7$ pN) predicts surprisingly small negative velocities that exhibit shallow minima 30–40% below F_S. In fact, Carter & Cross[41] have now observed such behavior; it can be linked theoretically to substep geometry.[104]

Experiments on myosin V presented in Figure 6 reported[66] mean dwell times measured before forward steps.[16,22] These were analyzed using

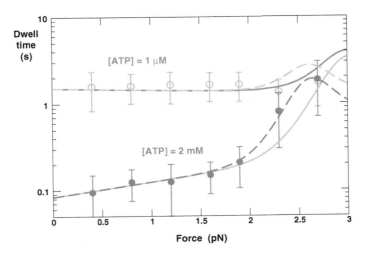

Figure 6. Mean dwell times of myosin V as a function of external load at different ATP concentrations. The symbols correspond to the experimental data of Mehta et al.[16] The solid lines are theoretical predictions from a discrete stochastic model with a fixed step length, whereas the dashed curves allow for variable step size. Figure adapted from Ref. 66 with permission from Elsevier.

$N = 2$ expressions for the mean forward cycle time,[66,67,105] which involves a random-walk first-passage calculation,[103] namely,

$$\tau_+^0 = (u_0 + u_1 + w_0 + w_1)/(u_0 w_0 + u_1 w_1). \tag{7}$$

Furthermore, as illustrated by the fits in Figure 6, one could allow for an observed distribution in step sizes.[66] A striking prediction[66] was the presence of a substep of magnitude $d_0^x = 13\text{–}14$ nm. Recent experiments[32] have indeed unambiguously revealed such a substep although of slightly smaller size, $d_0^x \simeq 11$ nm. The difference might indicate the need to allow for vertical displacements[69,70] of the point of attachment.[b] Even more intriguing, the data appear to demand a branching reaction because not all full steps of mean size $d \simeq 36$ nm display the substep.[32,33]

Available work and efficiency

The free energy, ΔG, available to a motor to do work via the hydrolysis of ATP (or other nucleotides) may be found from biochemical studies. Under the physiological conditions normally used for in vitro studies, Howard[4] concluded that $|\Delta G_{\text{ATP}}| \lesssim 25\ k_B T$. The maximum force a motor taking

---

[b] In connection with myosin V, attention should be drawn to more ambitious mechanoelastic structural models recently advanced.[75–77]

a step d can exert is $F_{\max} = \Delta G/d$ which, because $k_B T \simeq 4.1$ pN nm, yields approximately 2.8 pN for actin-based motors. For myosin V this corresponds closely to the observed stall force, F_S. From this perspective, therefore, myosin V operates at close to 100% efficiency. The same holds for the rotary $F_0 F_1$-ATPase motor.[18] For microtubule tracks, however, one finds $F_{\max} \lesssim 10\text{--}13$ pN, which significantly exceeds the observed stall forces of 7–8.5 pN for kinesin and dynein.[41,46] Understanding this 30% loss of efficiency at a molecular level is a major unsolved problem.

Back steps and dwell times

In the absence of back steps, or when their fractional occurrence, say, $\pi_- = 1 - \pi_+$, is negligible, as at low loads and high [ATP], the full cycle dwell time τ_+^0 given in Eq. (7) is essentially the reciprocal of the velocity.[66,67] However, when stall is approached, the ratio π_+/π_- must fall rapidly; at stall, back steps balance forward steps, so $\pi_+/\pi_- = 1$. Then for F_x beyond F_S, as observed by Carter & Cross,[41] the ratio falls rapidly to zero. Furthermore, the theory shows, perhaps surprisingly, that the full-cycle dwell times, τ_+^0 and τ_-^0, prior to forward and backward steps are equal (notwithstanding that the probabilities of moving forward, backward, or dissociating are quite distinct).[67] Appropriate observations confirm this fairly well.[23,41,c]

However, an important subtlety emerges when the observed back-step fraction is nonnegligible.[105] Specifically, although a principal step typically of magnitude close to d is readily seen, small mechanochemical substeps (e.g., less than 1 nm) escape detection. Then, for example, a motor may execute a major forward step (e.g., between states 2_l and 3_l in an $N = 4$ description) but fail to complete a full cycle (say, to $0_{l+1} \equiv 4_l$) before making a back step. When one allows for such hidden substeps,[105] one finds, in particular, that in the near-stall exponential fit

$$\pi_+/\pi_- \approx \exp[(F_x - F_S)d^*/k_B T], \qquad (8)$$

valid for $(F_x - F_S)$ small;[41,104–106] the putative effective step size, d^*, is significantly smaller — by a factor of $\sim \frac{1}{2}$ for kinesin[41] — than d, which is the full-cycle prediction. Developing the appropriate first-passage theory[104,105] further demonstrates the value of recording and analyzing more detailed

---

[c]As a result, interpretations suggesting that ATP is, hence, hydrolyzed on back steps, or not synthesized, and so on, are unconvincing.[67,69]

statistics such as π_{++} and π_{--}, the fraction of forward steps following a forward step and vice versa, and, correspondingly, the conditional dwell times τ_{++} and τ_{+-} etc.[105–107]

Multimotor complexes

In considering dimeric processive motors, one may regard the cooperative coupling of the two heads as a lead problem in the general issue of interacting molecular motors,[68,76,77,79,82] and in mitosis, homotetrameric kinesin plays a role.[3,84] Again, in living cells more than one motor may bind to a single vesicle and cooperate or, as in the case of a kinesin and a dynein on the same microtubule, even compete.[108] Other motor proteins, such as RecBCD helicases, consist of distinct domains that individually have enzymatic activity.[1,2,30] Indeed, separated RecB and RecD domains consume ATP and unwind DNA;[30] but single-molecule experiments find that the RecBCD cluster moves faster than the separate components.[30] A simple, discrete stochastic model provides insight into these facts by invoking a coupling energy of order $6k_BT$; it predicts that as the velocity is increased, the fluctuations are reduced.[68] Groups of motor and multimotor complexes also produce many interesting phenomena of collective behavior, such as flagellar beating, and chromosome and spindle oscillations, that are important for biological systems.[80]

A concluding remark

As illustrated by the issues raised above, such as the inefficient stall forces of kinesin and dynein, and yet others unmentioned, such as the mechanism of the cylindrical viral capsid-packing motors,[21] there are many open problems concerning motor proteins and their dynamics in which theory may provide predictions and conceptual insights. Furthermore, the field is constantly growing, with the discovery of new biological nanomachines, the refinement of established techniques, and the development of novel experimental approaches. One may anticipate exciting progress on all fronts.

SUMMARY POINTS

1. Motor proteins are special enzyme molecules that transform chemical energy into mechanical work. They exist in many forms.

2. Current experimental techniques enable one to study the dynamics of a single molecular motor with high spatial and time resolution.
3. The modeling of motor-protein dynamics has utilized two approaches: continuum ratchet pictures and discrete stochastic or chemical-kinetic descriptions.
4. Discrete stochastic models can account for available experimental observations, such as the load and [ATP] dependence of mean motor velocities, and they provide a flexible theoretical framework for understanding motor-protein mechanisms.

ACKNOWLEDGMENTS

We have appreciated interactions and correspondence with many colleagues, especially Howard Berg, Richard M. Berry, Steven M. Block., N. J. Carter, R. A. Cross, Sebastian Doniach, Yale E. Goldman, Hideo Higuchi, Jonathon Howard, S. Ishiwata, Frank Jülicher, Matthew J. Lang, Alex Mogilner, Justin Molloy, George Oster, Paul Selvin, Hong Qian, Edwin W. Taylor, S. Uemura, and Claudia Veigel. Ronald D. Vale and Toshio Yanagida kindly allowed us to reproduce graphic materials from their papers. Denis Tsygankov and Martin Lindén commented helpfully on a draft manuscript. A. B. K. acknowledges support from the Welch Foundation (grant C-1559), the Alfred P. Sloan Foundation (grant BR-4418), the NSF (grant CHE 02-37105), and a Hamill Innovation Award. This report was prepared in part (by A. B. K.) at the Kavli Institute for Theoretical Physics at UCSB, where it received partial support from NSF Grant PHY 99-07949. M. E. F. has received support from the National Science Foundation via grant CHE 03-01101.

References

1. H. A. Lodish, A. Berk, S. L. Zipursky and P. Matsudaira, *Molecular Cell Biology*, New York: Scientific American, 1084 pp, 4th ed. (1999).
2. B. Alberts, A. Johnson, J. Lewis, M. Raff, K. Roberts and P. Walter, *Molecular Biology of the Cell*, New York: Garland Science, 1463 pp, 4th ed. (2002).
3. D. Bray, *Cell Movements: From Molecules to Motility*, New York: Garland, 372 pp, 2nd ed. (2001).
4. J. Howard, *Mechanics of Motor Proteins and the Cytoskeleton*, Sunderland, MA: Sinauer Assoc., 367 pp (2001).
5. M. Schliwa, ed. *Molecular Motors*, Weinheim, Germ.: Wiley-VCH, 577 pp (2003).

6. R. D. Vale, The molecular motor toolbox for intracellular transport, *Cell* **112**, 467–80 (2003).
7. H. C. Berg, *E. coli in Motion*, New York: AIP Press/Springer-Verlag, 133 pp (2004).
8. K. Svoboda and S. M. Block, Force and velocity measured for single kinesin molecules, *Cell* **77**, 773–84 (1994).
9. F. Gittes, E. Meyhofer, S. Baek and J. Howard, Directional loading of the kinesin motor molecule as it buckles a microtubule, *Biophys. J.* **70**, 418–29 (1996).
10. H. Higuchi, E. Muto, Y. Inoue and T. Yanagida, Kinetics of force generation by single kinesin molecules activated by laser photolysis of caged ATP, *Proc. Natl. Acad. Sci. USA* **94**, 4395–400 (1997).
11. J. Gelles and R. Landick, RNA polymerase as a molecular motor, *Cell* **93**, 13–16 (1998).
12. K. Visscher, M. J. Schnitzer and S. M. Block, Single kinesin molecules studied with a molecular force clamp, *Nature* **400**, 184–89 (1999).
13. S. Rice, A. W. Lin, D. Safer, C. L. Hart, N. Naber et al., A structural change in the kinesin motor protein that drives motility, *Nature* **402**, 778–84 (1999).
14. R. M. Berry and J. P. Armitage, The bacterial flagella motor, *Adv. Microb. Phys.* **41**, 292–337 (1999).
15. E. M. De La Cruz, A. L. Wells, S. S. Rosenfeld, E. M. Ostap and H. L. Sweeney, The kinetic mechanism of myosin V, *Proc. Natl. Acad. Sci. USA* **96**, 13726–31 (1999).
16. A. D. Mehta, R. S. Rock, M. Rief, J. A. Spudich, M. S. Mooseker and R. E. Cheney, Myosin-V is a processive actin-based motor, *Nature* **400**, 590–93 (1999).
17. M. J. Schnitzer, K. Visscher and S. M. Block, Force production by single kinesin motors, *Nat. Cell. Biol.* **2**, 718–23 (2000).
18. K. Adachi, R. Yasuda, H. Noji, H. Itoh, Y. Harada et al., Stepping rotation of F_1-ATPase visualized through angle-resolved single-fluorophore imaging, *Proc. Natl. Acad. Sci. USA* **97**, 7243–47 (2000).
19. T. R. Strick, J. F. Allemand and D. Bensimon, Stress-induced structural transitions in DNA and proteins, *Annu. Rev. Biophys. Biomol. Struct.* **29**, 523–43 (2000).
20. E. M. De La Cruz, E. M. Ostap and H. L. Sweeney, Kinetic mechanism and regulation of myosin VI, *J. Biol. Chem.* **276**, 32373–81 (2001).
21. D. E. Smith, S. J. Tans, S. B. Smith, S. Grimes, D. L. Anderson and C. Bustamante, The bacteriophage ϕ29 portal motor can package DNA against a large internal force, *Nature* **413**, 748–52 (2001).
22. A. Mehta, Myosin learns to walk, *J. Cell Sci.* **114**, 1981–98 (2001).
23. M. Nishiyama, H. Higuchi and Y. Yanagida, Chemomechanical coupling of the forward and backward steps of single kinesin molecules, *Nat. Cell Biol.* **4**, 790–97 (2002).
24. K. Kaseda, H. Higuchi and K. Hirose, Coordination of kinesin's two heads studied with mutant heterodimers, *Proc. Natl. Acad. Sci. USA* **99**, 16058–63 (2002).
25. M. J. Lang, C. L. Asbury, J. W. Shaevitz and S. M. Block, An automated two-dimensional optical force clamp for single molecule studies, *Biophys. J.* **83**, 491–501 (2002).
26. S. M. Block, C. L. Asbury, J. W. Shaevitz and M. J. Lang, Probing the kinesin reaction cycle with a 2D optical force clamp, *Proc. Natl. Acad. Sci. USA* **100**, 2351–56 (2003).
27. C. L. Asbury, A. N. Fehr and S. M. Block, Kinesin moves by an asymmetric hand-over-hand mechanism, *Science* **302**, 2130–34 (2003).
28. A. Yildiz, M. Tomishige, R. D. Vale and P. R. Selvin, Kinesin walks hand-over-hand, *Science* **302**, 676–78 (2003).
29. G. Charvin, D. Bensimon and V. Croquette, Single-molecule study of DNA unlinking by eukaryoric and prokaryotic type-II topoisomerases, *Proc. Natl. Acad. Sci. USA* **100**, 9820–25 (2003).

30. A. F. Taylor and G. R. Smith, RecBCD enzyme is a DNA helicase with fast and slow motors of opposite polarity, *Nature* **423**, 889–93 (2003).
31. J. N. Forkey, M. E. Quinlan, M. A. Shaw, J. E. Corrie and Y. E. Goldman, Three-dimensional structural dynamics of myosin V by single-molecule fluorescence polarization, *Nature* **422**, 399–404 (2003).
32. S. Uemura, H. Higuchi, A. O. Olivares, E. M. De La Cruz and S. Ishiwata, Mechanochemical coupling of two substeps in a single myosin V motor, *Nat. Struct. Mol. Biol.* **9**, 877–83 (2004).
33. J. E. Baker, E. B. Krementsova, G. G. Kennedy, A. Armstrong, K. M. Trybus and D. M. Warshaw, Myosin V processivity: multiple kinetic pathways for head-to-head coordination, *Proc. Natl. Acad. Sci. USA* **101**, 5542–46 (2004).
34. T. T. Perkins, H. W. Li, R. V. Dalal, J. Gelles and S. M. Block, Forward and reverse motion of RecBCD molecules on DNA, *Biophys. J.* **86**, 1640–48 (2004).
35. B. Maier, M. Koomey and M. P. Sheetz, A force-dependent switch reverses type IV pilus retraction, *Proc. Natl. Acad. Sci. USA* **101**, 10961–66 (2004).
36. H. Itoh, A. Takahashi, K. Adachi, H. Noji, R. Yasuda et al., Mechanically driven ATP synthesis by F_1-ATPase, *Nature* **427**, 465–68 (2004).
37. G. E. Snyder, T. Sakamoto, J. A. Hammer, J. R. Sellers and P. R. Selvin, Nanometer localization of single green fluorescent proteins: evidence that myosin V walks hand-over-hand via telemark configuration, *Biophys. J.* **87**, 1776–83 (2004).
38. A. Yildiz, H. Park, D. Safer, Z. Yang, L. Q. Chen et al., Myosin VI steps via a hand-over-hand mechanism with its lever arm undergoing fluctuations when attached to actin, *J. Biol. Chem.* **279**, 37223–26 (2004).
39. K. Oiwa and H. Sakakibara, Recent progress in dynein structure and mechanism, *Curr. Opin. Cell Biol.* **17**, 98–103 (2005).
40. T. J. Purcell, H. L. Sweeney and J. A. Spudich, A force-dependent state controls the coordination of processive myosin V, *Proc. Natl. Acad. Sci. USA* **102**, 13873–78 (2005).
41. N. J. Carter and R. A. Cross, Mechanics of the kinesin step, *Nature* **435**, 308–12 (2005).
42. Y. Sowa, A. D. Rowe, M. C. Leake, T. Yakushi, M. Homma et al., Direct observation of steps in rotation of the bacterial flagellar motor, *Nature* **437**, 916–19 (2005).
43. Y. Rondelez, G. Tresset, T. Nakashima, Y. Kato-Yamada, H. Fujita et al., Highly coupled ATP synthesis by F_1-ATPase single molecules, *Nature* **433**, 774–77 (2005).
44. C. Veigel, S. Schmitz, F. Wang and J. R. Sellers, Load-dependent kinetics of myosin-V can explain its high processivity, *Nat. Cell Biol.* **7**, 861–69 (2005).
45. M. T. Valentine, P. M. Fordyce, T. C. Krzysiak, S. P. Gilbert and S. M. Block, Individual dimers of the mitotic kinesin motor Eg5 step processively and support substantial loads in vitro, *Nat. Cell Biol.* **8**, 470–76 (2006).
46. S. Toba, T. M. Watanabe, L. Yamaguchi-Okimoto, Y. Y. Toyoshima and H. Higuchi, Overlapping hand-over-hand mechanism of single molecular motility of cytoplasmic dynein, *Proc. Natl. Acad. Sci. USA* **103**, 5741–45 (2006).
47. S. L. Reck-Peterson, A. Yildiz, A. P. Carter, A. Gennerich, N. Zhang and R. D. Vale, Single-molecule analysis of dynein processivity and stepping behavior, *Cell* **126**, 335–48 (2006).
48. S. Leibler and D. A. Huse, Porters versus rowers: a unified stochastic model of motor proteins, *J. Cell Biol.* **121**, 1356–68 (1993).
49. C. S. Peskin and G. Oster, Coordinated hydrolysis explains the mechanical behavior of kinesin, *Biophys. J.* **68**, S202–11 (1995).
50. F. Jülicher, A. Ajdari and J. Prost, Modeling molecular motors, *Rev. Mod. Phys.* **69**, 1269–81 (1997).
51. T. C. Elston and G. Oster, Protein turbines I: the bacterial flagellar motor, *Biophys. J.* **73**, 703–21 (1997).

52. H. Y. Wang, T. Elston, A. Mogilner and G. Oster, Force generation in RNA polymerase, *Biophys. J.* **74**, 1186–202 (1998).
53. A. B. Kolomeisky and B. Widom, A simplified "ratchet" model of molecular motors, *J. Stat. Phys.* **93**, 633–45 (1998).
54. M. E. Fisher and A. B. Kolomeisky, The force exerted by a molecular motor, *Proc. Natl. Acad. Sci. USA* **96**, 6597–602 (1999).
55. M. E. Fisher and A. B. Kolomeisky, Molecular motors and the forces they exert, *Phys. A* **274**, 241–66 (1999).
56. R. Lipowsky, Universal aspects of the chemomechanical coupling for molecular motors, *Phys. Rev. Lett.* **85**, 4401–4 (2000).
57. D. Keller and C. Bustamante, The mechanochemistry of molecular motors, *Biophys. J.* **78**, 541–56 (2000).
58. R. M. Berry, Theories of rotary motors, *Philos. Trans. R. Soc. London Ser. B* **355**, 503–9 (2000).
59. A. B. Kolomeisky and M. E. Fisher, Periodic sequential kinetic models with jumping, branching and deaths, *Phys. A* **279**, 1–20; Erratum *ibid* **284**, 496 (2000).
60. A. B. Kolomeisky and M. E. Fisher, Extended kinetic models with waiting-time distributions: exact results, *J. Chem. Phys.* **113**, 10867–77 (2000).
61. A. B. Kolomeisky, Exact results for parallel-chain kinetic models of biological transport, *J. Chem. Phys.* **115**, 7253–59 (2001).
62. M. E. Fisher and A. B. Kolomeisky, Simple mechanochemistry describes the dynamics of kinesin molecules, *Proc. Natl. Acad. Sci. USA* **98**, 7748–53 (2001).
63. C. Bustamante, D. Keller and G. Oster, The physics of molecular motors, *Acc. Chem. Res.* **34**, 412–20 (2001).
64. A. Mogilner, A. J. Fisher and R. J. Baskin, Structural changes in the neck linker of kinesin explain the load dependence of the motor's mechanical cycle, *J. Theor. Biol.* **211**, 143–57 (2001).
65. P. Reimann, Brownian motors: noisy transport far from equilibrium, *Phys. Rep.* **361**, 57–265 (2002).
66. A. B. Kolomeisky and M. E. Fisher, A simple kinetic model describes the processivity of myosin-V, *Biophys. J.* **84**, 1642–50 (2003).
67. A. B. Kolomeisky, E. B. Stukalin and A. A. Popov, Understanding mechanochemical coupling in kinesins using first-passage-time processes, *Phys. Rev. E* **71**, 031902 (2005).
68. E. B. Stukalin, H. Phillips and A. B. Kolomeisky, Coupling of two motor proteins: A new motor can move faster, *Phys. Rev. Lett.* **94**, 238101 (2005).
69. M. E. Fisher and Y. C. Kim, Kinesin crouches to sprint but resists pushing, *Proc. Natl. Acad. Sci. USA* **102**, 16209–14 (2005).
70. Y. C. Kim and M. E. Fisher, Vectorial loading of processive motor proteins: implementing a landscape picture, *J. Phys. Condens. Matter* **17**, S3821–38 (2005).
71. H. Qian, Cycle kinetics, steady state thermodynamics and motors: a paradigm for living matter physics, *J. Phys. Condens. Matter* **17**, S3783–94 (2005).
72. J. Xing, J. C. Liao and G. Oster, Making ATP, *Proc. Natl. Acad. Sci. USA* **102**, 16536–46 (2005).
73. R. Kanada and K. Sasaki, Theoretical model for motility and processivity of two-headed molecular motors, *Phys. Rev. E* **67**, 061917 (2003).
74. J. Xing, H. Wang and G. Oster, From continuum Fokker-Planck models to discrete kinetic models, *Biophys. J.* **89**, 1551–63 (2005).
75. P. Xie, S. X. Dou and P. Y. Wang, Model for processive movements of myosin V and myosin VI, *Chin. Phys. J.* **14**, 744–52 (2005).
76. G. Lan and S. X. Sun, Dynamics of myosin V processivity, *Biophys. J.* **88**, 999–1008 (2005).

77. A. Vilfan, Elastic lever-arm model for myosin V, *Biophys. J.* **88**, 3792–805 (2005).
78. A. B. Kolomeisky and H. Phillips III, Dynamic properties of motor proteins with two subunits, *J. Phys. Condens. Matter* **17**, S3887–99 (2005).
79. S. Klumpp and R. Lipowsky, Cooperative cargo transport by several molecular motors, *Proc. Natl. Acad. Sci. USA* **102**, 17284–89 (2005).
80. S. W. Grill, K. Kruse and F. Jülicher, Theory of mitotic spindle oscillations, *Phys. Rev. Lett.* **94**, 108104 (2005).
81. E. B. Stukalin and A. B. Kolomeisky, Transport of single molecules along periodic parallel lattices with coupling, *J. Chem. Phys.* **124**, 204901 (2006).
82. O. Campas, Y. Kafri, K. B. Zeldovich, J. Casademunt and J.-F. Joanny, Collective dynamics of interacting motor proteins, *Phys. Rev. Lett.* **97**, 038101 (2006).
83. K. I. Skau, R. B. Hoyle and M. S. Turner, A kinetic model describing the processivity of myosin-V, *Biophys. J.* **91**, 2475–89 (2006).
84. R. D. Vale and R. J. Fletterick, The design plan of kinesin motors, *Annu. Rev. Cell Dev. Biol.* **13**, 745–77 (1997).
85. F. Kozielski, S. Sack, A. Marx, M. Thormählen, Schönbrunn et al., The crystal structure of dimeric kinesin and implications for microtubule-dependent motility, *Cell* **91**, 985–94 (1997).
86. M. Tomishige, D. R. Klopfenstein and R. D. Vale, Conversion of Unc104/KIF1A kinesin into a processive motor after dimerization, *Science* **297**, 2263–67 (2002).
87. E. M. De La Cruz and E. M. Ostap, Relating biochemistry and function in the myosin superfamily, *Curr. Opin. Cell Biol.* **16**, 1–7 (2004).
88. A. Sadhu and E. W. Taylor, A kinetic study of the kinesin ATPase, *J. Biol. Chem.* **267**, 11352–59 (1992).
89. Y. Z. Ma and E. W. Taylor, Mechanism of microtubule kinesin ATPase, *Biochemistry* **34**, 13242–51 (1995).
90. M. L. Moyer, S. P. Gilbert and K. A. Johnson, Pathway of ATP hydrolysis by monomeric and dimeric kinesin, *Biochemistry* **37**, 800–13 (1998).
91. A. L. Lucius and T. L. Lohman, Effects of temperature and ATP on the kinetic mechanism and kinetic step-size for *E. coli* RecBCD helicase-catalyzed DNA unwinding, *J. Mol. Biol.* **339**, 751–71 (2004).
92. J. C. Cochran, J. E. Gatial, T. M. Kapoor and S. P. Gilbert, Monastrol inhibition of the mitotic kinesin Eg5, *J. Biol. Chem.* **280**, 12658–67 (2005).
93. M. J. Schnitzer, S. M. Block, Kinesin hydrolyzes one ATP per 8-nm step, *Nature* **388**, 386–90 (1997).
94. W. Hua, E. C. Young, M. L. Fleming and J. Gelles, Coupling of kinesin steps to ATP hydrolysis, *Nature* **388**, 390–93 (1997).
95. D. L. Coy, M. Wagenbach and J. Howard, Kinesin takes one 8-nm step for each ATP that it hydrolyzes, *J. Biol. Chem.* **274**, 3667–71 (1999).
96. Z. Koza, General relation between drift velocity and dispersion of a molecular motor, *Acta Phys. Polon. B* **33**, 1025–30 (2002).
97. D. D. Hackney, The tethered motor domain of a kinesin-microtubule complex catalyzes reversible synthesis of bound ATP, *Proc. Natl. Acad. Sci. USA* **102**, 18338–43 (2005).
98. B. Derrida, Velocity and diffusion constant of a periodic one-dimensional hopping model, *J. Stat. Phys.* **31**, 433–50 (1983).
99. K. C. Neuman, O. A. Saleh, T. Lionnet, G. Lia, J.-F. Allemand et al., Statistical determination of the step size of molecular motors, *J. Phys. Condens. Matter* **17**, S3811–20 (2005).

100. Y. Okada and N. Hirokawa, Mechanism of the single-headed processivity: diffusional anchoring between the K-loop of kinesin and the C terminus of tubulin, *Proc. Natl. Acad. Sci. USA* **97**, 640–45 (2000).
101. M. Guthold, X. Zhu, C. Rivetti, G. Yang, N. H. Thomson et al., Direct observation of one-dimensional diffusion and transcription by *Escherichia coli* RNA polymerase, *Biophys. J.* **77**, 2284–94 (1999).
102. U. Landman, E. W. Montroll and M. Schlesinger, Random walks and generalized master equations with internal degrees of freedom, *Proc. Natl. Acad. Sci. USA* **74**, 430–33 (1977).
103. N. G. van Kampen, *Stochastic Processes in Physics and Chemistry*, New York: Elsevier, 465 pp, 2nd ed. (1992).
104. D. Tsygankov and M. E. Fisher, Mechanoenzymes under superstall and large assisting loads reveal structural features, *Proc. Natl. Acad. Sci. USA* **104**, 19321 (2007).
105. D. Tsygankov, M. Lindén and M. E. Fisher, Back-stepping, hidden substeps, and conditional dwell times in molecular motors, *Phys. Rev. E.* **75**, 021909 (2007).
106. Y.-X. Zhang and M. E. Fisher, Measuring the limping of processive motor proteins, *J. Stat. Phys.* **142**, 1218 (2011).
107. D. Tsygankov and M. E. Fisher, Kinetic models for mechanoenzymes: structural aspects under large loads, *J. Chem. Phys.* **128**, 015102 1–12 (2008).
108. Y.-X. Zhang and M. E. Fisher, Dynamics of the tug-of-war model for cellular transport, *Phys. Rev. E* **82**, 011923 1–14 (2010).

9

Renormalization Group Theory, the Epsilon Expansion and Ken Wilson as I knew Him*

Michael E. Fisher

Institute for Physical Science and Technology,
University of Maryland, College Park, Maryland 20742-8510, USA

THE tasks posed for renormalization group theory (RGT) within statistical physics by critical phenomena theory in the 1960's are set out briefly in contradistinction to quantum field theory (QFT), which was the origin for Ken Wilson's concerns. Kadanoff's 1966 block spin scaling picture and its difficulties are presented; Wilson's early vision of flows is described from the author's perspective. How Wilson's subsequent breakthrough ideas, published in 1971, led to the epsilon expansion and the resulting clarity is related. Concluding sections complete the general picture of flows in a space of Hamiltonians, universality and scaling. The article represents a 40% condensation (but with added items) of an earlier account: *Rev. Mod. Phys.* **70**, 653–681 (1998).

Keywords: Statistical physics; critical phenomena; renormalization group theory; epsilon expansion; flows in space of Hamiltonians.

Foreword

In March 1996 the Departments of Philosophy and of Physics at Boston University cosponsored a two-day Colloquium "On the Foundations of Quantum Field Theory." But in the full title, this was preceded by the phrase "A Historical Examination and Philosophical Reflections," which set

---

*Reprinted from *Int. J. Mod. Phys. B*, **29**, 1530006 (2015).

the aims of the meeting. The participants were mainly high energy physicists, experts in field theories and interested philosophers of science.[a] I was called on to speak, essentially in a service role, presumably because I had witnessed and had some hand in the development of renormalization group concepts and because I had played a role in applications where these ideas really mattered. A version of my talk was published in *Rev. Mod. Phys.* (Fisher, 1998)[b] and, with some revisions, in the Proceedings of the Colloquium (Cao, 1999) edited by the prime organizer Tian Yu Cao.

On 15 June 2013 my former Cornell colleague and friend, Kenneth Geddes Wilson, who had played such a major role in the conception, formulation and initial applications of renormalization group theory died unexpectedly a week after his 77th birthday. A Memorial Symposium was held in his honor at Cornell University on 16 November that year; it was my sad but rewarding duty to speak on the occasion to a general and varied audience. For this memorial volume, however, I felt it would be more appropriate to edit and, in places, to expand my earlier talk to especially stress the role that I, myself, saw Ken Wilson play in conceiving and formulating the renormalization group approach for critical phenomena and in its effective application *via* the so-called *epsilon expansion*. This is what I hope readers will appreciate in the following text and figures.

1. Introduction

It is held by some that the "Renormalization Group" (RG) — or, better, renormalization group*s* or, let us say, *Renormalization Group Theory* (or RGT) is "one of the underlying ideas in the theoretical structure of Quantum Field Theory (or QFT)." That belief suggests the potential value of a historical and conceptual account of RG theory and the ideas and sources from which it grew, as viewed from the perspective of statistical mechanics and condensed matter physics. Especially pertinent are the roots in the theory of critical phenomena.

The proposition just stated regarding the significance of RG theory for Quantum Field Theory (or QFT, for short) is open to debate even though

---

[a]The proceedings of the conference were published under the title *Conceptual Foundations of Quantum Field Theory* (Cao, 1999): for details see the references collected in the Selected Bibliography.

[b]References in this form or as a name in the text followed by a date in parentheses will be found below in the Selected Bibliography.

experts in QFT have certainly invoked RG ideas. Indeed, one may ask: How far is some concept only instrumental? How far is it crucial? It is surely true in physics that when we have ideas and pictures that are extremely useful, they acquire elements of reality in and of themselves. But, philosophically, it is instructive to look at the degree to which such objects are purely instrumental — merely useful tools — and the extent to which physicists seriously suppose they embody an essence of reality. Certainly, many parts of physics are well established and long precede RG ideas. Among these is statistical mechanics itself, a theory *not* reduced and, in a deep sense, *not* directly reducible to lower, more fundamental levels without the introduction of specific, new postulates.

Furthermore, statistical mechanics has reached a stage where it is well posed mathematically; many of the basic theorems (although by no means all)[c] have been proved with full rigor. In that context, I believe it is possible to view the RG as merely an instrument or a computational device. On the other hand, at one extreme, one might say: "Well, the partition function itself is really just a combinatorial device." But most practitioners tend to think of it (and especially its logarithm, the free energy) as rather more basic!

Now my aim here is not to instruct those who understand these matters well.[d] Rather, I hope to convey to nonexperts and, in particular, to any with a philosophical interest, a little more about what Renormalization Group Theory is[e] — at least in the eyes of some of those who have earned a living by using it! One hopes such information may be useful to those who might want to discuss its implications and significance or assess how it fits into QFT in particular or into physics more broadly.

Whence came Renormalization Group Theory? This is a good question to start with.[f] I will try to respond, sketching the foundations of

---

[c] See, *e.g.* Fisher on p. 165 of (Cao, 1999).

[d] Such as the field theorists D. Gross and R. Shankar (see Cao, 1999 and Shankar, 1994). Note also Bagnuls and Bervillier (1997).

[e] It is worthwhile to stress, at the outset, what a "RG" is *not*! Although in many applications the particular RG employed may be invertible, and so constitute a continuous or discrete, group of transformations, it is, in general, only a *semigroup*. In other words a RG is not necessarily invertible and, hence, cannot be 'run backwards' without ambiguity: In short it is *not* a "group." This misuse of mathematical terminology may be tolerated since these aspects play, at best, a small role in RG theory. The point will be returned to later.

[f] Five influential reviews antedating RG concepts are Domb (1960), Fisher (1965, 1967b), Kadanoff *et al.* (1967) and Stanley (1971). Early reviews of RG developments are provided by Wilson and Kogut (1974) and Fisher (1974b): see also Wilson (1983) and Fisher (1983).

RG theory in the *critical exponent relations* and crucial *scaling concepts* of Leo P. Kadanoff, Benjamin Widom and myself developed in 1963–1966[g] — among, of course, other important workers, particularly Cyril Domb[h] and his group at King's College London, of which, originally, I was a member, George A. Baker, Jr., whose introduction of Padé approximant techniques proved so fruitful in gaining quantitative knowledge,[i] and Valery L. Pokrovskii and A. Z. Patashinskii in the Soviet Union who were, perhaps, the first to bring field-theoretic perspectives to bear.[j] Especially I will say something of the genesis of the full RG concept — the systematic integrating out of appropriate degrees of freedom and the resulting RG flows — in the inspired work of Ken Wilson[k] when he was a colleague of Ben Widom and myself at Cornell University in 1965–1972. One must point also to the general, clarifying formulation of RG theory by Franz J. Wegner (1972a) when he was associated with Leo Kadanoff at Brown University: their focus on *relevant*, *irrelevant* and *marginal* 'operators' (or perturbations) played a central role.[l]

2. The Task for Renormalization Group Theory

Let us, at this point, step back briefly by highlighting, from the viewpoint of statistical physics, what it is one would wish RG theory to accomplish. First and foremost, (A) it should explain the ubiquity of power laws at and near critical points.

To be more explicit, consider gas–liquid criticality in single-component fluids and the closely anologous situation of an anisotropic, single-axis ferromagnet. As the temperature, T, approaches the critical point value, T_c, it is convenient to introduce the reduced temperature variable:

$$t \equiv (T - T_c)/T_c \to 0\pm . \tag{1}$$

Then, measuring the specific heat (at constant volume for a fluid or in zero magnetic field, $H = 0$, for a ferromagnet, *etc.*) one finds a power-law divergence to infinity of the form:

$$C(T) \approx A^\pm/|t|^\alpha, \quad \text{as} \quad t \to 0\pm . \tag{2}$$

---

[g]See Essam and Fisher (1963), Widom (1965a,b), Kadanoff (1966) and Fisher (1967a).
[h]Note Domb (1960), Domb and Hunter (1965) and the account in Domb (1996).
[i]See Baker (1961) and the overview in Baker (1990).
[j]The original paper is Patashinskii and Pokrovskii (1966); their text (1979), which includes a chapter on RGT, appeared in Russian around 1975 but did not then discuss RGT.
[k]Wilson (1971a,b) which he later described within a QFT context in Wilson (1983).
[l]Note the reviews by Kadanoff (1976) and Wegner (1976).

The amplitudes A^+ and A^- are *non*universal depending on the system (as, of course, is T_c); but their dimensionless ratio, A^+/A^-, and the characteristic *critical exponent* (or index), α are both found to be *universal*. (In fact, for bulk ($d = 3$)-dimensional systems one has $A^+/A^- \simeq 0.52$ and $\alpha \simeq 0.11$; but for $d = 2$, as predicted by Onsager (1944), $|t|^{-\alpha}$ is replaced by $\log|t|$ while $A^+/A^- = 1$. The power law (2) is markedly different from the 'classical' (or 'traditional' or 'mean-field') prediction of a mere jump in specific heat, $\Delta C = C_c^- - C_c^+ > 0$, for all d.

Next, one should recognize, following Landau[m] and Ginzburg,[n] the existence, at points **r** in d-dimensional Euclidean space, of a locally defined *order parameter*, say $\Psi(\mathbf{r})$. In a fluid this might well be the fluctuating density $\rho(\mathbf{r})$; in a ferromagnet it might be the local magnetization, $M(\mathbf{r})$, or spin variable, $S(\mathbf{r})$.

The order parameter and its exponent are directly revealed in the critical region of a fluid via the shape of the coexistence curve. This is described by

$$\Delta\rho \equiv \tfrac{1}{2}[\rho_{\text{liq}}(T) - \rho_{\text{gas}}(T)] \approx B|t|^\beta, \quad \text{as} \quad t \to 0-, \qquad (3)$$

in which the amplitude B is *non*universal but the *critical exponent* β, takes a *universal* value close to $\beta \simeq 0.325$ for $d = 3$: see, *e.g.*, Heller and Benedek (1962). For $d = 2$ the result is $\beta = \tfrac{1}{8}$ (Onsager, 1949; Yang, 1952; Kim and Chan, 1984); but both values differ significantly from the classical value $\beta = \tfrac{1}{2}$ predicted, *e.g.*, by the van der Waals equation.

By the same token, the spontaneous magnetization, $M_0(T)$, of an anisotropic ferromagnet with $M = \pm M_0(T)$, varies as $B|t|^\beta$ (the $+$ or $-$ depending on whether the field, H, below T_c, approaches 0 from above or below).

In terms of the order parameter one may also define the basic two-point or *pair correlation function*, $G(\mathbf{r})$, via

$$G(\mathbf{r}; T, \cdots) \equiv \langle \Delta\Psi(\mathbf{r})\Delta\Psi(\mathbf{0}) \rangle, \qquad (4)$$

where $\Delta\Psi(\mathbf{r}) = \Psi(\mathbf{r}) - \langle\Psi\rangle$, while the angular brackets $\langle\cdot\rangle$ denote a statistical average over the thermal (and, if relevant, quantal) fluctuations. Physically, $G(\mathbf{r})$ is important because it provides a direct measure of the development of order as criticality is approached. Indeed, the correlation length, $\xi(T)$, specifies the scale (or range) of the decay of $G(\mathbf{r})$ — typically

---

[m] See Landau and Lifshitz (1958), especially Sec. 135.
[n] In particular for the theory of superconductivity: see V. L. Ginzburg and L. D. Landau, *Zh. Eksp. Teor. Fiz.* **20**, 1064 (1959); and, for a personal historical account, V. L. Ginzburg, *Phys.-Usp.* **40**, 407–432 (1997).

exponential in character[o] relative to its long-range limit. The power law:

$$\xi(T) \approx \xi_0^{\pm}/|t|^{\nu}, \quad \text{as} \quad t \to 0\pm, \tag{5}$$

then describes the divergence of the correlation length; the exponent ν is close to 0.63 for fluids, *etc.*, when $d = 3$ (while for $d = 2$ one has[p] $\nu = 1$). It is worth remarking that in quantum field theory, the inverse correlation length ξ^{-1} is basically equivalent to the *renormalized mass* of the field $\psi(\mathbf{r})$.

Precisely at the critical point itself, however, an exponential decay is replaced, rather generally, by a power-law decay, namely,

$$G_c(\mathbf{r}) \approx D/r^{d-2+\eta}, \quad \text{as} \quad r \to \infty. \tag{6}$$

The critical exponent η, sometimes referred to as the *dimensional anomaly*, vanishes identically in all classical theories! However, it takes the value $\eta = \frac{1}{4}$ in $d = 2$ dimensions[q] while being nonzero and close to 0.036 for ($d = 3$)-dimensional fluids and anisotropic ferromagnets.[r]

Another central quantity is the divergent isothermal compressibility $\chi(T)$ (for a fluid) or isothermal susceptibility, $\chi(T) \propto (\partial M/\partial H)_T$ (for a ferromagnet). For this function, we write

$$\chi(T) \approx C^{\pm}/|t|^{\gamma}, \quad \text{as} \quad t \to 0\pm, \tag{7}$$

and find the universal value $\gamma \simeq 1.24$ for ($d = 3$)-dimensional fluid systems and anisotropic ferromagnets (while $\gamma = 1\frac{3}{4}$ for $d = 2$).[s] The classical value is simply $\gamma = 1$.

It is important to realize that there are other *universality classes* known theoretically although only a few are found experimentally.[t] Nevertheless, distinct universality classes yield distinct, observably different exponent values. Indeed, one of the early successes of RG theory was delineating and sharpening our grasp of various important universality classes: this will be illustrated briefly below. Furthermore, if and when a control variable, such as the pressure or a magnetic field, can change the universality class, it becomes important to know the value of the corresponding *crossover exponent* ϕ.[u] Its value, positive or negative, speaks to the *relevance* or *irrelevance* of the controlled variable.

---

[o] See Onsager (1944) and, *e.g.*, Fisher (1962, 1964).
[p] As shown by Onsager (1944).
[q] As can be shown from the work of Kaufman and Onsager (1949); see Fisher (1959) and also Fisher (1965, Sec. 29; 1967b, Sec. 6.2) and Fisher and Burford (1967).
[r] See, *e.g.*, Fisher and Burford (1967), Fisher (1983), Baker (1990) and Domb (1996).
[s] First advanced for $d = 2$ by Fisher (1959).
[t] See, *e.g.*, the survey in Fisher (1974a) and Aharony (1976).
[u] Fisher and Pfeuty (1972), Wegner (1972b).

In demanding, as we agreed, that RG theory should explain, in the context of critical phenomena, "the ubiquity of universal power laws," it may be helpful to compare the issue with the challenge to atomic physics of explaining the ubiquity of sharp spectral lines. Quantum mechanics responds, crudely speaking, by saying: "Well, (a) there is some wave — or a *wave function* ψ — needed to describe electrons in atoms, and (b) to fit a wave into a confined space the wavelength must be quantized: hence (c) only certain definite energy levels are allowed and, thence, (d) there are sharp, spectral transitions between them!"

Of course, that is far from being the whole story in quantum mechanics; but I believe it captures an important essence. Neither is the first RG response the whole story: but, *to anticipate* (A), in Wilson's conception RG theory crudely says: "Well, (a) there is a *flow* in some *space*, $\mathbb{H}$, *of Hamiltonians* (or "coupling constants"); (b) the critical point of a system is associated with a *fixed point* (or stationary point) of that flow; (c) the flow operator — technically the *RG transformation*,[v] $\mathbb{R}$ — can be *linearized* about that fixed point; and (d) typically, such a linear operator (as in quantum mechanics) has a spectrum of discrete, but nontrivial eigenvalues, say λ_k; then (e) each (asymptotically independent) exponential term in the flow varies as $e^{\lambda_k l}$, where l is the *flow* (or renormalization) *parameter*, and corresponds to a physical power law, say $|t|^{\phi_k}$, with critical exponent ϕ_k proportional to the eigenvalue λ_k." How one may find suitable transformations $\mathbb{R}$ and why the flows matter, are the subjects for the following chapters of our story.

Within this picture, distinct fixed points may correspond to distinct universality classes of distinct character (which are often blessed with names such as: Ising, XY, Heisenberg, *etc.*).

Just as quantum mechanics does much more than explain sharp spectral lines, so RG theory should also explain, at least in principle, (B) the values of the leading thermodynamic and correlation exponents, $\alpha, \beta, \gamma, \delta, \nu, \eta$ and ω (most already mentioned above) and (C) clarify why and how the classical values are in error, including the existence of borderline dimensionalities, like $d_\times = 4$, above which classical theories become valid. Beyond the

---

[v] As explained in more detail in Secs. 5 and 6 below, a specific renormalization transformation, say $\mathbb{R}_b$, acts on some 'initial' Hamiltonian $\mathcal{H}^{(0)}$ in the space $\mathbb{H}$ to transform it into a new Hamiltonian, $\mathcal{H}^{(1)}$. Under repeated operation of $\mathbb{R}_b$ the initial Hamiltonian "flows" into a sequence $\mathcal{H}^{(l)}$ ($l = 1, 2, \ldots$) corresponding to the iterated RG transformation $\mathbb{R}_b \cdots \mathbb{R}_b$ (l times) which, in turn, specifies a new transformation $\mathbb{R}_{b^l}$. These "products" of repeated RG operations serve to define a *semigroup* of transformations that, in general, does *not* actually give rise to a group: see footnote e above and the discussion below in Sec. 5 associated with Eq. (23).

leading exponents, one wants (D) the correction-to-scaling exponent θ (and, ideally, the higher-order correction exponents) and, especially, (E) one needs a method to compute crossover exponents, ϕ, to check for the relevance or irrelevance of a multitude of possible perturbations. Two central issues, of course, are (F) the understanding of *universality* with nontrivial exponents and (G) a derivation of *scaling*. The establishment of scaling leads directly to exponent relations such as

$$\alpha + 2\beta + \gamma = 2 \quad \text{and} \quad (2 - \eta)\nu = \gamma, \quad \text{etc}. \tag{8}$$

But much more is implied in terms of *data collapse*, the existence of *scaling functions, etc.*[w]

And, more subtly, one wants (H) to understand the *breakdown* of universality and scaling in certain circumstances — one might recall continuous spectra in quantum mechanics — and (J) to handle effectively logarithmic and more exotic dependences on temperature, *etc.*[x]

An important further requirement as regards to condensed matter physics is that RG theory should be firmly related to the science of statistical mechanics as perfected by Gibbs. Certainly, there is no need and should be no desire, to replace standard statistical mechanics as a basis for describing equilibrium phenomena in pure, homogeneous systems.[y] Accordingly, it is appropriate to summarize briefly the demands of statistical mechanics in a way suitable for describing the formulation of RG transformations.

We may start by supposing that one has a set of microscopic, fluctuating, mechanical variables: in QFT these would be the various quantum fields, $\psi(\mathbf{r})$, defined — one supposes — at all points in a Euclidean (or Minkowski) space. In statistical physics we will, rather, suppose that in a physical system of volume V there are N discrete "degrees of freedom." For classical fluid systems one would normally use the coordinates $\mathbf{r}_1, \mathbf{r}_2, \cdots, \mathbf{r}_N$ of the

---

[w]Historically, scaling concepts grew slowly, starting with exponent relations (Fisher, 1959, 1962, 1964; Buckingham and Gunton, 1969; Essam and Fisher, 1963; Rushbrooke, 1963; and Griffiths, 1965, 1972). Most transparent was Widom (1965a,b) but note also Domb and Hunter (1965); Kadanoff (1966); Kadanoff *et al.* (1967); Fisher (1967, 1971, 1974a); and Stanley (1971, Chaps. 11, 12). For *corrections-to-scaling* and the related concepts of *irrelevance, marginality* and *relevance* one must cite Wegner (1972, 1976); see also Kadanoff (1976) and Fisher (1974, 1983).

[x]See, *e.g.*, Ahlers *et al.* (1975), Aharony (1973) and Kadanoff and Wegner (1971).

[y]One may, however, raise legitimate concerns about the adequacy of customary statistical mechanics when it comes to the analysis of random or impure systems, such as, *e.g.*, "vortex-glass superconductors" (Koch *et al.*, 1984) — or in applications to systems far from equilibrium or in metastable or steady states — *e.g.*, in fluid turbulence, in sandpiles and earthquakes, *etc*. And the use of RG ideas in chaotic mechanics and various other topics listed by Benfatto and Gallavotti (1995) Chap. 1, clearly does *not* require a statistical mechanical basis.

constituent particles. However, it is simpler mathematically — and the analogies with QFT are closer — if we consider here a set of "*spins*" s_x (which could be vectors, tensors, operators, *etc.*) associated with discrete lattice sites located at uniformly spaced points x. If the lattice spacing is a, one can take $V = Na^d$ and the density of degrees of freedom in d spatial dimensions is $N/V = a^{-d}$.

In terms of the basic variables s_x, one can form various "local operators" (or "physical densities" or "observables") like the local magnetization and energy densities,

$$M_x = \mu_B s_x, \quad \mathcal{E}_x = -\tfrac{1}{2} J \sum_\delta s_x s_{x+\delta}, \quad \cdots, \qquad (9)$$

(where μ_B and J are fixed coefficients while δ runs over the nearest-neighbor lattice vectors). A physical system of interest is then specified by its *Hamiltonian* $\mathcal{H}[\{s_x\}]$ — or energy function, as in mechanics — which is usually just a spatially uniform sum of local operators. The crucial function is the *reduced Hamiltonian*:

$$\bar{\mathcal{H}}[s; t, h, \cdots, h_j, \cdots] = -\mathcal{H}[\{s_x\}; \cdots, h_j, \cdots]/k_B T, \qquad (10)$$

where s denotes the set of all the microscopic spins s_x while $t, h, \ldots, h_j, \ldots$ are various "*thermodynamic fields*" (in QFT — the coupling constants). We may suppose that one or more of the thermodynamic fields, in particular the temperature, can be controlled directly by the experimenter [see Eq. (1)]; but others may be "given" since they will, for example, embody details of the physical system that are "fixed by nature."

Normally in condensed matter physics one thus focuses on some specific form of $\bar{\mathcal{H}}$ with at most two or three variable parameters — the Ising model is one such particularly simple form with just two variables, t, the reduced temperature, and $h = \mu_B H/k_B T$, the reduced field. An important feature of Wilson's approach, however, is to regard any such "physical Hamiltonian" as merely specifying a subspace (spanned, say, by "coordinates" t and h) in a very large space of possible (reduced) Hamiltonians, $\mathbb{H}$: see the schematic illustration in Figure 1. This change in perspective proves crucial to the proper formulation of a renormalization group: in principle, it enters also in QFT although in practice, it is usually given little attention.

Granted a microscopic Hamiltonian, statistical mechanics promises to tell one the thermodynamic properties of the corresponding macroscopic system! First one must compute the partition function:

$$Z_N[\bar{\mathcal{H}}] = \text{Tr}_N^s \{e^{\bar{\mathcal{H}}[s]}\}, \qquad (11)$$

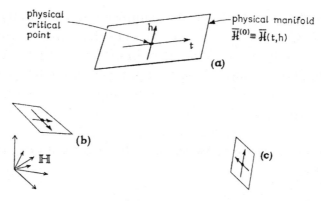

Figure 1. Schematic illustration of the space of Hamiltonians, $\mathbb{H}$, having, in general, infinitely many dimensions (or coordinate axes). A particular physical system or model representing, say, the ferromagnet, iron, is specified by its reduced Hamiltonian $\bar{\mathcal{H}}(t,h)$, with $t = (T - T_c)/T_c$ and $h = \mu_B H/k_B T$ defined for *that* system: but in $\mathbb{H}$ this Hamiltonian specifies only a submanifold — the physical manifold, labelled (a), that is parametrized by the 'local coordinates' t and h. Other submanifolds, (b) and (c), located elsewhere in $\mathbb{H}$, depict the physical manifolds for Hamiltonians corresponding to other particular physical systems, say, the ferromagnets nickel and gadolinium, *etc.*

where the *trace operation*, $\text{Tr}_N^s\{\cdot\}$, denotes a summation or integration[z] over all the possible values of all the N spin variables s_x in the system of volume V. The *Boltzmann factor*, $\exp(\bar{\mathcal{H}}[s])$, measures, of course, the probability of observing the microstate specified by the set of values $\{s_x\}$ in an equilibrium ensemble at temperature T. Then the thermodynamics follow from the total free energy density, which is given by[aa]

$$f[\bar{\mathcal{H}}] \equiv f(t, h, \cdots, h_j, \cdots) = \lim_{N,V \to \infty} V^{-1} \log Z_N[\bar{\mathcal{H}}]; \qquad (12)$$

this includes a *singular part* $f_s[\bar{\mathcal{H}}]$ near a critical point of interest as well as smooth *background terms* which are analytic (in $t, h, \ldots$) through the critical point. Correlation functions are defined similarly in standard manner.

To the degree that one can actually perform the trace operation in Eq. (11) for a particular model system and take the "thermodynamic limit" in Eq. (12) one will obtain the precise critical exponents, scaling functions and so on. This was Onsager's (1944) route in solving the $d = 2$, spin-$\tfrac{1}{2}$ Ising models in zero magnetic field. At first sight one then has no need of RG theory. That surmise, however, turns out to be far from the truth. The

---

[z]Here, for simplicity, we suppose the s_x are classical, commuting variables. If they are operator-valued then, in the standard way, the trace must be defined as a sum or integral over diagonal matrix elements computed with a complete basis set of N-variable states.

[aa]In Eq. (12), we have explicitly indicated the thermodynamic limit in which N and V become infinite maintaining the ratio $V/N = a^d$ fixed: in QFT this corresponds to an infinite system with an ultraviolet lattice cutoff.

issue is "simply" one of understanding! (Should one ever achieve truly high precision in simulating critical systems on a computer — a prospect which now seems closer than in the past — the same problem would remain.) In short, while one knows for sure that $\alpha = 0$ (log), $\beta = \frac{1}{8}$, $\gamma = 1\frac{3}{4}$, $\nu = 1, \eta = \frac{1}{4}, \cdots$ for the planar Ising models one does not know *why* the exponents have these values or *why* they satisfy the exponent relations Eq. (8) or why the scaling laws are obeyed. Indeed, the seemingly inevitable mathematical complexities of solving even such physically oversimplified models exactly[bb] serve to conceal almost all traces of general, underlying mechanisms and principles that might "explain" the results. Thus, it comes to pass that even a rather crude and approximate solution of a two-dimensional Ising model by a real-space RG method can be truly instructive.[cc]

3. Kadanoff's Scaling Picture

The 12 months from late-1965 through 1966 saw the clear formulation of scaling for the thermodynamic properties in the critical region and the fuller appreciation of scaling for the correlation functions.[dd] One may highlight Widom's (1965) approach since it was the most direct and phenomenological — a bold, new thermodynamic hypothesis was advanced by generalizing a particular feature of the classical theories. But Domb and Hunter (1965) reached essentially the same conclusion for the thermodynamics based on analytic and series-expansion considerations, as did Patashinskii and Pokrovskii (1966) using a more microscopic formulation that brought out the relations to the full set of correlation functions (of all orders).[ee]

Kadanoff (1966), however, derived scaling by introducing a completely new concept, namely, the *mapping* of a critical or near-critical system onto itself by a reduction in the effective number of degrees of freedom.[ff] This

---

[bb]As expounded in the monograph by Rodney Baxter (1982).
[cc]See Niemeijer and van Leeuwen (1976) and Burkhardt and van Leeuwen (1982) and Wilson (1975, 1983) for discussion of real-space RG methods.
[dd]Note the remarks made above in footnote w (on page 330). One might also recall, in this respect, earlier work (Fisher, 1959, 1962, 1964) restricted (in the application to ferromagnets) to zero magnetic field.
[ee]It was later seen, furthermore, (Kiang and Stauffer, 1970; Fisher, 1971, Sec. 4.4) that thermodynamic scaling with general exponents (but particular forms of scaling function) was embodied in the "droplet model" partition function advanced by Essam and Fisher (1963) from which the exponent relations $\alpha' + 2\beta + \gamma' = 2$, etc., were originally derived.
[ff]Novelty is always relative! From a historical perspective one should recall a suggestive contribution by M. J. Buckingham, presented in April 1965, in which he proposed a division of a lattice system into cells of geometrically increasing size, $L_n = b^n L_0$, with controlled intercell couplings. This led him to

paper attracted much favorable notice since, beyond obtaining all the scaling properties, it seemed to lay out a direct route to the actual *calculation* of critical properties. On closer examination, however, the implied program seemed — as I will explain briefly — to run rapidly into insuperable difficulties and interest faded. In retrospect, however, Kadanoff's scaling picture embodied important features eventually seen to be basic to Wilson's conception of the full RG. Accordingly, it is appropriate to present a sketch of Kadanoff's seminal ideas.

For simplicity, consider with Kadanoff (1966), a lattice of spacing a (and dimensionality $d > 1$) with $S = \frac{1}{2}$ Ising spins s_x which, by definition, take only the values $+1$ or -1. Spins on nearest-neighbor sites are coupled by an energy parameter or coupling constant, $J > 0$, which favors parallel alignment [see, *e.g.*, Eq. (9) above]. Thus, at low temperatures the majority of the spins point "up" ($s_x = +1$) or, alternatively, "down" ($s_x = -1$); in other words, there will be a spontaneous magnetization, $M_0(T)$, which decreases when T rises until it vanishes at the critical temperature $T_c > 0$: recall paragraph after Eq. (3).

Now divide the lattice up into (disjoint) blocks, of dimensions $L \times L \times \cdots \times L$ with $L = ba$ so that each block contains b^d spins. Then associate with each block, say $\mathcal{B}_{x'}$ centered at point x', a new, effective *block spin*, $s'_{x'}$: see Figure 2. If, finally, we *rescale* all spatial coordinates according to

$$\mathbf{x} \Rightarrow \mathbf{x}' = \mathbf{x}/b, \qquad (13)$$

the new lattice of block spins $s'_{x'}$ looks just like the original lattice of spins s_x. Note, in particular, the density of degrees of freedom is unchanged: see Figure 2.

But if this appearance is to be more than superficial one must be able to relate the new or "renormalized" coupling J' between the block spins

---

propose "the *existence* of an asymptotic 'lattice problem' such that the description of the nth order in terms of the $(n-1)$th is the same as that of the $(n+1)$th in terms of the nth." This is practically a description of "scaling" or "self-similarity" as we recognize it today. Unfortunately, however, Buckingham failed to draw any significant, correct conclusions from his conception and his paper seemed to have little influence despite its presentation at the notable international conference on *Phenomena in the Neighborhood of Critical Points* organized by M. S. Green (with G. B. Benedek, E. W. Montroll, C. J. Pings and the author) and held at the National Bureau of Standards, then in Washington, D. C. The Proceedings, complete with discussion remarks, were published, in December 1966, under the editorship of Green and J. V. Sengers (1966). Nearly all the presentations addressed the rapidly accumulating experimental evidence, but many well-known theorists from a range of disciplines attended including P. W. Anderson, P. Debye, C. de Dominicis, C. Domb, S. F. Edwards, P. C. Hohenberg, K. Kawasaki, J. S. Langer, E. Lieb, W. Marshall, P. C. Martin, T. Matsubara, E. W. Montroll, O. K. Rice, J. S. Rowlinson, G. S. Rushbrooke, L. Tisza, G. E. Uhlenbeck and C. N. Yang; but B. Widom, L. P. Kadanoff and K. G. Wilson are *not* listed among the participants.

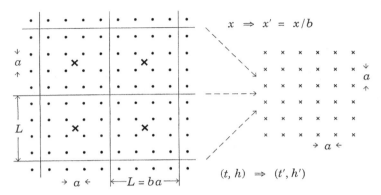

Figure 2. A lattice of spacing a of Ising spins $s_x = \pm 1$ (in $d = 2$ dimensions) marked by solid dots, divided up into Kadanoff blocks or cells of dimensions $(L = ba) \times (L = ba)$ each containing a block spin $s'_{x'} = \pm 1$, indicated by a cross. After a rescaling, $\mathbf{x} \Rightarrow \mathbf{x}' = \mathbf{x}/b$, the lattice of block spins appears identical with the original lattice. However, one supposes that the temperature t, and magnetic field h, of the original lattice can be renormalized to yield appropriate values, t' and h', for the rescaled, block-spin lattice: see text. In this illustration the spatial rescaling factor is $b = 4$.

to the original coupling J, or, equivalently, the renormalized temperature deviation t' to the original value t. Likewise one must relate the new, renormalized magnetic field h' to the original field h.

To this end, Kadanoff supposes that b is large but less than the ratio, ξ/a, of the *correlation length*, $\xi(t,h)$, to the lattice spacing a; since ξ diverges at criticality — see Eq. (5) — this allows, asymptotically, for b to be chosen *arbitrarily*. Then Kadanoff argues that the total coupling of the magnetic field h to a block of b^d spins is equivalent to a coupling to b^d average spins:

$$\bar{s}_{\mathbf{x}'} \equiv b^{-d} \sum_{\mathbf{x} \in \mathcal{B}_{\mathbf{x}'}} s_{\mathbf{x}} \cong \zeta(b) s'_{\mathbf{x}'} \tag{14}$$

where the sum runs over all the sites $\mathbf{x}$ in the block $\mathcal{B}_{\mathbf{x}'}$, while the "asymptotic equivalence" to the new, Ising block spin $s'_{\mathbf{x}'}$ entails, Kadanoff proposes, some "spin rescaling or renormalization factor" $\zeta(b)$. Introducing a similar thermal renormalization factor, $\theta(b)$, leads to the *recursion relations*:

$$t' \approx \theta(b) t \quad \text{and} \quad h' \approx b^d \zeta(b) h. \tag{15}$$

Correspondingly, the basic correlation function — compare with Eq. (4) — should renormalize as

$$G(\mathbf{x}; t, h) \equiv \langle s_0 s_{\mathbf{x}} \rangle \approx \zeta^2(b) G(\mathbf{x}'; t', h'). \tag{16}$$

In summary, under a spatial scale transformation and the integration out of all but a fraction b^{-d} of the original spins, the system asymptotically *maps back into itself* although at a renormalized temperature and field! However, the map is *complete* in the sense that *all* the statistical properties should be related by similarity.

But how should one choose — or, better, determine — the renormalization factors ζ and θ? Let us consider the basic relation Eq. (16) *at* criticality, so that $t = h = 0$ and, by Eq. (15), $t' = h' = 0$. Then, if we accept the observation/expectation Eq. (6) of a power law decay, i.e., $G_c(\mathbf{x}) \sim 1/|\mathbf{x}|^{d-2+\eta}$, one soon finds that $\zeta(b)$ must be just a power of b. It is natural, following Kadanoff (1966), then to propose the forms:

$$\zeta(b) = b^{-\omega} \quad \text{and} \quad \theta(b) = b^\lambda, \tag{17}$$

where the two exponents ω and λ characterize the critical point under study while b is an essentially unrestricted *scaling parameter*.

By capitalizing on the freedom to choose b as $t, h \to 0$, or, more-or-less equivalently, by *iterating* the recursion relations Eqs. (15) and (16), one can, with some further work, show that the previous exponent relations, Eqs. (8), must hold. But, beyond that, the full scaling laws also follow. Thus, one may write the correlation function in zero field ($h = 0$) as

$$G(\mathbf{r}; T) \approx \frac{D}{r^{d-2+\eta}} \mathcal{G}\left(\frac{r}{\xi(T)}\right), \tag{18}$$

where for consistency with Eq. (6), the *scaling function*, $\mathcal{G}(x)$, satisfies the normalization condition $\mathcal{G}(0) = 1$. Integrating $\mathbf{r}$ over all space yields the compressibility/susceptibility $\chi(T)$ and, thence, one may derive the relation $\gamma = (2 - \eta)\nu$, stated in Eqs. (8).

Similarly, one can derive scaling for the singular part of the free energy (introduced in Eq. (12) above). If, for the sake of some extra generality we introduce another field, say, $g = P/k_BT$, where *e.g.*, P denotes the pressure, the scaling hypothesis may be written:

$$f_s(t, h, g) \approx |t|^{2-\alpha} \mathcal{F}\left(\frac{h}{|t|^\Delta}, \frac{g}{|t|^\phi}\right), \tag{19}$$

where α is the specific heat exponent, introduced in Eq. (2), while the new exponent Δ, which determines "how h scales with t," is simply given by $\Delta = \beta + \gamma = \beta\delta$, where the exponent δ describes the order parameter variation *at $t = 0$* via $\langle\Psi\rangle \sim h^{1/\delta}$. The crossover exponent, ϕ, was mentioned in Sec. 2, shortly before Eqs. (8), but is not, at first sight, directly relevant for Kadanoff's analysis.

Of course, all the exponents are now determined by ω and λ in Eqs. (17); thus one finds $\nu = 1/\lambda$ and $\beta = \omega\nu$. Furthermore, the analysis leads to new exponent relations, namely, the so-called *hyperscaling laws*[gg] which explicitly involve the spatial dimensionality d: most notable is[hh]

$$d\nu = 2 - \alpha. \qquad (20)$$

Although this relation must be questioned for $d > 4$, Kadanoff's scaling picture is greatly strengthened by the fact that it holds *exactly* for the $d = 2$ Ising model! And likewise for all other exactly soluble models when $d < 4$.[ii]

Historically, the careful numerical studies of the $d = 3$ Ising models by series expansions[jj] for many years suggested a small but significant deviation from Eq. (20) as allowed by pure scaling phenomenolgy.[kk] But, in later years, the accumulating weight of evidence, when critically reviewed, convinced even the most cautious skeptics of the validity of Eq. (20) in three dimensions![ll]

Nevertheless, all is not roses! Unlike the previous exponent relations (all being independent of d) hyperscaling fails for the classical theories unless $d = 4$. And since one knows (rigorously for certain models) that the classical exponent values are valid for $d > 4$, it follows that hyperscaling cannot be generally valid. Thus, something is certainly missing from Kadanoff's picture. Now, thanks to RG insights, we know that the breakdown of hyperscaling is to be understood via the second argument in the "fuller" scaling form Eq. (19): when d exceeds the appropriate borderline dimension, $d_\times$, a "dangerous irrelevant variable" appears and must be allowed for.[mm] In essence one finds that the scaling function limit $\mathcal{F}(y, z \to 0)$, previously accepted without question, is no longer well defined but, rather, diverges as a power of z: asymptotic scaling survives but $d^* \equiv (2 - \alpha)/\nu$ sticks at the value 4 for $d > d_\times = 4$.

However, the issue of hyperscaling was *not* the main road block to the analytic development of Kadanoff's picture. The principal difficulties arose

---

[gg]See (Fisher, 1974a) where the special character of the hyperscaling relations is stressed.
[hh]See Kadanoff (1966), Widom (1965a) and Stell (1965, unpublished, quoted in Fisher, 1969). The relation $\delta = (d + 2 - \eta)/(d - 2 + \eta)$ is also worthy of note.
[ii]See, e.g., Fisher (1983) and, for the details of the exactly solved models, Baxter (1982).
[jj]For accounts of series expansion techniques and their important role see: Domb (1960, 1996), Baker (1961, 1990), Essam and Fisher (1963), Fisher (1965, 1967b) and Stanley (1971).
[kk]As expounded systematically in (Fisher, 1974a) with hindsight enlightened by RGT.
[ll]See Fisher and Chen (1985) and Baker and Kawashima (1995, 1996).
[mm]See Fisher in Gunton and Green (1974, p. 66) where a "dangerous irrelevant variable" is characterized as a "hidden relevant variable;" and Fisher (1983, Appendix D).

in explaining the *power-law* nature of the rescaling factors in Eqs. (15)–(17) and, in particular, in justifying the idea of a *single*, effective, renormalized coupling J' between adjacent block spins, say $s'_{x'}$ and $s'_{x'+\delta'}$. Thus the interface between two adjacent $L \times L \times L$ blocks (taking $d = 3$ as an example) separates two block faces each containing b^2, strongly interacting, original lattice spins s_x. Well below T_c all these spins are frozen, "up" or "down," and a single effective coupling could well suffice; but at and above T_c these spins must fluctuate on many scales and a single effective-spin coupling seems inadequate to represent the inherent complexities.[nn]

One may note, also that Kadanoff's picture, like the scaling hypothesis itself, provides no real hints as to the origins of universality: the rescaling exponents ω and λ in Eqs. (17) might well change from one system to another. Wilson's (1971a) conception of the RG answered *both* the problem of the "lost microscopic details" of the original spin lattice *and* provided a natural explanation of universality.

4. Wilson's Quest

Now because this account has a historical perspective, and since I was Ken Wilson's colleague at Cornell for some 20 years, I will say something about how his search for a deeper understanding of Quantum Field Theory (QFT) led him to formulate Renormalization Group Theory (RGT) as we know it today. The first remark to make is that Ken Wilson was a markedly independent and original thinker as well as being a rather private and reserved person. Second, in his 1975 article, in *Rev. Mod. Phys.*, from which I have already quoted, Ken Wilson gave his own considered overview of RGT which, in my judgement, still stands well today. In 1982 he received the Nobel Prize and in his Nobel lecture, published in 1983, he devotes a section to "Some History Prior to 1971" in which he recounts his personal scientific odyssey.

He explains that as a student at Caltech in 1956–1960, he failed to avoid "the default for the most promising graduate students (which) was to enter elementary-particle theory." There he learned of the 1954 paper by Gell-

---

[nn]In hindsight, we know this difficulty is profound: in general, it is *impossible* to find an adequate single coupling. However, for certain special models it does prove possible and Kadanoff's picture goes through: see Nelson and Fisher (1975) and Fisher (1983). Further, in defense of Kadanoff, the condition $b \ll \xi/a$ was supposed to "freeze" the original spins in each block sufficiently well to justify their replacement by a simple block spin.

Mann and Low "which was the principal inspiration for (his) own work prior to Kadanoff's (1966) formulation of the scaling hypothesis." By 1963, Ken Wilson had resolved to pursue QFTs as applied to the strong interactions. Prior to summer 1966 he heard Ben Widom present his scaling equation-of-state in a seminar at Cornell "but was puzzled by the absence of any theoretical basis for the form Widom wrote down." Later, in summer 1966, on studying Onsager's solution of the Ising model in the reformulation of Lieb, Schultz and Mattis,[oo] Wilson became aware of analogies with field theory and realized the applicability of his own earlier RG ideas (developed for a truncated version of fixed-source meson theory[pp]) to critical phenomena. This gave him a scaling picture but he discovered that he "had been scooped by Leo Kadanoff." Thereafter Ken Wilson amalgamated his thinking about field theories on a lattice and critical phenomena learning, in particular, about Euclidean QFT[qq] and its close relation to the transfer matrix method in statistical mechanics — the basis of Onsager's (1944) solution.

That same summer of 1966 I joined Ben Widom at Cornell and we jointly ran an open and rather wide-ranging seminar loosely centered on statistical mechanics. Needless to say, the understanding of critical phenomena and of the then new scaling theories was a topic of much interest. Ken Wilson frequently attended and, perhaps partially through that route, soon learned a lot about critical phenomena. He was, in particular, interested in the series expansion and extrapolation methods for estimating critical temperatures, exponents, amplitudes, *etc.*, for lattice models that had been pioneered by Cyril Domb and the King's College, London group. This approach is, incidentally, still one of the most reliable and precise routes available for estimating critical parameters.[rr] At that time I, myself, was completing a paper on work with a London University student, Robert J. Burford, using high-temperature series expansions to study in detail the correlation functions and scattering behavior of the two- and three-dimensional Ising models.[ss] Our theoretical analysis had already brought out some of the analogies with

---
[oo] See Schultz *et al.* (1964).
[pp] See Wilson (1983).
[qq] As stressed by Symanzik (1966) the Euclidean formulation of QFT makes more transparent the connections to statistical mechanics. Note, however, that in his 1966 article Symanzik did not delineate the special connections to critical phenomena *per se* that were gaining increasingly wide recognition; see, *e.g.*, Patashinskii and Pokrovskii (1966), Fisher (1969, Sec. 12) and the remarks below concerning Fisher and Burford (1967).
[rr] See the reviews Domb (1960), Fisher (1965, 1967b) and Stanley (1971).
[ss] Fisher and Burford (1967).

field theory revealed by the transfer matrix approach. Ken himself undertook large-scale series expansion calculations in order to learn and understand the techniques. Indeed, relying on the powerful computer programs Ken Wilson developed and kindly made available to us, another one of my students, Howard B. Tarko, extended the series analysis of the Ising correlation functions to temperatures below T_c and to all values of the magnetic field.[tt] Our results lasted fairly well and many of them were only later revised and improved.[uu]

Typically, then, Ken Wilson's approach was always "hands on" and his great expertise with computers was ever at hand to check his ideas and focus his thinking.[vv] From time to time Ken would intimate to Ben Widom or myself that he might be ready to tell us where his thinking about the central problem of explaining scaling had got to. Of course, we were eager to hear him speak at our seminar although his talks were frequently hard to grasp. From one of his earlier talks and the discussion afterwards, however, I carried away a powerful and vivid picture of *flows* — flows in a large space. And the point was that at the initiation of the flow, when the "time" or "flow parameter" l was small, two nearby points would travel close together: see Figure 3. But as the flow developed a point could be reached — a bifurcation point (and hence, as one later realized, a stationary or fixed point of the flow) — beyond which the two originally close points could separate and, as l increased, diverge to vastly different destinations: see Figure 3. At the time, I vaguely understood this as indicative of how a sharp, nonanalytic phase transition could grow from smooth analytic initial data.[ww]

But it was a long time before I understood the nature of the space — the space $\mathbb{H}$ of Hamiltonians — and the *mechanism* generating the flow, that is, a renormalization group transformation. Nowadays, when one looks at Figure 3, one sees the locus of initial points, $l = 0$, as identifying the manifold corresponding to the original or 'bare' Hamiltonian (see Figure 1) while the trajectory leading to the bifurcation point represents a locus of critical points; the two distinct destinations for $l \to \infty$ then typically, correspond to

---

[tt] Tarko and Fisher (1975).
[uu] See, *e.g.*, Zinn and Fisher (1996), Zinn, Lai and Fisher (1996), Butera and Comi (1997), Guida and Zinn-Justin (1998), Campostrini *et al.* (2002) and El-Showk *et al.* (2014a) who exploit conformal symmetry, and references therein.
[vv] See his remarks in Wilson (1983) on p. 591, column 1.
[ww] See the (later) introductory remarks in Wilson (1971a) related to Figure 1 there.

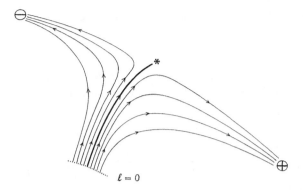

Figure 3. A "vision" of flows in some large space inspired by a seminar of K. G. Wilson in the period 1967–1970. The idea conveyed is that initially close, smoothly connected points at the start of the flow — the locus $l = 0$ — can eventually separate and run to far distant regions representing very different "final" physical states: the essence of a phase transition. In modern terms the flow is in the space $\mathbb{H}$ of Hamiltonians; the intersection of the separatrix, shown bolder, with the initial locus ($l = 0$) represents the physical critical point; $*$ denotes the controlling fixed point, while $\oplus$ and $\ominus$ represent asymptotic high-T, disordered states, and low-T, ordered states, respectively.

a high-temperature fully disordered system and to a low-temperature fully ordered system: see Figure 3.

In 1969 word reached Cornell that two Italian theorists, C. Di Castro and G. Jona-Lasinio, were claiming[xx] that the "multiplicative renormalization group," as expounded in the field-theory text by Bogoliubov and Shirkov (1959), could provide "a microscopic foundation" for the scaling laws (which, by then, were well established phenomenologically). The formalism and content of the field-theoretic renormalization group was totally unfamiliar to most critical-phenomena theorists: but the prospect of a microscopic derivation was clearly exciting! However, the articles[yy] proved hard to interpret as regards concrete progress and results. Nevertheless, the impression is sometimes conveyed that Wilson's final breakthrough was somehow anticipated by Di Castro and Jona-Lasinio.[zz]

Such an impression would, I believe, be quite misleading. Indeed, Di Castro was invited to visit Cornell where he presented his ideas in a seminar that was listened to attentively. Again I have a vivid memory: walking to lunch at the Statler Inn after the seminar I checked my own impressions with Ken Wilson by asking: "Well, did he really say anything new?" (By

---

[xx]The first published article was Di Castro and Jona-Lasinio (1969).
[yy]See the later review by Di Castro and Jona-Lasinio (1976) for references to their writings in the period 1969–1972 prior to Wilson's 1971 papers and the ϵ-expansion in 1972.
[zz]See, for example, Benfatto and Gallavotti (1995) on p. 96 in *A Brief Historical Note*, which is claimed to represent only the authors' personal "cultural evolution through the subject."

"new" I meant some fresh insight or technique that carried the field forward.) The conclusion of our conversation was "No." The point was simply that none of the problems then outstanding — see the "tasks" outlined above (in Sec. 2) — had been solved or came under effective attack. In fairness, I must point out that the retrospective review by Di Castro and Jona-Lasinio themselves (1976) is reasonably well balanced: one accepted a scaling hypothesis and injected that as an ansatz into a general formalism; then certain insights and interesting features emerged; but, in reality, only scaling theory had been performed; and, in the end, as Di Castro and Jona-Lasinio say: "Still one did not see how to perform explicit calculations." Incidentally, it is also interesting to note Wilson's sharp criticism[aaa] of the account presented by Bogoliubov and Shirkov (1959) of the original renormalization group ideas of Stueckelberg and Petermann (who, in 1953, coined the phrase "groupes de normalization") and of Gell-Mann and Low (1954).

One more personal anecdote may be permissible here. In August 1973, I was invited to present a tutorial seminar on renormalization group theory while visiting the Aspen Center for Physics. Ken Wilson's thesis advisor, Murray Gell-Mann, was in the audience. In the discussion period after the seminar Gell-Mann expressed his appreciation for the theoretical structure created by his famous student that I had set out in its generality, and he asked: "But tell me, what has all that got to do with the work Francis Low and I did so many years ago?"[bbb] In response, I explained the connecting thread and the far-reaching intellectual inspiration: certainly there is a thread but — to echo my previous comments — I believe that its length is comparable to that reaching from Maxwell, Boltzmann and ideal gases to Gibbs' general conception of ensembles, partition functions, and their manifold interrelations.

5. The Construction of Renormalization Group Transformations: the Epsilon Expansion

In telling my story I have purposefully incorporated a large dose of hindsight by emphasizing the importance of viewing a particular physical system — or its reduced Hamiltonian, $\bar{\mathcal{H}}(t, h, \ldots)$: see Eq. (10) — as

---

[aaa]See, especially, Wilson (1975) on p. 796, column 1 and footnote 10 in Wilson (1971a).
[bbb]That is, in Gell-Mann and Low (1954).

specifying only a relatively small manifold in a large space, $\mathbb{H}$, of possible Hamiltonians. But why is that more than a mere formality? One learns the answer as soon as, following Wilson (1975, 1983), one attempts to implement Kadanoff's scaling description in some concrete, computational way. In Kadanoff's picture (in common with the Gell-Mann–Low, Callan–Symanzik, and general quantum field theory viewpoints) one *assumes* that after a "rescaling" or "renormalization" the new, renormalized Hamiltonian (or, in QFT, the Lagrangean) has the *identical form* except for the renormalization of a single parameter (or coupling constant) or — as in Kadanoff's picture — of at most a small *fixed* number, like the temperature t and field h. That assumption is the dangerous and, unless one is especially lucky,[ccc] the *generally false* step! Wilson (1975, p. 592) has described his "liberation" from this straight jacket and how the freedom gained opened the door to the systematic design of renormalization group transformations.

To explain, we may state matters as follows: Gibbs' prescription for calculating the partition function — see Eq. (11) — tells us to sum (or to integrate) over the allowed values of *all* the N spin variables s_x. But this is very difficult! Let us, instead, adopt a strategy of "divide and conquer," by separating the set $\{s_x\}$ of N spins into two groups: first, $\{s_x^<\}$, consisting of $N' = N/b^d$ spins which we will leave as untouched fluctuating variables; and, second, $\{s_x^>\}$ consisting of the remaining $N - N'$ spin variables over which we will integrate (or sum) so that they drop out of the problem. If we draw inspiration from Kadanoff's (or Buckingham's[ddd]) block picture we might reasonably choose to integrate over all but one central spin in each block of b^d spins. This process, which Kadanoff has dubbed "decimation" (after the Roman military practice), preserves translational invariance and clearly represents a concrete form of "coarse graining" (which, in earlier days, was typically cited as a way to derive, "in principle," mesoscopic or Landau–Ginzburg descriptions).

Now, after taking our partial trace we must be left with some new, *effective Hamiltonian*, say, $\mathcal{H}_{\text{eff}}[s^<]$, involving only the preserved, unintegrated spins. On reflection one realizes that, in order to be faithful to the original physics, such an effective Hamiltonian must be defined via its Boltzmann factor: recalling our brief outline of statistical mechanics, that leads directly

---

[ccc] See footnote nn above and Nelson and Fisher (1975) and Fisher (1983).
[ddd] Recall footnote ff on page 333.

to the explicit formula

$$e^{\bar{\mathcal{H}}_{\text{eff}}[s^<]} = \text{Tr}_{N-N'}^{s^>}\{e^{\bar{\mathcal{H}}_{\text{eff}}[s^< \cup s^>]}\}, \tag{21}$$

where the 'union,' $s^< \cup s^>$, simply stands for the full set of original spins $s \equiv \{s_\mathbf{x}\}$. By a spatial rescaling, as in Eq. (13), and a relabelling, namely, $s_\mathbf{x}^< \Rightarrow s'_{\mathbf{x}'}$, we obtain the "renormalized Hamiltonian," $\bar{\mathcal{H}}'[s'] \equiv \bar{\mathcal{H}}_{\text{eff}}[s^<]$. Formally, then, we have succeeded in defining an *explicit renormalization transformation*. We will write

$$\bar{\mathcal{H}}'[s'] = \mathbb{R}_b\{\bar{\mathcal{H}}[s]\}, \tag{22}$$

where we have elected to keep track of the spatial rescaling factor, b, as a subscript on the renormalization group operator $\mathbb{R}$.

Note that if we complete the Gibbsian prescription by taking the trace over the renormalized spins we simply get back to the desired partition function, $Z_N[\bar{\mathcal{H}}]$. (The formal derivation for those who might be interested is set out in the footnote below.[eee]) Thus nothing has been lost: the renormalized Hamiltonian retains all the thermodynamic information. On the other hand, experience suggests that, rather than try to compute Z_N directly from $\bar{\mathcal{H}}'$, it will prove more fruitful to *iterate* the transformation so obtaining a sequence, $\bar{\mathcal{H}}^{(l)}$, of renormalized Hamiltonians, namely,

$$\bar{\mathcal{H}}^{(l)} = \mathbb{R}_b[\bar{\mathcal{H}}^{(l-1)}] = \mathbb{R}_{b^l}[\bar{\mathcal{H}}], \tag{23}$$

with $\bar{\mathcal{H}}^{(0)} \equiv \bar{\mathcal{H}}$, $\bar{\mathcal{H}}^{(1)} = \bar{\mathcal{H}}'$. It is these iterations that give rise to the *semigroup* character of the renormalization group transformation.[fff]

But now comes the crux: thanks to the rescaling and relabelling, the microscopic variables $\{s'_{\mathbf{x}'}\}$ are, indeed, completely equivalent to the original spins $\{s_\mathbf{x}\}$. However, when one proceeds to determine the nature of

---

[eee]We start with the definition Eq. (21) and recall Eq. (11) to obtain

$$Z_{N'}[\bar{\mathcal{H}}'] \equiv \text{Tr}_{N'}^{s'}\{e^{\bar{\mathcal{H}}[s']}\} = \text{Tr}_{N'}^{s^<}\{e^{\bar{\mathcal{H}}_{\text{eff}}[s^<]}\} = \text{Tr}_{N'}^{s^<}\text{Tr}_{N-N'}^{s^>}\{e^{\bar{\mathcal{H}}[s^< \cup s^>]}\}$$
$$= \text{Tr}_N^s\{e^{\bar{\mathcal{H}}[s]}\} = Z_N[\bar{\mathcal{H}}],$$

from which the free energy $f[\bar{\mathcal{H}}]$ follows via Eq. (12).

[fff]Thus successive decimations with scaling factors b_1 and b_2 yield the quite general relation:

$$\mathbb{R}_{b_2}\mathbb{R}_{b_1} = \mathbb{R}_{b_2 b_1},$$

which essentially defines a unitary *semigroup* of transformations. See footnotes e and v above (on pages 325 and 329), and the formal algebraic definition in MacLane and Birkhoff (1967): a unitary semigroup (or 'monoid') is a set M of elements, $u, v, w, x, \cdots$ with a binary operation, $xy = w \in M$, which is associative, so $v(wx) = (vw)x$, and has a unit u, obeying $ux = xu = x$ (for all $x \in M$) — in RGT, the unit transformation corresponds simply to $b = 1$. Hille (1948) and Riesz and Sz.-Nagy (1955) describe semigroups within a continuum, functional analysis context and discuss the existence of an infinitesimal generator when the flow parameter l is defined for continuous values $l \geq 0$: see Eq. (33) below and Wilson's (1971a) introductory discussion.

$\bar{\mathcal{H}}_{\text{eff}}$, and thence of $\bar{\mathcal{H}}'$, by using the formula (21), one soon discovers that one *cannot* expect the original form of $\bar{\mathcal{H}}$ to be reproduced in $\bar{\mathcal{H}}_{\text{eff}}$. Consider, for concreteness, an initial Hamiltonian, $\bar{\mathcal{H}}$, that describes Ising spins ($s_x = \pm 1$) on a square lattice in zero magnetic field with just nearest-neighbor interactions of coupling strength $K_1 = J_1/k_B T$: in the most conservative Kadanoff picture there must be *some* definite recursion relation for the renormalized coupling, say, $K_1' = \mathcal{T}_1(K_1)$, embodied in a definite function $\mathcal{T}_1(\cdot)$. But, in fact, one finds that $\bar{\mathcal{H}}_{\text{eff}}$ must actually contain *further* nonvanishing spin couplings, K_2, between second-neighbor spins, K_3, between third-neighbors and so on up to *indefinitely* high orders. Worse still, four-spin coupling terms like $K_{\square 1} s_{x_1} s_{x_2} s_{x_3} s_{x_4}$ appear in $\bar{\mathcal{H}}_{\text{eff}}$, again for *all* possible arrangements of the four spins! And also six-spin couplings, eight-spin couplings,.... Indeed, for any given set Q of $2m$ Ising spins on the lattice (and its translational equivalents), a nonvanishing coupling constant, K_Q', is generated and appears in $\bar{\mathcal{H}}'$!

The only saving grace is that further iteration of the decimation transformation Eq. (21) cannot (in zero field) lead to anything worse! In other words the space $\mathbb{H}_{\text{Is}}$ of Ising spin Hamiltonians in zero field may be specified by the infinite set $\{K_Q\}$, of all possible spin couplings, and is *closed* under the decimation transformation Eq. (21). Formally, one can thus describe $\mathbb{R}_b$ by the full set of *recursion relations*:

$$K_P' = \mathcal{T}_P(\{K_Q\}), \quad \text{(for all } P\text{)}. \tag{24}$$

Clearly, this answers our previous questions as to what becomes of the complicated across-the-faces-of-the-block interactions in the original Kadanoff picture: they actually carry the renormalized Hamiltonian *out* of the (too small) manifold of nearest-neighbor Ising models and introduce (infinitely many) further couplings. The resulting situation is portrayed schematically in Figure 4: the renormalized manifold for $\bar{\mathcal{H}}'(t', h')$ generally has no overlap with the original manifold. Further iterations, and *continuous* [see Eq. (33) below] as against discrete renormalization group transformations, are suggested by the flow lines or "trajectories" also shown in Figure 4. We will return to some of the details of these below.

In practice, the naive decimation transformation specified by Eq. (21) generally fails as a foundation for useful calculations.[ggg] Indeed, the design of effective renormalization group transformations turns out to be an art

---

[ggg]See Kadanoff and Niemeijer in Gunton and Green (1974), Niemeijer and van Leeuwen (1976) and Fisher (1983).

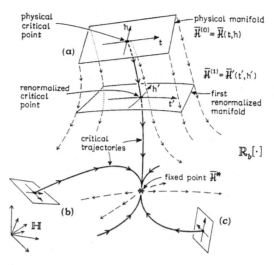

Figure 4. A depiction of the space of Hamiltonians $\mathbb{H}$ — compare with Figure 1 — showing initial or physical manifolds [labelled (a), (b), ..., as in Figure 1] and the flows induced by repeated application of a discrete renormalization group transformation $\mathbb{R}_b$ with a spatial rescaling factor b (or induced by a corresponding continuous or differential renormalization group). Critical trajectories are shown bold: they all terminate, in the region of $\mathbb{H}$ shown here, at a fixed point $\bar{\mathcal{H}}^*$. The full space contains, in general, other nontrivial, critical fixed points, describing multicritical points and distinct critical-point universality classes; in addition, trivial fixed points, including high-temperature "sinks" with no outflowing or relevant trajectories, typically appear. *Lines of fixed points* and other more complex structures may arise and, indeed, play a crucial role in certain problems. [After Fisher (1983).]

more than a science: there is no standard recipe! Nevertheless, there are guidelines: the general philosophy enunciated by Wilson and expounded well, for example, in a lecture by Shankar treating fermionic systems,[hhh] is to attempt to *eliminate* first those microscopic variables or degrees of freedom of "least direct importance" to the macroscopic phenomenon under study, while *retaining* those of most importance. In the case of ferromagnetic or gas–liquid critical points, the phenomena of most significance take place on long length scales — the correlation length, ξ, diverges; the critical point correlations, $G_c(\mathbf{r})$, decay slowly at long-distances; long-range order sets in below T_c.

Thus in his first, breakthrough articles in 1971, Wilson used an ingenious "phase-space cell" decomposition for *continuously variable scalar spins* (as against ± 1 Ising spins) to treat a lattice Landau–Ginzburg model with a general, single-spin or 'on-site' potential $V(s_\mathbf{x})$ acting on each spin, $-\infty < s_\mathbf{x} < \infty$. Blocks of cells of the smallest spatial extent were averaged over to

---

[hhh] See R. Shankar in Cao (1999) and Shankar (1994).

obtain a single, renormalized cell of twice the linear size (so that $b = 2$). By making sufficiently many simplifying approximations (that, incidentally, imposed $\eta = 0$), Wilson obtained an explicit *nonlinear, integral recursion relation* that transformed the l-times renormalized potential, $V^{(l)}(\cdot)$, into $V^{(l+1)}(\cdot)$. This recursion relation could be handled by computer and led to a *specific numerical estimate* for the exponents γ and ν for $d = 3$ dimensions, namely, $\gamma = 2\nu \simeq 1.217$, that were *quite different* from the classical values (and from the results of any previously soluble models like the *spherical model* [iii]). On seeing those results I knew that a major barrier to progress had been overcome!

I returned from a year's sabbatic leave at Stanford University in the summer of 1971, by which time Ken Wilson's two basic papers were in print. Shortly afterwards, in September, again while walking to lunch as I recall, Ken Wilson discussed his latest results from the nonlinear recursion relation with me. Analytical expressions could be obtained by expanding $V^{(l)}(s)$ in a power series:

$$V^{(l)}(s) = r_l s^2 + u_l s^4 + v_l s^6 + \cdots . \tag{25}$$

If truncated at quadratic order one had a soluble model — the Gaussian model (or free-field theory) — and the recursion relation certainly worked *exactly* for that! But to have a nontrivial model, one had to start not only with r_0 (as, essentially, the temperature variable) but, as a minimum, one also had to include $u_0 > 0$: the model then corresponded to the well-known $\lambda \varphi^4$ field theory. Although one might, thus, initially set $v_0 = w_0 = \cdots = 0$, all these higher order terms were immediately generated under renormalization; furthermore, there was no reason for u_0 to be small and, for this reason and others, the standard field-theoretic perturbation theories were ineffective.

Now, I had had a long-standing interest in the effects of *the spatial dimensionality* d on singular behavior in various contexts:[jjj] so that issue was raised for Ken's recursion relation. Indeed, d appeared simply as an explicit parameter.[kkk] It then became clear that $d = 4$ was a special case in which

---

[iii] The spherical model, which is fully soluble, was devised by Berlin and Kac (1952); see also Montroll and Berlin (1951) and Helfand and Langer (1967). For accounts of the critical behavior of the spherical model, see Fisher (1966a), where long-range forces were also considered, and, *e.g.*, Stanley (1971), Baxter (1982) and Fisher (1983).

[jjj] Fisher and Sykes (1959), Fisher and Gaunt (1964), Fisher (1966a,b; 1967c; 1972). See also Helfand and Langer (1967) p. 438, column 2.

[kkk] See the recursion relations (4.29) and (4.30) in Wilson (1971b). Note that the "2" in Eq. (4.30) should, in general, be replaced by b.

the leading order corrections to the Gaussian model vanished. Furthermore, above $d = 4$ dimensions classical behavior reappeared in a natural way (since the parameters $u_0, v_0, \cdots$ all then became irrelevant). These facts fitted in nicely with the known special role of $d = 4$ in other situations.[lll]

For $d = 3$, however, one seemed to need the infinite set of coefficients in Eq. (25) which all coupled together under renormalization. But I suggested that maybe one could treat the dimensional deviation, $\epsilon = 4 - d$, as a small, *nonintegral parameter* in analyzing the recursion relations for $d < 4$. Ken soon showed this was effective!

Indeed, using his analysis (Wilson, 1971b) and supposing the initial values, r_0 and u_0, are of order ϵ the recursion relations[kkk] become

$$r_{l+1} = 4[r_l + 3u_l(1 + r_l)^{-1} - 9u_l^2] + O(\epsilon^3), \tag{26}$$

$$u_{l+1} = (1 + \epsilon \ln 2)u_l - 9u_l^2 + O(\epsilon^3), \tag{27}$$

where we may recall the $\ln 2$ would, in general, be $\ln b$. A fixed point is then simply a solution $r_l = r^*$ and $u_l = u^*$ independent of l. From the second relation we see there are just two fixed points: the Gaussian fixed point $r^* = u^* = 0$ and a new, nontrivial fixed point[mmm]

$$u = u^* = \tfrac{1}{9}\epsilon \ln 2 + O(\epsilon^2), \tag{28}$$

$$r = r^* = -\tfrac{4}{9}\epsilon \ln 2 + O(\epsilon^2). \tag{29}$$

In $d = 4$ spatial dimensions when $\epsilon = 0$, these two fixed points coincide. Linearizing for small deviations about the new fixed point leads to

$$\gamma = 2\nu = 1 + \tfrac{1}{6}\epsilon + O(\epsilon^2). \tag{30}$$

Furthermore, to order ϵ the results prove to be independent of b and, as hoped, universal. The critical value for r (which is effectively T_c) turns out to be $r_c = -4u_0 + O(\epsilon^2)$; as expected, r_c depends on the single-spin potential $V(s) \equiv V^{(0)}(s)$ through its variation with u_0.

A paper, entitled by Ken "Critical Exponents in 3.99 Dimensions," was shortly written, submitted and published; it contained the first general expressions for nonclassical exponents.[nnn] In addition, the analysis was

---

[lll]See footnotes iii and jjj and Larkin and Khmel'nitskii (1969), especially Appendix 2.
[mmm]Sometimes, these days, known as the Wilson–Fisher fixed point or "line" or "family," especially when continued up to $\epsilon = 2$ or down to $d = 2$, the evidence suggesting it remains well defined: see, e.g., El-Showk et al. (2014b).
[nnn]See Wilson and Fisher (1972); Equations (9) and (10) are these relations. The first draft of this letter was written by Ken Wilson who graciously listed the authors in alphabetical order.

extended to XY [or $O(n = 2)$] models with planar two-component spins and to a version of Baxter's eight-vertex model.[ooo]

It transpired, however, that the perturbation parameter ϵ provided more — namely, a systematic way of ordering the infinite set of discrete recursion relations not only for the expansion coefficients of $V^{(l)}(s)$ but also for further multispin terms in the appropriate full space $\mathbb{H}$, involving spatial gradients, etc.[ppp] With that facility in hand, the previous approximations entailed in the phase-space cell analysis could be dispensed with; it became essential, however,[ppp] to then rescale the spins with the factor $b^{l(d-2+\eta)/2}$, where the exponent η had to be determined. Thus it could be established *exactly*, that $\eta = \frac{1}{54}\epsilon^2$, subject to corrections of order ϵ^3, while the result (30) for γ was correct to order ϵ.

In summary, Ken Wilson had seen that, by employing momenta or wave vectors $\mathbf{q}$ and labelling the spin variables, now re-expressed in Fourier space, as $\hat{s}_\mathbf{q}$, he could precisely implement his *momentum-shell renormalization group*[qqq] — subsequently one of the most-widely exploited tools in critical phenomena studies![rrr]

In essence the momentum-shell transformation is like decimation[sss] except that the division of the variables in Eq. (21) is made in momentum space: for ferromagnetic or gas-liquid type critical points the set $\{\hat{s}_\mathbf{q}^<\}$ contains those 'long-wavelength' or 'low-momentum' variables satisfying $|\mathbf{q}| \leq q_\Lambda/b$, where $q_\Lambda = \pi/a$ is the (ultraviolet) momentum cut-off implied by the lattice structure. Conversely, the 'short-wavelength', 'high-momentum' spin components $\{\hat{s}_\mathbf{q}^>\}$ having wave vectors lying in the momentum-space *shell*: $q_\Lambda/b < |\mathbf{q}| < q_\Lambda$, are integrated out. The spatial

---

[ooo]See Baxter (1971, 1982). The four-vertex model was reformulated by Kadanoff and Wegner (1971) as an energy–energy-type coupling; see also Wu (1971).
[ppp]See Eq. (18) in Wilson and Fisher (1972) and the subsequent text.
[qqq]Wilson (1972), entitled: "Feynman-Graph Expansion for Critical Exponents."
[rrr]See, *e.g.*, Brézin, Wallace and Wilson (1972), Wilson and Kogut (1974), the reviews Brézin, Le Guillou and Zinn-Justin (1976), and Wallace (1976), and the texts Amit (1978) and Itzykson and Drouffe (1983), etc.
[sss]A considerably more general form of RG transformation can be written as

$$\exp(\bar{\mathcal{H}}'[s']) = \text{Tr}_N^s\{\mathcal{R}_{N',N}(s';s)\exp(\bar{\mathcal{H}}[s])\},$$

where the trace is taken over the full set of original spins s. The $N' = N/b^d$ renormalized spins $\{s'\}$ are introduced via the RG kernel $\mathcal{R}_{N',N}(s';s)$ which incorporates spatial and spin rescalings, *etc.*, and which should satisfy a trace condition to ensure the partition-function-preserving property (see footnote eee on page 344) which leads to the crucial free-energy flow equation: see Eq. (38) below. The decimation transformation, the momentum-shell RG and other transformations can be written in this form.

rescaling now takes the form:

$$\mathbf{q} \Rightarrow \mathbf{q}' = b\mathbf{q}, \qquad (31)$$

as follows from Eq. (13); but in analogy to $\zeta(b)$ in Eq. (14), a *nontrivial spin rescaling factor* ("multiplicative-wavefunction renormalization" in QFT) is introduced via

$$\hat{s}_\mathbf{q} \Rightarrow \hat{s}'_{\mathbf{q}'} = \hat{s}_\mathbf{q}/\hat{c}[b, \bar{\mathcal{H}}] \ . \qquad (32)$$

The crucially important rescaling factor $\hat{c}$ takes the form $b^{d-\omega}$ and must be *tuned* in the critical region of interest [which leads to $\omega = \frac{1}{2}(d - 2 + \eta)$: compare with Eqs. (4) and (6)]. It is also worth mentioning that by letting $b \to 1+$, one can derive a *differential* or continuous flow RG and rewrite the recursion relation Eq. (22) as[ttt]

$$\frac{d}{dl}\bar{\mathcal{H}} = \mathbb{B}\left[\bar{\mathcal{H}}\right] \ . \qquad (33)$$

Such continuous flows are illustrated in Figures 3 and 4. (If it happens that $\bar{\mathcal{H}}$ can be represented, in general only approximately, by a single coupling constant, say, g, then $\mathbb{B}$ reduces to the so-called beta-function $\beta(g)$ of QFT.)

For deriving ϵ expansions on the basis of the momentum-shell RG, Feynman-graph perturbative techniques as developed for QFT prove very effective. (See footnotes qqq & rrr on the previous page.) They enter basically because one can take $u_0 = O(\epsilon)$ as small and they play a role both in efficiently organizing the calculation and in performing the essential integrals (particularly for systems with simple propagators and vertices).[uuu] Capitalizing on his field-theoretic expertise, Ken obtained, in only a few weeks after submitting the first article, *exact expansions* for the exponents ν, γ and ϕ to order ϵ^2 (and, by scaling, for all other exponents).[vvv] Furthermore, the anomalous dimension — defined in Eq. (6) close to the beginning of our story — was calculated exactly to order ϵ^3. Here, with $\epsilon = 4 - d > 0$, is this striking result:

$$\eta = \frac{(n+2)}{2(n+8)^2}\epsilon^2 + \frac{(n+2)}{2(n+8)^2}\left[\frac{6(3n+14)}{(n+8)^2} - \frac{1}{4}\right]\epsilon^3 + O(\epsilon^4), \qquad (34)$$

[ttt]See Wilson (1971a) and footnote fff above (on page 344): in this form the RG semigroup can typically be extended to an Abelian group (MacLane and Birkhoff, 1967); but as already stressed, this fact plays a negligible role.

[uuu]Nickel (1978). Nevertheless, many more complex situations arise in condensed matter physics for which the formal application of graphical techniques without an adequate understanding of the appropriate RG structure can lead one seriously astray.

[vvv]Note that Wilson (1972) was received on 1 December 1971 while Wilson and Fisher (1972) carries a receipt date of 11 October 1971. Fisher and Pfeuty (1972) meanwhile submitted an order ϵ treatment of the anisotropic n-vector model on 17 November 1971 finding $\phi = 1 + n\epsilon/2(n+8)$. The first introductory textbook regarding the RG, Pfeuty and Toulouse (1975), was sent to press in August 1974; the graduate text by Ma (1976a) soon followed.

where the symmetry parameter n denotes the number of components of the microscopic spin vectors, $\mathbf{s_x} \equiv (s_\mathbf{x}^\mu)_{\mu=1,\ldots,n}$, so that one has just $n=1$ for Ising spins.[www] For completeness we quote also the results for γ and ϕ while as regards ν, the relation $(2-\eta)\nu = \gamma$ should be recalled.

$$\gamma = 1 + \frac{n+2}{2(n+8)}\epsilon + \frac{(n+2)(n^2+22n+52)}{4(n+8)^3}\epsilon^2 + O(\epsilon^3), \quad (35)$$

$$\phi = 1 + \frac{n\epsilon}{2(n+8)} + \frac{n(n^2+24n+68)}{4(n+8)^3}\epsilon^2 + O(\epsilon^3). \quad (36)$$

In fact these ϵ expansions have been extended to order ϵ^5 (and ϵ^6 for η).[xxx] Beyond that, Ken's analysis inspired the development of many further related expansions for critical exponents and other universal parameters.[yyy]

6. Flows, Fixed Points, Universality and Scaling

To complete my story — and to fill in a few logical gaps over which we have jumped — I should explain how Wilson's construction of RG transformations in the space $\mathbb{H}$ enables RGT to accomplish the "tasks" set out above in Sec. 2. As illustrated in Figure 4, the recursive application of an RG transformation $\mathbb{R}_b$ induces a *flow* in the space of Hamiltonians, $\mathbb{H}$. Then one observes that "sensible," "reasonable," or, better, "well-designed" RG transformations are *smooth*, so that points in the original physical manifold, $\bar{\mathcal{H}}^{(0)} = \bar{\mathcal{H}}(t,h)$, that are close, say in temperature, remain so in $\bar{\mathcal{H}}^{(1)} \equiv \bar{\mathcal{H}}'$, i.e., under renormalization, and likewise as the flow parameter l increases, in $\bar{\mathcal{H}}^{(l)}$. Notice, incidentally, that since the spatial scale renormalizes via $\mathbf{x} \Rightarrow \mathbf{x}' = b^l \mathbf{x}$ one may regard

$$l = \log_b(|\mathbf{x}'|/|\mathbf{x}|), \quad (37)$$

[www] See, e.g., Domb and Sykes (1962), Fisher (1967b, 1974b, 1983), Kadanoff et al. (1967), Stanley (1971), Fisher and Pfeuty (1972), Pfeuty and Toulouse (1975), Aharony (1976), LeGuillou and Zinn-Justin (1977) and Patashinskii and Pokrovskii (1979). As discovered by Stanley (1968) the limit $n \to \infty$ corresponds precisely to the spherical model: see footnote iii above (on page 347).

[xxx] See Gorishny, Larin, and Tkachov (1984) for the initial result later corrected in Kleinert et al. (1991) but to little practical consequence for the summation procedures of Le Guillou and Zinn-Justin (1987), as shown by Guida and Zinn-Justin (1998).

[yyy] Especial mention should be made first of $1/n$ expansions, where n is the number of components of the vector order parameter (Abe, 1972, 1973; Fisher, Ma and Nickel, 1972; Suzuki, 1972), and see Fisher (1974) and Ma (1976a); second, of coupling-constant expansions in fixed dimension: see Parisi (1973, 1974); Baker, Nickel, Green and Meiron (1976); Le Guillou and Zinn-Justin (1977); Baker, Nickel and Meiron (1978). Later Wilson himself undertook large scale Monte Carlo RG calculations for the simple-cubic Ising model (Pawley et al. 1984). For other problems, modified *dimensionality expansions* have been made by writing $d = 8 - \epsilon, 6 - \epsilon, 4 + \frac{1}{2}m - \epsilon$ ($m = 1, 2, \ldots$), $3 - \epsilon, 2 + \epsilon$ and $1 + \epsilon$.

as measuring, logarithmically, the scale on which the system is being described — one might speak of the *scale dependence of parameters*; but note that, in general, the *form* of the Hamiltonian is also changing as the "scale" is changed or l increases. Thus, a partially renormalized Hamiltonian can be expected to take on a more-or-less generic, mesoscopic form: hence it represents an appropriate candidate to give meaning to a Landau–Ginzburg or, now, adding Wilson's name, a Landau–Ginzburg–Wilson (LGW) effective Hamiltonian.

Thanks to the smoothness of the RG transformation, if one knows the free energy $f_l \equiv f[\bar{\mathcal{H}}^{(l)}]$ at the lth stage of renormalization, then one knows the *original* free energy $f[\bar{\mathcal{H}}]$ and its critical behavior: explicitly one has[zzz]

$$f(t, h, \cdots) \equiv f[\bar{\mathcal{H}}] = b^{-dl} f[\bar{\mathcal{H}}^{(l)}] \equiv b^{-dl} f_l(t^{(l)}, h^{(l)}, \cdots). \tag{38}$$

Furthermore, the smoothness implies that all the universal critical properties are preserved under renormalization. Similarly one finds[aaaa] that the critical point of $\bar{\mathcal{H}}^{(0)} \equiv \bar{\mathcal{H}}$ maps on to that of $\bar{\mathcal{H}}^{(1)} \equiv \bar{\mathcal{H}}'$, and so on, as illustrated by the bold flow lines in Figure 4. Thus, it is instructive to follow the *critical trajectories* in $\mathbb{H}$, i.e., those RG flow lines that emanate from a physical critical point. In principle, the topology of these trajectories could be enormously complicated and even chaotic: in practice, however, for a well-designed or "apt" RG transformation, one most frequently finds that the critical flows terminate — or, more accurately, come to an asymptotic halt — at a *fixed point* $\bar{\mathcal{H}}^*$, of the RG: see Figure 4. Such a fixed point is defined, via Eqs. (22) or (33), simply by

$$\mathbb{R}_b[\bar{\mathcal{H}}^*] = \bar{\mathcal{H}}^* \quad \text{or} \quad \mathbb{B}[\bar{\mathcal{H}}^*] = 0. \tag{39}$$

One then searches for fixed-point solutions: the role of the fixed-point equation is, indeed, roughly similar to that of Schrödinger's equation $H\psi = E\psi$, for stationary states ψ_k of energy E_k in quantum mechanics (where ψ now denotes a wave function of appropriate variables).

Why are the fixed points so important? Some, in fact, are *not*, being merely *trivial*, corresponding to *no interactions* or to *all spins frozen*, etc. But the *non*trivial fixed points represent critical states; furthermore, the nature of their criticality, and of the free energy in their neighborhood, must, as explained, be *identical* to that of all those distinct Hamiltonians whose critical

---

[zzz] Recall the partition-function-preserving property set out in footnote[eee] above, which actually implies the basic relation Eq. (38).
[aaaa] See Wilson (1971a), Wilson and Kogut (1974) and Fisher (1983).

trajectories converge to the same fixed point! In other words, a particular fixed point *defines* a *universality class* of critical behavior which "governs" or "attracts" all those systems whose critical points eventually map onto it: see Figure 4.

Here, then we at last have the natural explanation of *universality*: systems of quite different physical character may, nevertheless, belong to the domain of attraction of the *same* fixed point $\bar{\mathcal{H}}^*$ in $\mathbb{H}$. The distinct sets of inflowing trajectories reflect their varying physical content of associated irrelevant variables and the corresponding nonuniversal rates of approach to the asymptotic power laws dicated by $\bar{\mathcal{H}}^*$.

From each physical critical fixed point, there flow at least two "unstable" or outgoing trajectories. These correspond to one or more *relevant* variables, specifically, for the case illustrated in Figure 4, to the temperature or thermal field, t, and the magnetic or ordering field, h. See also Figure 3. If there are further relevant trajectories then, as mentioned above, one can expect *crossover* to different critical behavior. In the space $\mathbb{H}$, such trajectories will then typically lead to distinct fixed points describing (in general) completely new universality classes.[bbbb]

But what about *power laws* and *scaling*? The answer to this question was already sketched in Sec. 2; but we will recapitulate here, giving a few more technical details. However, trusting readers or those uninterested in the analysis may skip to the paragraph containing Eq. (46)!

That said, one must start by noting that the smoothness of a well-designed RG transformation means that it can always be expanded locally — to at least some degree — in a Taylor series.[cccc] It is worth stressing that it is this very property that fails for free energies in a critical region: to regain this ability, the large space of Hamiltonians is crucial. Near a fixed point satisfying Eq. (39) we can, therefore, rather generally expect to be able to *linearize* by writing

$$\mathbb{R}_b[\bar{\mathcal{H}}^* + gQ] = \bar{\mathcal{H}}^* + g\mathbb{L}_b Q + o(g) \qquad (40)$$

[bbbb] A skeptical reader may ask: "But what if no fixed points are found? This can well mean, as it has frequently meant in the past, simply that the chosen RG transformation was poorly designed or "not apt." On the other hand, a fixed point represents only the simplest kind of asymptotic flow behavior: other types of asymptotic flow may well be identified and translated into physical terms. Indeed, near certain types of trivial fixed point, such procedures, long ago indicated by Wilson (1971a, Wilson and Kogut, 1974), *must* be implemented: see, *e.g.*, Fisher and Huse (1985). Limit cycles may also arise and be dealt with as shown by Glazek and Wilson (2002).

[cccc] See Wilson (1971a), Wilson and Kogut (1974), Fisher (1974b), Wegner (1972, 1976) and Kadanoff (1976).

as $g \to 0$, or in differential form,

$$\frac{d}{dl}(\bar{\mathcal{H}}^* + gQ) = g\mathbb{B}_1 Q + o(g). \tag{41}$$

Now $\mathbb{L}_b$ and $\mathbb{B}_1$ are *linear operators* (albeit acting in a large space $\mathbb{H}$). As such we can seek eigenvalues and corresponding "eigenoperators," say Q_k (which basic or "critical operators" will be partial Hamiltonians). Thus, in parallel to quantum mechanics, we may write

$$\mathbb{L}_b Q_k = \Lambda_k(b) Q_k \quad \text{or} \quad \mathbb{B}_1 Q_k = \lambda_k Q_k, \tag{42}$$

where, in fact, (by the semigroup property) the eigenvalues must be related by $\Lambda_k(b) = b^{\lambda_k}$. As in any such linear problem, knowing the spectrum of eigenvalues and eigenoperators or, at least, its dominant parts, tells one much of what one needs to know. Reasonably, the Q_k should form a basis for a general expansion:

$$\bar{\mathcal{H}} \cong \bar{\mathcal{H}}^* + \sum_{k \geq 1} g_k Q_k. \tag{43}$$

Physically, the expansion coefficient g_k ($\equiv g_k^{(0)}$) then represents the thermodynamic field[dddd] conjugate to the critical operator Q_k which, in turn, will often be close to some combination of *local* operators. Indeed, in a characteristic critical-point problem one finds two *relevant operators*, say Q_1 and Q_2 with $\lambda_1, \lambda_2 > 0$. Invariably, one of these operators can, say by its symmetry, be identified with the local energy density, $Q_1(\mathbf{r}) \cong E(\mathbf{r})$, so that $g_1 \cong t$ is the thermal field; the second then characterizes the order parameter, $Q_2(\mathbf{r}) \cong \Psi(\mathbf{r})$ with field $g_2 \cong h$. Under renormalization each g_k varies simply as $g_k^{(l)} \approx b^{\lambda_k l} g_k^{(0)}$.

Finally,[cccc] one examines the flow equation (38) for the free energy and for the correlations. The essential point is that the degree of renormalization, b^l, can be *chosen* as large as one wishes. When $t \to 0$, i.e., in the critical region which it is our aim to understand, a good choice proves to be $b^l = 1/|t|^{1/\lambda_1}$, which clearly diverges to ∞. One then finds that Eq. (38) leads to the *basic scaling relation* Eq. (19) which we will rewrite here in greater generality as

$$f_s(t, h, \cdots, g_j, \cdots) \approx |t|^{2-\alpha} \mathcal{F}\left(\frac{h}{|t|^{\Delta}}, \cdots, \frac{g_j}{|t|^{\phi_j}}, \cdots\right). \tag{44}$$

[dddd] Reduced, as expected, by the factor $1/k_B T$.

This is the essential result: thus it is easy to see that it leads to the "collapse" of equation-of-state data,[eeee] etc.

Now, however, the critical exponents can be expressed directly in terms of the RG eigenexponents λ_k (for the fixed point in question). Specifically one finds

$$2 - \alpha = \frac{d}{\lambda_1}, \quad \Delta = \beta + \gamma = \beta\delta = \frac{\lambda_2}{\lambda_1}, \quad \phi_j = \frac{\lambda_j}{\lambda_1} \quad \text{and} \quad \nu = \frac{1}{\lambda_1}. \quad (45)$$

Then, as already mentioned, the sign of a given ϕ_j and, hence, of the corresponding λ_j determines the *relevance* (for $\lambda_j > 0$), *marginality* (for $\lambda_j = 0$), or *irrelevance* (for $\lambda_j < 0$) of the corresponding critical operator Q_j (or "perturbation") and of its conjugate field g_j: this field might, but for most values of j will *not*, be under direct experimental control. As explained previously, all exponent relations (8), (20), *etc.*, follow from scaling, while the first and last of the Eqs. (45) yield the dimensionality-dependent *hyperscaling relation* Eq. (20).

When there are no marginal variables and the least negative ϕ_j is larger than unity in magnitude, a simple scaling description will usually work well and the Kadanoff picture almost applies. When there are *no* relevant variables and only one or a few *marginal variables*, field-theoretic perturbative techniques of the Gell-Mann–Low (1954), Callan–Symanzik[ffff] or so-called "parquet diagram" varieties[gggg] may well suffice (assuming the dominating fixed point is sufficiently simple to be well understood). There may then be little incentive for specifically invoking general RGT. From the perspective of high energy physics this seems to be the current situation in QFT and it applies also in some condensed matter problems.[hhhh]

Finally, for the set of critical operators $Q_i, Q_j, \cdots$ it is appropriate to follow Wilson (1969) and Kadanoff (1969) and define their *dimensions*[iiii] Δ_i,

---

[eeee]See Vicentini-Missoni (1972), Domb (1966) Figures 1.10 and 6.6, Fisher (1967b) Figure 18, pp. 710–712.

[ffff]See Wilson (1975), Brézin *et al.* (1976), Amit (1978) and Itzykson and Drouffe (1989).

[gggg]As used effectively by Larkin and Khmel'nitskii (1969).

[hhhh]See, *e.g.*, the case of dipolar Ising-type ferromagnets in $d = 3$ dimensions investigated experimentally by Ahlers, Kornblit and Guggenheim (1975) following theoretical work by Larkin and Khmel'nitskii (1969), Fisher and Aharony (1973) and Aharony (see 1976, Sec. 4E).

[iiii]Clearly the dimensions, Δ_j, should not be confused with the *gap exponent*, $\Delta = \beta + \gamma$, appearing in Eqs. (44) and (45), or earlier in Eq. (19), which specifies how h scales with t. The notation Δ_j, or Δ_Ψ, etc. originally employed by Polyakov (1970) seems currently favored: see El-Showk *et al.* (2014). Wilson (1969) originally used d_j while Kadanoff (1969) initially employed ν_Ψ, etc., but later adopted $x_j = d - y_j$ or $x_h = d - y_h$, etc. (Kadanoff, 1976). Note, however, that the exponent ω introduced in Eq. (17) for Kadanoff's spin rescaling factor is, in fact, just Δ_2. In the present treatment the critical operator dimensions are simply related to the eigenvalues λ_j defined in Eq. (42).

$\Delta_j, \cdots$ via the correlation functions evaluated *at* criticality ($t = h = 0$), i.e., recalling Eq. (4),

$$G^c_{ij}(\mathbf{r}_1, \mathbf{r}_2) = \langle \Delta Q_i(\mathbf{r}_1) \Delta Q_j(\mathbf{r}_2) \rangle \sim |\mathbf{r}_1 - \mathbf{r}_2|^{-(\Delta_i + \Delta_j)} \qquad (46)$$

for $|\mathbf{r}_1 - \mathbf{r}_2| \gg a$. It then becomes not unnatural to enquire about the nature and, more specifically, the dimensions of products of operators like $Q_i(\mathbf{r}_1) Q_j(\mathbf{r}_2)$, where $\mathbf{r}_1$ and $\mathbf{r}_2$ are fairly *close* together. This then leads to the formulation of appropriate *operator algebras* — as independently proposed by Wilson (1969) and by Kadanoff (1969), while no doubt also in the mind of Polyakov (1969). In more concrete terms an algebra can be specified by its *operator product expansion* in the form:

$$Q_i(\mathbf{r}_1) Q_j(\mathbf{r}_2) = \sum_k A_{ij,k}(\mathbf{r}_1 - \mathbf{r}_2) Q_k\left(\frac{\mathbf{r}_1 + \mathbf{r}_2}{2}\right), \qquad (47)$$

in which the numerical coefficients $A_{ij,k}$ specify the structure of the algebra.

Perhaps rather optimistically, this raised the hope — see Kadanoff (1969, 1971), Polyakov (1969) — that understanding the character of the algebra — subject, as it had to be, to various symmetries arising naturally in the problem at hand — might suffice to determine explicitly (or, at least, more directly) the actual values of the basic critical dimensions. If so, then all the critical exponents could be found! Thus, noting[iiii] $\Delta_1 = d - \lambda_1$, $\Delta_2 = d - \lambda_2$, *etc.*, the results (45) of our formulation may be rewritten, in terms *only* of the critical dimensions and the spatial dimensionality, as

$$\alpha = \frac{d - 2\Delta_1}{d - \Delta_1}, \quad \beta = \frac{\Delta_2}{d - \Delta_1}, \quad \gamma = \frac{d - 2\Delta_2}{d - \Delta_1}, \quad \delta = \frac{d}{\Delta_2} - 1, \quad \phi_j = \frac{d - \Delta_j}{d - \Delta_1} \qquad (48)$$

and, especially,

$$\nu = 1/(d - \Delta_1) \quad \text{and} \quad \eta = 2(\Delta_2 + 1) - d, \qquad (49)$$

from which, following Kadanoff (1976), we have

$$\Delta_1 = d\frac{1 - \alpha}{2 - \alpha} \quad \text{and} \quad \Delta_2 = \frac{d}{\delta + 1} = \frac{1}{2}(d - 2 + \eta). \qquad (50)$$

However, a further hypothesis regarding the algebraic character underlying the set of operators in fact proved necessary to fulfill the hope of fixing the exponent values. This was the concept of *conformal covariance*, initially proposed by Polyakov (1970, 1974) and, in essence, fully verified for the two-dimensional Ising model. More recently this has led to a line of research for Ising-type models in $d = 3$ dimensions that promises most accurate values for the exponents: see El-Showk *et al.* (2014) who found $\Delta_1 = 1.41267(13)$ and $\Delta_2 = 0.518154(15)$.

7. Conclusion

My tale is now told: following Wilson's 1971 papers and the introduction of the ϵ-expansion in 1972 the significance of the renormalization group approach in statistical mechanics was soon widely recognized[jjjj] and exploited by many authors interested in critical and multicritical phenomena and in other problems in the broad area of condensed matter physics, physical chemistry and beyond. Some of these successes have already been mentioned in order to emphasize, in particular, those features of the full RG that are of general significance in the wide range of problems lying beyond the confines of quantum field theory.[kkkk]

A further issue, originally motivated by Ken Wilson, is the relevance of renormalization group concepts to quantum field theory. I have addressed that only in various peripheral ways. Insofar as I am no expert in quantum field theory, that is not inappropriate; but perhaps one may step aside a moment and look at QFT from the general philosophical perspective of

---

[jjjj]Footnote ff (on page 333) drew attention to the first international conference on critical phenomena organized by Melville S. Green and held in Washington D. C. in April 1965. Eight years later, in late May 1973, Mel Green, with an organizing committee of J. D. Gunton, L. P. Kadanoff, K. Kawasaki, K. G. Wilson and the author, mounted another conference to review the progress in theory of the previous decade. The meeting was held in a Temple University Conference Center in rural Pennsylvania. The proceedings (Gunton and Green, 1974) entitled *Renormalization Group in Critical Phenomena and Quantum Field Theory*, are now mainly of historical interest. The discussions were recorded in full but most papers only in abstract or outline form. Whereas in the 1965 conference the overwhelming number of talks concerned experiments, now only J. M. H. (Anneke) Levelt Sengers and Guenter Ahlers spoke to review experimental findings in the light of theory. Theoretical talks were presented, in order, by P. C. Martin, Wilson, Fisher, Kadanoff, B. I. Halperin, E. Abrahams, Niemeijer (with van Leeuwen), Wegner, Green, Suzuki, Fisher and Wegner (again), E. K. Riedel, D. J. Bergman (with Y. Imry and D. Amit), M. Wortis, Symanzik, Di Castro, Wilson (again), G. Mack, G. Dell-Antonio, J. Zinn-Justin, G. Parisi, E. Brézin, P. C. Hohenberg (with Halperin and S.-K. Ma) and A. Aharony. Sadly, there were no participants from Russia, then the Soviet Union, but others included R. Abe, G. A. Baker, Jr., T. Burkhardt, R. B. Griffiths, T. Lubensky, D. R. Nelson, E. Siggia, H. E. Stanley, D. Stauffer, M. J. Stephen, B. Widom and A. Zee. As the lists of names and participants illustrates, many active young theorists had been attracted to the area, had made significant contributions, and were to make more in subsequent years.

[kkkk]Some reviews that may be cited to illustrate applications are Fisher (1974b), Wilson (1975), Wallace (1976), Aharony (1976), Pokrovskii and Patashinskii (1979), Nelson (1983) and Creswick *et al.* (1992). Beyond these, attention should be drawn to the notable article by Hohenberg and Halperin (1977) that reviews dynamic critical phenomena, and to many articles on further topics in the Domb and Lebowitz series *Phase Transitions and Critical Phenomena*, Vols. 7–20 (Academic, London, 1983–2001).

understanding complex, interacting systems. Then, I would claim, statistical mechanics is a central science of great intellectual significance — as just one reminder, the concepts of "spin-glasses" and the theoretical and computational methods developed to analyze them (such as "simulated annealing") have proved of interest in physiology for the study of neuronal networks and in operations research for solving hard combinatorial problems. In that view, even if one focuses only on the physical sciences, the land of statistical physics is broad, with many hills and dales, valleys and peaks to explore that are of relevance to the real world and to our ways of thinking about it. Within that land one may find an island, surrounded by water: but these days, more and broader bridges happily span the waters and communicate with the mainland! That island is devoted to what was "particle physics" and is now "high energy physics" or, more generally, to the deepest lying and, in that sense, the "most fundamental" aspects of physics. The reigning theory on the island is surely quantum field theory — the magnificent set of ideas and techniques that inspired the symposium that led to this article.[llll] Those laboring on the island have built most impressive skyscrapers reaching to the heavens!

Nevertheless, from the global viewpoint of statistical physics — where many degrees of freedom, the ever-present fluctuations and the diverse spatial and temporal scales pose the central problems — quantum field theory may be regarded as describing a rather special set of statistical mechanical models. As regards to applications they have been largely restricted to $d = 4$ spatial dimensions [more physically, of course to $(3 + 1)$ dimensions] although in subsequent decades *string theory* dramatically changed that! The practitioners of QFT insist on the preeminence of some special symmetry groups, the Poincaré group, $SU(3)$ and so on, which are not all so "natural" at first sight — even though the role of guage theories as a unifying theme in modeling nature has been particularly impressive. But, of course, we know these special features of QFT are not matters of choice — rather, they are forced on us by our explorations of Nature itself. Indeed, as far as experiment tells us, there is only one high energy physics; whereas, by contrast, the ingenuity of chemists, materials scientists and of Life itself, offers a much broader, multifaceted and varied panorama of systems to explore both conceptually and in the laboratory.

[llll] See the Foreword above and Cao (1999).

From this global standpoint, renormalization group theory represents a theoretical tool of depth and power. It first flowered luxuriantly in condensed matter physics, especially in the study of critical phenomena. But it is ubiquitous because of its potential for linking physical behavior across disparate scales; its ideas and techniques will continue to play a vital role in those situations where the fluctuations on many different physical scales truly interact. But it provides a valuable perspective — through concepts such as 'universality,' 'relevance,' 'marginality' and 'irrelevance,' even when scales are well separated! One can reasonably debate how vital renormalization group concepts are for quantum field theory itself. Certain aspects of the full theory do seem important because Nature teaches us, and particle physicists have learned, that quantum field theory is, indeed, one of those theories in which the different scales are connected together in nontrivial ways via the intrinsic quantum-mechanical fluctuations. However, in current quantum field theory, only certain facets of quantum field theory play a pivotal role.[mmmm] High energy physics did not have to be the way it is! But, even if it were quite different, we would still need quantum field theory in its fullest generality in condensed matter physics and, one suspects, in future scientific endeavors.

Acknowledgments

Thanks are due to Professor Emeritus Alfred I. Tauber, Director (until 2010) and Professor Tian Yu Cao of the Center for Philosophy and History of Science at Boston University for their part in arranging the original symposium on quantum field theory and for inviting me to speak. Daniel M. Zuckerman kindly assisted with the Bibliography. Comments on the initial draft manuscript (based on my talk at the symposium) from colleagues and friends Stephen G. Brush, N. David Mermin, R. Shankar, David J. Wallace, B. Widom and Kenneth G. Wilson were much appreciated. Stimulating interactions over many years have been enjoyed with Leo P. Kadanoff (including specifically for this article) and, but on fewer occasions, with Valery Pokrovskii.

[mmmm] It is interesting to look back and read in Gunton and Green (1973) pp. 157–160, Wilson's thoughts in May 1973 regarding the "Field Theoretic Implications of the Renormalization Group" at a point just before the ideas of *asymptotic freedom* became clarified for non-Abelian gauge theory by Gross and Wilczek (1973) and Politzer (1973).

Selected Bibliography

The reader is cautioned that this article is not intended as a systematic review of renormalization group theory and its origins. Likewise, this bibliography makes no claims of completeness; however, it includes those contributions of most significance in the personal views of the author (mainly prior to 1999). The reviews of critical phenomena and RG theory cited in footnote f (on page 325 above) contain many additional references. Further review articles appear in the series *Phase Transitions and Critical Phenomena*, edited by C. Domb and M. S. Green (later replaced by J. L. Lebowitz) and published by Academic Press, London (1972–2001): some are listed below. Introductory accounts in an informal lecture style are presented in Fisher (1965, 1983).

References

Abe, R. "Expansion of a Critical Exponent in Inverse Powers of Spin Dimensionality," *Prog. Theor. Phys.* **48**, 1414–1415 (1972).

Abe, R. "Expansion of a Critical Exponent in Inverse Powers of Spin Dimensionality," *Prog. Theor. Phys.* **49**, 113–128 (1973).

Aharony, A. "Critical Behavior of Magnets with Dipolar Interactions. V. Uniaxial Magnets in d Dimensions," *Phys. Rev. B* **8**, 3363–3370; (1974), erratum *ibid.* **9**, 3946 (1973).

Aharony, A. "Dependence of Universal Critical Behavior on Symmetry and Range of Interaction," in *Phase Transitions and Critical Phenomena*, eds. C. Domb and M. S. Green, Vol. 6 (Academic, London, 1976), pp. 357–424.

Ahlers, G., A. Kornblit and H. J. Guggenheim "Logarithmic Corrections to the Landau Specific Heat near the Curie Temperature of the Dipolar Ising Ferromagnet LiTbF$_4$," *Phys. Rev. Lett.* **34**, 1227–1230 (1975).

Amit, D. J. *Field Theory, the Renormalization Group and Critical Phenomena* (McGraw-Hill, London); see also (1993), the expanded Revised 2nd edn. (World Scientific, Singapore, 1978).

Bagnuls, C. and C. Bervillier "Field-Theoretic Techniques in the Study of Critical Phenomena," *J. Phys. Stud.* **1**, 366–382 (1997).

Baker, G. A., Jr. "Application of the Padé Approximant Method to the Investigation of some Magnetic Properties of the Ising Model," *Phys. Rev.* **124**, 768–774 (1961).

Baker, G. A., Jr. *Quantitative Theory of Critical Phenomena* (Academic, San Diego, 1990).

Baker, G. A., Jr. and N. Kawashima "Renormalized Coupling Constant for the Three-Dimensional Ising Model," *Phys. Rev. Lett.* **75**, 994–997 (1995).

Baker, G. A., Jr. and N. Kawashima "Reply to Comment by J.-K. Kim," *Phys. Rev. Lett.* **76**, 2403 (1996).

Baker, G. A., Jr., B. G. Nickel, M. S. Green and D. I. Meiron "Ising Model Critical Indices in Three Dimensions from the Callan–Symanzik Equation," *Phys. Rev. Lett.* **36**, 1351–1354 (1976).

Baker, G. A., Jr., B. G. Nickel and D. I. Meiron "Critical Indices from Perturbation Analysis of the Callan–Symanzik Equation," *Phys. Rev. B* **17**, 1365–1374 (1978).

Baxter, R. J. "Eight-vertex Model in Lattice Statistics," *Phys. Rev. Lett.* **26**, 832–833 (1971).

Baxter, R. J. *Exactly Solved Models in Statistical Mechanics* (Academic, London, 1982).

Benfatto, G. and G. Gallavotti *Renormalization Group*, Physics Notes, eds. P. W. Anderson, A. S. Wightman and S. B. Treiman (Princeton University, Princeton, 1995).

Berlin, T. H. and M. Kac "The Spherical Model of a Ferromagnet," *Phys. Rev.* **86**, 821–831 (1952).

Bogoliubov, N. N. and D. V. Shirkov *Introduction to the Theory of Quantized Fields* (Interscience, New York), Chap. 8 (1959).

Brézin, E., J. C. Le Guillou and J. Zinn-Justin "Field Theoretical Approach to Critical Phenomena," in *Phase Transitions and Critical Phenomena*, eds. C. Domb and M. S. Green, Vol. 6 (Academic, London), pp. 125–247 (1976).

Brézin, E., D. J. Wallace and K. G. Wilson "Feynman-Graph Expansion for the Equation of State near the Critical Point," *Phys. Rev. B* **7**, 232–239 (1972).

Buckingham, M. J. and J. D. Gunton "Correlations at the Critical Point of the Ising Model," *Phys. Rev.* **178**, 848–853 (1969).

Burkhardt, T. W. and J. M. J. van Leeuwen (eds.) *Real Space Renormalization* (Springer, Berlin, 1982).

Butera, P. and M. Comi "N-Vector Spin Models on the Simple-Cubic and Body-Centered-Cubic Lattices: A Study of the Critical Behavior of the Susceptibility and of the Correlation Length by High-Temperature Series Extended to Order β^{21}," *Phys. Rev. B* **56**, 8212–8240 (1997).

Campostrini, M., A. Pelissetto, P. Rossi and E. Vicari "25th Order High-Temperature Expansion Results for Three-Dimensional Ising-like Systems on the Simple Cubic Lattice," *Phys. Rev. E* **65**, 066127 (2002).

Cao, T. Y. (ed.) *Conceptual Foundations of Quantum Field Theory* (Cambridge University Press, Cambridge, 1999).

Creswick, R. J., H. A. Farach and C. P. Poole, Jr. *Introduction to Renormalization Group Methods in Physics* (John Wiley and Sons, New York, 1992).

Di Castro, C. and G. Jona-Lasinio "On the Microscopic Foundation of Scaling Laws," *Phys. Lett. A* **29**, 322–323 (1969).

Di Castro and Jona-Lasinio "The Renormalization Group Approach to Critical Phenomena," in *Phase Transitions and Critical Phenomena*, eds. C. Domb and M. S. Green, Vol. 6 (Academic, London), pp. 507–558 (1976).

Domb, C. "On the Theory of Cooperative Phenomena in Crystals," *Adv. Phys. (Philos. Mag. Suppl.)* **9**, 149–361 (1960).

Domb, C. *The Critical Point: A Historical Introduction to the Modern Theory of Critical Phenomena* (Taylor and Francis, London, 1996).

Domb, C. and D. L. Hunter "On the Critical Behavior of Ferromagnets," *Proc. Phys. Soc. London* **86**, 1147–1151 (1965).

Domb, C. and M. F. Sykes "Effect of Change of Spin on the Critical Properties of the Ising and Heisenberg Models," *Phys. Rev.* **128**, 168–173 (1962).

El-Showk, S., M. F. Paulos, D. Poland, S. Rychkova, D. Simmons-Duffin and A. Vichi "Solving the 3D Ising Model with the Conformal Bootstrap II. c-Minimization and Precise Critical Exponents," *J. Stat. Phys.* **157**, 869–914 (2014a).

El-Showk, S., M. F. Paulos, D. Poland, S. Rychkova, D. Simmons-Duffin and A. Vichi "Conformal Field Theories in Fractional Dimensions," *Phys. Rev. Lett.* **112**, 141601 (2014b).

Essam, J. W. and M. E. Fisher "Padé Approximant Studies of the Lattice Gas and Ising Ferromagnet below the Critical Point," *J. Chem. Phys.* **38**, 802–812 (1963).

Fisher, D. S. and D. A. Huse "Wetting Transitions: A Functional Renormalization Group Approach," *Phys. Rev. B* **32**, 247–256 (1985).

Fisher, M. E. "The Susceptibility of the Plane Ising Model," *Physica* **25**, 521–524 (1959).
Fisher, M. E. "On the Theory of Critical Point Density Fluctuations," *Physica* **28**, 172–180 (1962).
Fisher, M. E. "Correlation Functions and the Critical Region of Simple Fluids," *J. Math. Phys.* **5**, 944–962 (1964).
Fisher, M. E. "The Nature of Critical Points," in *Lectures in Theoretical Physics*, Vol. 107 (University of Colorado Press, Boulder), pp. 1–159 (1965).
Fisher, M. E. "Notes, Definitions and Formulas for Critical Point Singularities," in *Critical Phenomena*, (eds.) M. S. Green and J. V. Sengers (National Bureau of Standards, Washington DC, 1966a).
Fisher, M. E. "Shape of a Self-Avoiding Walk or Polymer Chain," *J. Chem. Phys.* **44**, 616–622 (1966b).
Fisher, M. E. "The Theory of Condensation and the Critical Point," *Physics* **3**, 255–283 (1967a).
Fisher, M. E. "The Theory of Equilibrium Critical Phenomena," *Rep. Prog. Phys.* **30**, 615–731 (1967b).
Fisher, M. E. "Critical Temperatures of Anisotropic Ising Lattices II. General Upper Bounds," *Phys. Rev.* **162**, 480–485 (1967c).
Fisher, M. E. "Phase Transitions and Critical Phenomena," in *Contemporary Physics, Proc. Int. Symp.*, Trieste, 7–28 June 1968, Vol. 1 (International Atomic Energy Agency, Vienna, 1969), pp. 19–46.
Fisher, M. E. "The Theory of Critical Point Singularities," in *Critical Phenomena, Proc. 1970 Enrico Fermi Int. Sch. Phys.*, Course No. 51, Varenna, Italy, ed. M. S. Green (Academic, New York, 1971), pp. 1–99.
Fisher, M. E. "Phase Transitions, Symmetry and Dimensionality," in *Essays in Physics*, Vol. 4 (Academic, London, 1972), pp. 43–89.
Fisher, M. E. "General Scaling Theory for Critical Points," in *Proc. Nobel Symp. XXIV*, Aspenäsgården, Lerum, Sweden, June 1973, *Collective Properties of Physical Systems*, eds. B. Lundqvist and S. Lundqvist (Academic, New York, 1974a), pp. 16–37.
Fisher, M. E. "The Renormalization Group in the Theory of Critical Behavior," *Rev. Mod. Phys.* **46**, 597–616 (1974b).
Fisher, M. E. "Scaling, Universality and Renormalization Group Theory," in *Critical Phenomena*, ed. F. J. W. Hahne, Lecture Notes in Physics, Vol. 186 (Springer, Berlin, 1983), pp. 1–139.
Fisher, M. E. "Renormalization Group Theory: Its Basis and Formulation in Statistical Physics," *Rev. Mod. Phys.* **70**, 653–681; reprinted with revisions in Cao (1999), Part IV, Chap. 8, pp. 89–135 (1998).
Fisher, M. E. and A. Aharony "Dipolar Interactions at Ferromagnetic Critical Points," *Phys. Rev. Lett.* **30**, 559–562 (1973).
Fisher, M. E. and R. J. Burford "Theory of Critical Point Scattering and Correlations. I. The Ising model," *Phys. Rev.* **156**, 583–622 (1967).
Fisher, M. E. and J.-H. Chen "The Validity of Hyperscaling in Three Dimensions for Scalar Spin Systems," *J. Phys.* **46**, 1645–1654 (1985).
Fisher, M. E. and D. S. Gaunt "Ising Model and Self-Avoiding Walks on Hypercubical Lattices and 'High Density' Expansions," *Phys. Rev.* **133**, A224–A239 (1964).
Fisher, M. E., S.-K. Ma and B. G. Nickel "Critical Exponents for Long-Range Interactions," *Phys. Rev. Lett.* **29**, 917–920 (1972).
Fisher, M. E. and P. Pfeuty "Critical Behavior of the Anisotropic n-Vector Model," *Phys. Rev. B* **6**, 1889–1891 (1972).

Fisher, M. E. and M. F. Sykes "Excluded-Volume Problem and the Ising Model of Ferromagnetism," *Phys. Rev.* **114**, 45–58 (1959).

Gell-Mann, M. and F. E. Low "Quantum Electrodynamics at Small Distances," *Phys. Rev.* **95**, 1300–1312 (1954).

Glazek, S. D. and K. G. Wilson "Limit Cycles in Quantum Theories," *Phys. Rev. Lett.* **89**, 230401 (2002).

Gorishny, S. G., S. A. Larin and F. V. Tkachov "ϵ-Expansion for Critical Exponents: The $O(\epsilon^5)$ Approximation," *Phys. Lett. A* **101**, 120–123 (1984).

Green, M. S. and J. V. Sengers (eds.) *Critical Phenomena: Proc. Conf.* Washington, D.C. April 1965, NBS. Misc. Pub. **273** (US Govt. Printing Off., Washington, 1966).

Griffiths, R. B. "Thermodynamic Inequality near the Critical Point for Ferromagnets and Fluids," *Phys. Rev. Lett.* **14**, 623–624 (1965).

Griffiths, R. B. "Rigorous Results and Theorems," in *Phase Transitions and Critical Phenomena* eds. C. Domb and M. S. Green, Vol. 1 (Academic, London, 1972), pp. 7–109.

Gross, D. and F. Wilczek "Ultraviolet Behavior of NonAbelian Gauge Theories," *Phys. Rev. Lett.* **30**, 1343–1346 (1973).

Guida, R. and J. Zinn-Justin "Critical Exponents of the N-Vector Model," *J. Phys. A, Math. Gen.* **31**, 8103–8121 (1998).

Gunton, J. D. and M. S. Green (eds.) in Renormalization Group in *Critical Phenomena and Quantum Field Theory: Proc. Conf.* Chestnut Hill, Pennsylvania, 29–31 May 1973 (Temple University, Philadelphia, 1974).

Helfand, E. and J. S. Langer "Critical Correlations in the Ising Model," *Phys. Rev.* **160**, 437–450 (1967).

Heller, P. and G. B. Benedek "Nuclear Magnetic Resonance in MnF_2 near the Critical Point," *Phys. Rev. Lett.* **8**, 428–432 (1962).

Hille, E. *Functional Analysis and SemiGroups* (American Mathematical Society, New York, 1948).

Hohenberg, P. C. and B. I. Halperin "Theory of Dynamic Critical Phenomena," *Rev. Mod. Phys.* **49**, 435–479 (1977).

Itzykson, D. and J.-M. Drouffe *Statistical Field Theory* (Cambridge University Press, Cambridge, 1989).

Kadanoff, L. P. "Scaling Laws for Ising Models near T_c," *Physics* **2**, 263–272 (1966).

Kadanoff, L. P. "Correlations along a Line in the Two-Dimensional Ising Model," *Phys. Rev.* **188**, 859–863 (1969a).

Kadanoff, L. P. "Operator Algebra and the Determination of Critical Indices," *Phys. Rev. Lett.* **23**, 1430–1433 (1969b).

Kadanoff, L. P. "Scaling, Universality and Operator Algebras," in *Phase Transitions and Critical Phenomena*, eds. C. Domb and M. S. Green, Vol. 5A (Academic, London, 1976), pp. 1–34.

Kadanoff, L. P. and H. Ceva "Determination of an Operator Algebra for the Two-Dimensional Ising Model," *Phys. Rev. B* **3**, 3918–3919 (1971).

Kadanoff, L. P., W. Götze, D. Hamblen, R. Hecht, E. A. S. Lewis, V. V. Palciauskas, M. Rayl, J. Swift, D. Aspnes and J. Kane "Static Phenomena near Critical Points: Theory and Experiment," *Rev. Mod. Phys.* **39**, 395–431 (1967).

Kadanoff, L. P. and F. J. Wegner "Some Critical Properties of the Eight-Vertex Model," *Phys. Rev. B* **4**, 3989–3993 (1971).

Kaufman, B. and L. Onsager "Crystal Statistics II. Short-Range Order in a Binary Lattice," *Phys. Rev.* **76**, 1244–1252 (1949).

Kiang, C. S. and D. Stauffer "Application of Fisher's Droplet Model for the Liquid–Gas Transition near T_c," *Z. Phys.* **235**, 130–139 (1970).

Kim, H. K. and M. H. W. Chan (1984) "Experimental Determination of a Two-Dimensional Liquid–Vapor Critical-Point Exponent," *Phys. Rev. Lett.* **53**, 170–173.

Kleinart, H., J. Neu, V. Schulte-Frohlinde, K. G. Chetyrkin and S. A. Larin "Five-Loop Renormalization Group Functions of $O(n)$-Symmetric ϕ^4-Theory and ϵ-Expansions of Critical Exponents up to ϵ^5," *Phys. Lett. B* **272**, 39–44 (1991).

Koch, R. H., V. Foglietti, W. J. Gallagher, G. Koren, A. Gupta and M. P. A. Fisher "Experimental Evidence for Vortex-Glass Superconductivity in Y-Ba-Cu-O," *Phys. Rev. Lett.* **63**, 1511–1514 (1989).

Landau, L. D. and E. M. Lifshitz *Statistical Physics*, Course of Theoretical Physics, Vol. 5 (Pergamon, London, 1958), Chap. 14.

Larkin, A. I. and D. E. Khmel'nitskii "Phase Transition in Uniaxial Ferroelectrics," *Zh. Eksp. Teor. Fiz.* **56**, 2087–2098; *Sov. Phys.-JETP* **29**, 1123–1128 (1969).

Le Guillou, J. C. and J. Zinn-Justin "Critical Exponents for the n-Vector Model in Three Dimensions from Field Theory," *Phys. Rev. Lett.* **39**, 95–98 (1977).

Le Guillou, J. C. and J. Zinn-Justin "Accurate Critical Exponents for Ising-like Systems in NonInteger Dimensions," *J. Phys.* **48**, 19–24 (1987).

Ma, S.-K. *Modern Theory of Critical Phenomena* (W. A. Benjamin, Reading, 1976a).

Ma, S.-K. "The $1/n$ Expansion," in *Phase Transitions and Critical Phenomena*, eds. C. Domb and M. S. Green, Vol. 6 (Academic, London), pp. 249–292 (1976b).

MacLane, S. and G. Birkhoff *Algebra* (MacMillan, New York), Chap. 2, Sec. 9; Chap. 3, Secs. 1, 2 (1967).

Montroll, E. W. and T. H. Berlin "An Analytical Approach to the Ising Problem," *Commun. Pure Appl. Math.* **4**, 23–30 (1951).

Nelson, D. R. and M. E. Fisher "Soluble Renormalization Groups and Scaling Fields for Low-Dimensional Spin Systems," *Ann. Phys.* **91**, 226–274 (1975).

Niemeijer, Th. and J. M. J. van Leeuwen "Renormalization Theory for Ising-like Systems," in *Phase Transitions and Critical Phenomena*, eds. C. Domb and M. S. Green, Vol. 6 (Academic, London, 1976), pp. 425–505.

Nickel, B. G. "Evaluation of Simple Feynman Graphs," *J. Math. Phys.* **19**, 542–548 (1978).

Onsager, L. "Crystal Statistics I. A Two-Dimensional Model with an Order-Disorder Transition," *Phys. Rev.* **62**, 117–149 (1944).

Onsager, L. "Discussion Remark Following a Paper by G. S. Rushbrooke at the International Conference on Statistical Mechanics in Florence," *Nuovo Cimento Suppl. Ser. 9* **6**, 261 (1949).

Parisi, G. "Perturbation Expansion of the Callan–Symanzik Equation," in Gunton and Green (1974), cited above (1973).

Parisi, G. "Large-Momentum Behavior and Analyticity in the Coupling Constant," *Nuovo Cimento A* **21**, 179–186 (1974).

Patashinskii, A. Z. and V. L. Pokrovskii "Behavior of Ordering Systems near the Transition Point," *Zh. Eksp. Teor. Fiz.* **50**, 439–447; *Sov. Phys.-JETP* **23**, 292–297 (1966).

Patashinskii, A. Z. and V. L. Pokrovskii *Fluctuation Theory of Phase Transitions* (Pergamon, Oxford, 1979).

Pawley, G. S., R. H. Swendsen, D. J. Wallace and K. G. Wilson "Monte Carlo Renormalization Group Calculations of Critical Behavior in the Simple-Cubic Ising Model," *Phys. Rev. B* **29**, 4030–4040 (1984).

Pfeuty, P. and G. Toulouse *Introduction au Group des Renormalisation et à ses Applications* (Univ. de Grenoble, Grenoble); translation, *Introduction to the Renormalization Group and to Critical Phenomena* (Wiley and Sons, London, 1977), (1975).

Politzer, H. D. "Reliable Perturbative Results for Strong Interactions?" *Phys. Rev. Lett.* **30**, 1346–1349 (1973).

Polyakov, A. M. "Properties of Long and Short Range Correlations in the Critical Region," *Zh. Eksp. Teor. Fiz.* **57**, 271–283; *Sov. Phys.-JETP* **30**, 151–157 (1969), (1970).

Polyakov, A. M. "Conformal Symmetry of Critical Fluctuations," *Zh. Eksp. Teor. Fiz. Pis'ma Red.* **12**, 538–541; *Sov. Phys. JETP-Lett.* **12**, 381–383 (1970).

Polyakov, A. M. "NonHamiltonian Approach to Conformal Quantum Field Theory," *Zh. Eksp. Teor. Fiz.* **66**, 23–42; *Sov. Phys.-JETP* **39**, 10–18 (1974).

Riesz, F. and B. Sz.-Nagy *Functional Analysis* (Frederick Ungar, New York), Chap. 10, Secs. 141, 142 (1955).

Rushbrooke, G. S. "On the Thermodynamics of the Critical Region for the Ising Problem," *J. Chem. Phys.* **39**, 842–843 (1963).

Schultz, T. D., D. C. Mattis and E. H. Lieb "Two-Dimensional Ising Model as a Soluble Problem of Many Fermions," *Rev. Mod. Phys.* **36**, 856–871 (1964).

Shankar, R. "Renormalization-Group Approach to Interacting Fermions," *Rev. Mod. Phys.* **66**, 129–192 (1994).

Stanley, H. E. "Spherical Model as the Limit of Infinite Spin Dimensionality," *Phys. Rev.* **176**, 718–722 (1968).

Stanley, H. E. *Introduction to Phase Transitions and Critical Phenomena* (Oxford University, New York, 1971).

Stell "Extension of the Ornstein–Zernike Theory of the Critical Region," *Phys. Rev. Lett.* **20**, 533–536 (1968).

Suzuki, M. "Critical Exponents and Scaling Relations for the Classical Vector Model with Long-Range Interactions," *Phys. Lett. A* **42**, 5–6 (1972).

Symanzik, K. "Euclidean Quantum Field Theory, I. Equations for a Scalar Model," *J. Math. Phys.* **7**, 510–525 (1966).

Tarko, H. B. and M. E. Fisher "Theory of Critical Point Scattering and Correlations. III. The Ising model below T_c and in a Field," *Phys. Rev. B* **11**, 1217–1253 (1975).

Vicentini-Missoni, M. "'Equilibrium Scaling in Fluids and Magnets," in *Phase Transitions and Critical Phenomena*, eds. C. Domb and M. S. Green, Vol. 2 (Academic, London, 1972), pp. 39–78.

Wallace, D. J. "The ϵ-Expansion for Exponents and the Equation of State in Isotropic Systems," in *Phase Transitions and Critical Phenomena*, eds. C. Domb and M. S. Green, Vol. 6 (Academic, London, 1976), pp. 293–356.

Wegner, F. J. "Corrections to Scaling Laws," *Phys. Rev. B* **5**, 4529–4536 (1972a).

Wegner, F. J. "Critical Exponents in Isotropic Spin Systems," *Phys. Rev. B* **6**, 1891–1893 (1972b).

Wegner, F. J. "The Critical State, General Aspects," in *Phase Transitions and Critical Phenomena*, eds. C. Domb and M. S. Green, Vol. 6 (Academic, London, 1976), pp. 7–124.

Widom, B. "Surface Tension and Molecular Correlations near the Critical Point," *J. Chem. Phys.* **43**, 3892–3897 (1965a).

Widom, B. "Equation of State in the Neighborhood of the Critical Point," *J. Chem. Phys.* **43**, 3898–3905 (1965b).

Wilson, K. G. "NonLagrangian Models of Current Algebra," *Phys. Rev.* **179**, 1499–1512 (1969).

Wilson, K. G. "Operator-Product Expansions and Anomalous Dimensions in the Thirring Model," *Phys. Rev. D* **2**, 1473–1477 (1970).

Wilson, K. G. "Renormalization Group and Critical Phenomena I. Renormalization Group and Kadanoff Scaling Picture," *Phys. Rev. B* **4**, 3174–3183 (1971a).

Wilson, K. G. "Renormalization Group and Critical Phenomena II. Phase-Space Cell Analysis of Critical Behavior," *Phys. Rev. B* **4**, 3184–3205 (1971b).

Wilson, K. G. "Feynman-Graph Expansion for Critical Exponents," *Phys. Rev. Lett.* **28**, 548–551 (1972).

Wilson, K. G. "The Renormalization Group: Critical Phenomena and the Kondo Problem," *Rev. Mod. Phys.* **47**, 773–840 (1975).

Wilson, K. G. "The Renormalization Group and Critical Phenomena" (1982, Nobel Prize Lecture), *Rev. Mod. Phys.* **55**, 583–600 (1983).

Wilson, K. G. and M. E. Fisher "Critical Exponents in 3.99 Dimensions," *Phys. Rev. Lett.* **28**, 240–243 (1972).

Wilson, K. G. and J. Kogut "The Renormalization Group and the ϵ Expansion," *Phys. Rep.* **12**, 75–200 (1974).

Wu, F. Y. "Ising Model with Four-Spin Interactions," *Phys. Rev. B* **4**, 2312–2314 (1971).

Yang, C. N. "The Spontaneous Magnetization of a Two-Dimensional Ising Model," *Phys. Rev.* **85**, 808–816 (1952).

Zinn, S.-Y. and M. E. Fisher "Universal Surface-Tension and Critical-Isotherm Amplitude Ratios in Three Dimensions," *Physica A* **226**, 168–180 (1996).

Zinn, S.-Y., S.-N. Lai and M. E. Fisher "Renormalized Coupling Constants and Related Amplitude Ratios for Ising Systems," *Phys. Rev. E* **54**, 1176–1182 (1996).

10

Statistical Physics in the Oeuvre of Chen Ning Yang*

Michael E. Fisher

Institute for Physical Science and Technology and Department of Physics, University of Maryland, College Park, MD 20742, USA

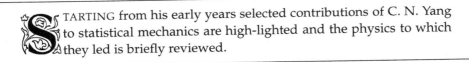

TARTING from his early years selected contributions of C. N. Yang to statistical mechanics are high-lighted and the physics to which they led is briefly reviewed.

1. First Publications

From his earliest days as a student Chen Ning Yang — later known to friends in the West as Frank Yang — had been aware of mathematics; his father, Wu-Chih Yang, was, indeed, a professor of the subject. Not surprisingly, perhaps, the first paper published by Cheng-Ning Yang (see Figure 1) appeared in the *Bulletin of the American Mathematical Society* in 1944. Nevertheless, as a graduate student at the Southwest Associated University in Kunming in 1942–44, he became intrigued by statistical mechanics. His advisor, Professor J. S. Wang (who had studied in Britain with R. H. Fowler in the 1930's) supervised his Master's thesis under the title: "*Contributions to the Statistical Theory of Order-Disorder Transitions.*"

Through his studies of order-disorder phenomena — one of the most notable examples being the transition in beta-brass which is an alloy of copper and zinc close to a 50:50 composition — the young Mr. Yang surely became

---

*Reprinted from *Int. J. Mod. Phys. B*, **29**, 1530013 (2015).

> **ON THE UNIQUENESS OF YOUNG'S DIFFERENTIALS**
>
> CHENG-NING YANG
>
> **Introduction.** The differential of a function of several variables may be defined in a variety of ways, of which the one given by Young[1] renders the best parallelism with the case of a single variable. Stated in the way given below, his definition is applicable to a function defined in a set S of points containing limiting points at which the function is to have differentials. The question of the uniqueness of the differentials, however, arises. In this paper we shall first define, and prove two theorems concerning, the "limiting directions" which describe the directional distribution of the points of S near a limiting point. Then we proceed to show that properties of these limiting directions determine whether the differential is unique or not.

Figure 1. C. N. Yang's first published paper based on Young's paper in *Proc. London Math. Soc.* **7**, 157 (1908). The article appeared in *Bull. Amer. Math. Soc.* **50**, 373–375 (1944). Note the spelling of "C. N."!.

familiar with X-ray scattering observations such as that illustrated in Figure 2. But here there is a bit of a cheat: the crystal studied in Figure 2 by Dan Shechtman and coworkers is actually Al_6Mn which, as might be concluded from the 10-fold symmetry, is actually a *quasicrystal*! Be that as it may, Yang's first paper in statistical physics was based on his Master's thesis

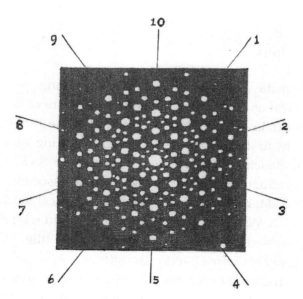

Figure 2. Scattering observations illustrating the evidence for order in a crystal that can undergo an order-disorder transition. [After D. Shectman, I. Blech, D. Gratias, and J. W. Cahn, *Phys. Rev. Lett.* **53**, 1951 (1984).]

THE JOURNAL OF CHEMICAL PHYSICS   VOLUME 13, NUMBER 2   FEBRUARY, 1945

A Generalization of the Quasi-Chemical Method in the Statistical Theory of Superlattices

C. N. YANG*

National Tsing Hua University, Kunming, China

(Received November 17, 1944)

The quasi-chemical method in the investigation of the equilibrium distribution of atoms in the pairs of neighboring sites in a superlattice is generalized by considering groups containing large numbers of sites. The generalized method may be used to obtain successive approximations of the free energy of the crystal. The labor of integration is avoided by the introduction of a Legendre transformation. In order to analyze the fundamental assumption underlying the method more closely, the number of arrangements of the atoms for given long-distance order is calculated and the hypothesis of the non-interference of local configurations discussed. The method is applied to the calculation of the free energy in the different approximations discussed in this paper, including Bethe's second approximation and a simple approximation for the face-centered cubic crystal Cu_3Au.

Figure 3. The second paper published. Note that the author is now identified as "C. N. Yang."

and published in *The Journal of Chemical Physics* in 1945: see Figure 3. This was submitted while he was a Research Fellow of the China Foundation at the National Tsing Hua University (which had been evacuated to Kunming owing to the Japan–China war, 1937–1945). Notable, and characteristic, is that the paper develops a new formulation of the quasi-chemical method: "which is capable of yielding successively higher approximations."

2. The Ising Model and Onsager

Doubtless already as a graduate student working on his Master's thesis, Yang had come across the Ising model of a ferromagnet. In its simplest form at each site, i, of a square lattice, in $d = 2$ dimensions (or a simple cubic lattice, if $d = 3$), there is a "spin variable," s_i, which takes only the values $+1$ or -1. Nearest neighbor spins, at sites i and j are coupled yielding a favorable energy $\Delta \mathcal{H} = -Js_is_j$ while the magnetic field, H, contributes $-mHs_i$ for each spin. To model a binary AB alloy, the values $s_i = +1$ or -1 are merely associated with an atom A or B.

Ernst Ising solved the model explicitly for $d = 1$ in 1924. Too soon, however, he became a refugee from Hitler's Germany. Much later he worked for almost three decades, 1948–1976, as a Professor of Physics in Bradley University in Peoria, Illinois. But to solve the Ising model for $d > 1$ proved intractable for many years. Furthermore, as we now know well, the available approximate solutions were quite misleading!

But Yang has related how one day in 1944–45, while still a Master's student, his normally quiet and reserved advisor, Professor J. S. Wang, became quite excited, almost agitated. It transpired that he had just learned that Lars Onsager (shown below with C. N. Yang in Figure 4) had exactly solved the square-lattice Ising model in zero magnetic field finding a logarithmic divergence of the specific heat at criticality! He told Yang about the paper: *Phys. Rev.* **65**, 117 (1944); but Yang, like so many others, could not see how to go further: no "strategic plan" seemed evident. A little while later, when already in Chicago, significant progress still eluded him.

Figure 4. Chen Ning Yang and Lars Onsager together much later in March 1965.

However, in November 1949, with a Chicago Ph.D. safely in hand, Yang had a chance conversation with J. M. Luttinger at the Institute for Advanced Study in Princeton. Through that he learned of Bruria Kaufman's simplification of Onsager's approach in terms of a system of anticommuting hermitian matrices. Kaufman's work[1] opened the door to understanding Onsager's solution and to an appreciation of how much had been learned about the Ising model on a square lattice, albeit in zero magnetic field.

In January 1951 Yang realized that the spontaneous magnetization, $M_0(T)$ as a function of the temperature T, could be derived from an off-diagonal matrix element between the two eigenvectors of the underlying

---
[1]See: B. Kaufman, *Phys. Rev.* **76**, 1232 (1949).

Reprinted from THE PHYSICAL REVIEW, Vol. 85, No. 5, 808–816, March 1, 1952.
Printed in U. S. A.

The Spontaneous Magnetization of a Two-Dimensional Ising Model

C. N. YANG
Institute for Advanced Study, Princeton, New Jersey
(Received September 18, 1951)

The spontaneous magnetization of a two-dimensional Ising model is calculated exactly. The result also gives the long-range order in the lattice.

— The "longest calculation in my career"

IT is the purpose of the present paper to calculate the spontaneous magnetization (i.e., the intensity of magnetization at zero external field) of a two-dimensional Ising model of a ferromagnet. Van der Waerden[1] and Ashkin and Lamb[2] had obtained a series expansion of the spontaneous magnetization that converges very rapidly at low temperatures. Near the critical temperature, however, their series expansion cannot be used. We shall here obtain a closed expression for the spontaneous magnetization by the matrix method

where

$$V_1 = \exp\{H^* \sum_1^n C_r\}, \quad (2)$$

and

$$V_2 = \exp\{H \sum_1^n s_r s_{r+1}\}. \quad (3)$$

H^* and H are given by

$$e^{-2H} = \tanh H^* = \exp[-(1/kT)(V_{\uparrow\downarrow} - V_{\uparrow\uparrow})]. \quad (4)$$

Figure 5. The 1951 article providing the exact solution of the spontaneous magnetization of the square-lattice Ising model and establishing the critical exponent $\beta = \frac{1}{8}$. Note the author's later remark! The basic transfer matrix is $\mathbf{V}_1\mathbf{V}_2$ where (in the notation of Bruria Kaufman[1] who used $H = J/k_B T$) $\mathbf{V}_1$ and $\mathbf{V}_2$ appear here in Eqs. (2) and (3).

transfer matrix that had the largest eigenvalues. Following that idea — which entailed months of hard labour — led to the submission in mid-September of the remarkable paper displayed in Figure 5. Following a series of "miraculous cancellations" (including elaborate factors denoted, with bold originality, **I**, **II**, **III**, and **IV**), the calculation led to the surprisingly simple expression

$$M_0(T) = \left[\frac{1+x^2}{(1-x^2)^2}(1-6x^2+x^4)^{\frac{1}{2}}\right]^{2\beta}, \quad (2.1)$$

where $x = \exp(-2J/k_B T)$, while k_B is Boltzmann's constant. The crucial and *universal critical exponent* proved to be $\beta = \frac{1}{8}$, in strong contrast with nearly all previous approximate theoretical treatments[2] which yielded $\beta = \frac{1}{2}$ — often called the "mean-field value."

The problem of a rectangular Ising model with distinct values of J_1 and J_2 — already broached by Onsager — was suggested to C. H. Chang (of the University of Washington, Seattle). His results (guided by Yang!) could be

---

[2]It is worth recalling, however, that Cyril Domb, using an extrapolation of the exact low-temperature series expansions for $M_0(T)$, had, in 1949, obtained the estimate $\beta \lesssim 0.16$ only 20% higher than Yang's exact result: see *Proc. Roy. Soc. A* **199**, 199–221 (1949).

expressed in terms of the modulus, k, of the crucial elliptic integral as
$$M_0(T) = (1 - k^2)^\beta, \qquad (2.2)$$
where, with $x_i = \exp(-2J_i/k_BT)$ for $i = 1, 2$,
$$k = \frac{2x_1}{1 - x_1^2} \frac{2x_2}{1 - x_2^2}. \qquad (2.3)$$

As expected, Chang found $\beta = \frac{1}{8}$ for all values of $n = J_2/J_1$; the striking behavior of $M_0(T; n)$ is shown in Figure 6.

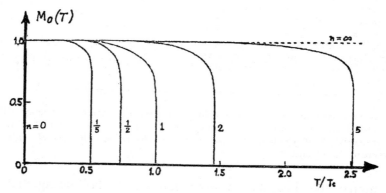

Figure 6. The spontaneous magnetization of a two-dimensional rectangular Ising model with vertical and horizontal couplings, J_1 and J_2, related by $n = J_2/J_1$. Note that T_c is the critical point of the square lattice with $J = J_1 = J_2$ and $x_c = \sqrt{2} - 1$. [After C. H. Chang, *Phys. Rev.* **88**, 1422 (1952).]

3. Lars Onsager and C. N. Yang

A year later Yang encountered Onsager at a 1953 *Tokyo–Kyoto International Conference*. But while Onsager spoke in his characteristic, somewhat vague style, punctuated by broad smiles, he did not address the Ising model. Thus it was only some twelve years later in March 1965 that Yang was able to learn, directly from Onsager himself, the reasons for all the unobvious commutator calculations in the 1944 paper. The background occasion, recorded in Figure 7, was the *University of Kentucky Centennial Conference on Phase Transitions* commemorating the founding of the university in 1865. While waiting after the meeting at the airport,[3] Onsager explained how, in succession, he had undertaken to diagonalize the transfer matrix by hand: first

---
[3] As related in Chen Ning Yang, *Selected Papers 1945–1980 With Commentary* (W.H. Freeman & Co., San Francisco, 1983).

Figure 7. Newly commisioned Kentucky Colonels with their host at the Conference on Phase Transitions.

the $2 \times \infty$ matrix, then the $3 \times \infty$, and so on! Eventually, by the $6 \times \infty$ case, he confirmed that the $64 = 2^6$ eigenvalues were (essentially) all of the form $\exp(\pm\gamma_1 \pm \cdots \pm \gamma_6)$. That suggested an underlying product algebra which, in turn, had led him to the elaborate structure of his original derivation.

One may wonder what brought Yang and Onsager together. It transpired that Onsager was an old friend of the host at the Conference, W. C. de Marcus, who approached him for advice. Thus, indeed, it was Onsager's approval that brought Yang to the meeting! He also suggested Professor

Mark Kac (by then some years at the Rockefeller University in New York) and, in addition, as a junior speaker the Author of this article. All four, as shown in Figure 7, were awarded a special honor of the *Commonwealth of Kentucky*, that is to say, we were commisioned as Kentucky Colonels. A document recording the author's commision is reproduced in Figure 8. In accepting the commisions on behalf of all four of us, Mark Kac expressed

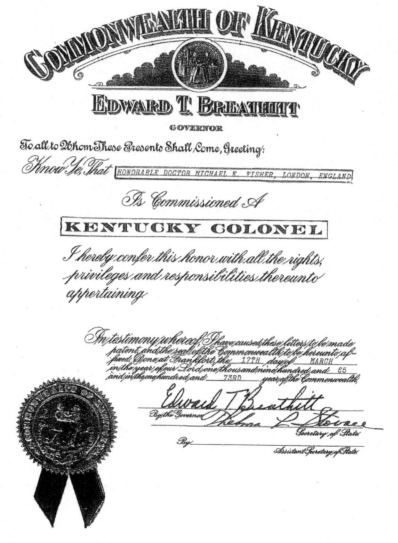

Figure 8. Document announcing the commissioning of Michael E. Fisher as a Kentucky Colonel on 17 March 1965 in the 173rd year of the Commonwealth of Kentucky.

his belief that this "Southern Honor" was particularly appropriate in that Lars Onsager came from Southern Norway, Frank Yang hailed from Southern China, Michael Fisher was brought up in Southern England, while he himself originated in Southern Poland!

4. Two Dimensions in the Real World

The critical exponent, $\beta = \frac{1}{8}$, found for the Ising model by Yang (and confirmed more broadly by Chang) was for a "manifestly artificial model" — as the Ising model was generally regarded in the middle of the last century. Beyond other aspects, the "two-dimensionality" added to the sense of artificiality. But that was no serious obstacle to another Chinese immigrant to the United States, namely, the talented experimentalist M. H. W. Chan. Moses, as he is known in the West, was born in November 1946 in Xi'an but moved with his family to Hong Kong in 1949 and came to America as a student, gaining his Ph.D. at Cornell University.

It was known that if one deposited a submonolayer of methane, CH_4, on graphite at a temperature below about 70 K it phase separated into a low density (or gaseous) "vapor" and a higher density "liquid." At first sight this appeared to be an ideal lattice gas and so, reasonably described by an Ising model. But previous attempts to measure the coexistence curve as a function of temperature and hence determine the exponent β experimentally had, at best, proved inconclusive. It was realized by Moses Chan, however, that the difficulty was closely associated with the steepness of the anticipated "two-dimensional" coexistence curve, clearly evident in Figure 6. He realized that an optimal route is to determine the phase diagram by carefully measuring the specific heat at a range of constant coverages, n: see the specific-heat traces in Figure 9. On crossing the coexistance curve at constant n from the two-phase (vapor + liquid) region below T_c to the supercritical fluid phase a clear break would be evident in temperature plots of the specific heat.

Using this approach, Chan (with his student H. K. Kim) was able, in 1984, to determine the coexistence curve displayed in Figure 10. The fitted value, $\beta = 0.127 \pm 0.02$, agrees remarkably well with Yang's result of $\beta = \frac{1}{8} = 0.125$ — now clearly seen as a "prediction" thirty two years before the observation! In celebration of these theoretical and experimental achievements the greeting shown in Figure 11 was displayed.

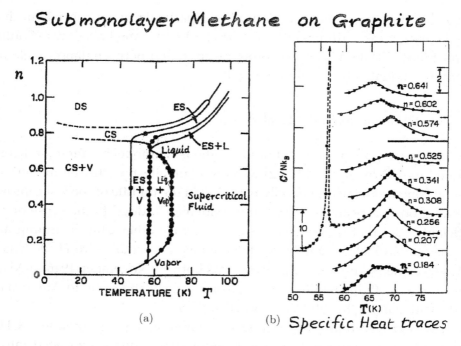

Figure 9. (a) The coverage, n, vs. temperature, T, phase diagram of methane deposited on graphite at low coverages as determined by (b) specific heat plots, C/Nk_B, versus T. [After H. K. Kim and M. H. W. Chan, *Phys. Rev. Lett.* **53**, 170–173 (1984).]

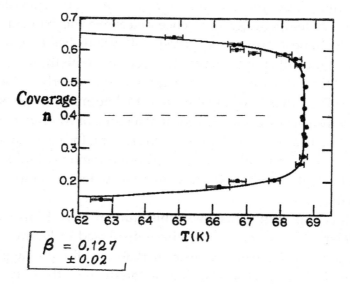

Figure 10. Coexistence curve for temperatures from 62 K to 69 K of submonolayer methane on graphite to determine the critical exponent β. [After H. K. Kim and M. H. W. Chan, *Phys. Rev. Lett.* **53**, 170–173 (1984).]

Figure 11. A good-will message from Moses Hung-Wai Chan to Professor Chen Ning Yang (as displayed on 28 May 2015 in the author's talk to the Conference at the Institute of Advanced Studies in Nanyang Technological University).

5. Condensation and Circles in the Complex Plane

In March 1982 a Sanibel Symposium was organized by Per-Olav Löwdin in Florida, in honor of Joseph Mayer (who died only a year later). In an interesting historical essay[4] C. N. Yang has related how Mayer, in 1937, approached the problem of *condensation* of a fluid from a vapor to a liquid. Mayer developed systematic cluster expansions that, as Yang observes, "started an analysis of the mathematics[5] and physics of such phenomena." In fact, Mayer expected some sort of mathematical singularity at the condensation point (with, as a consequence, the failure of any analytical continuation of an isotherm beyond condensation). But conceptual puzzles

---

[4]C. N. Yang, *Int. J. Quant. Chem.: Quantum Chemistry Symposium*, **16**, 21–24 (1982). See also pages 43–46 in Chen Ning Yang, *Selected Papers II With Commentaries* (World Scientific Publishing Co., Singapore, 2013).

[5]Among authors inspired to address the issues mathematically, Yang cites L. van Hove, *Physica* **15**, 951 (1949). Unhappily, however, as orginally pointed out by N. G. van Kampen and as I reported in *Arch. Ratl. Mech. Anal.* **17**, 377–410 (1964), the second (but unnumbered!) equation in the appendix of van Hove's article is deeply flawed. This fact actually invalidates van Hove's claim of a proof of the existence of a well defined *thermodynamic limit* of the statistical mechanical expression for the free energy of a system.

Statistical Theory of Equations of State and Phase Transitions. I. Theory of Condensation

C. N. YANG AND T. D. LEE

Institute for Advanced Study, Princeton, New Jersey

(Received March 31, 1952)

A theory of equations of state and phase transitions is developed that describes the condensed as well as the gas phases and the transition regions. The thermodynamic properties of an infinite sample are studied rigorously and Mayer's theory is re-examined.

[Handwritten annotations: "for Fluids"; "Go to the Complex Plane of Fugacity: $z = e^{-2H/kT} \propto e^{\mu/kT}$."]

Statistical Theory of Equations of State and Phase Transitions. II. Lattice Gas and Ising Model

T. D. LEE AND C. N. YANG

Institute for Advanced Study, Princeton, New Jersey

(Received March 31, 1952)

[Handwritten annotation: "Totally Equivalent!"]

The problems of an Ising model in a magnetic field and a lattice gas are proved mathematically equivalent. From this equivalence an example of a two-dimensional lattice gas is given for which the phase transition regions in the p–v diagram is exactly calculated.

A theorem is proved which states that under a class of general conditions the roots of the grand partition function always lie on a circle. Consequences of this theorem and its relation with practical approximation methods are discussed. All the known exact results about the two-dimensional square Ising lattice are summarized, and some new results are quoted.

Figure 12. The two-part 1952 article by C. N. Yang and T. D. Lee addressing the statistical mechanical theory of phase transitions. This work established the significance of the complex plane for understanding how a sharp first-order phase transition could appear in a large system.

remained: "How can the gas models 'know' when they have to coagulate to form a liquid or solid?" asked Born and Fuchs in 1938."[4] This was the topic that attracted Yang's attention in 1952. It led to a collaboration with T. D. Lee and to the two-part articles shown in Figure 12. Their recipe, as the figure states, was: "Go to the Complex Plane!"

As indicated in Figure 12, Yang and Lee chose to examine the complex plane of the fugacity z. For a ferromagnet subject to a magnetic field, H, this is simply the corresponding Boltzmann factor; but for a fluid the *chemical potential*, μ, is involved. This variable is much beloved by chemists but anathema to many physicists! However, in terms of z the grand canonical partition function, $\Xi(z;\Omega)$, for a finite domain Ω that contains N particles is merely a polynomial of degree N. Since $\Xi(z)$ must be real and positive for positive real z and T (or μ/T or H/T) all zeroes of the polynomial $\Xi(z)$ must lie in the complex plane (or on the negative real axis). As a result, knowing the distribution of the zeroes amounts to a full knowledge of the thermodynamics. Consequently if the zeroes approach the real axis as $N \to \infty$, a

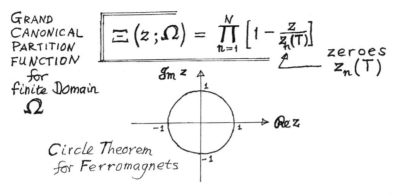

Figure 13. Illustration of the Lee–Yang circle theorem for lattice gases or for ferromagnetic Ising models where $z = \exp(2H/k_B T)$: all zeroes must lie on the unit circle. In the thermodynamic limit this implies that the only possible singularities in the free energy — indicative of a phase transition — must be located in zero magnetic field ($H = 0$) or, for a lattice gas, on a unique locus $\mu_\sigma(T)$.

singularity must appear in the thermodynamic limit at the corresponding value of μ and T.

In Part II of their article Lee and Yang explicitly considered the Ising model and the lattice gas and found them to be completely equivalent mathematically. But the great surprise was the *circle theorem*! Specifically, as illustrated in Figure 13, the zeroes of the grand canonical partition function for any pairwise ferromagnetic interactions whatsoever (of arbitrary range or structure, etc.) were proved to always lie on the unit circle in the complex z-plane. Equivalently, the zeroes are restricted to lie *on* the pure imaginary axis in the complex plane of μ, the chemical potential.

It is worth stressing that, building on the work of Yang and Lee, we now know that Mayer's vision of a singularity at condensation is, in fact, correct. At a first-order transition one must expect[6,7] an *essential singularity*. Thus, all the nth derivatives remain bounded but their behavior as $n \to \infty$ is insufficient to make the associated Taylor series converge.

For many years it was felt that the Lee–Yang zeroes must basically remain unobservable — even though the *density of zeroes*, $g(\theta)$ near the

---

[6] See M. E. Fisher, IUPAP Conf. Stat. Mech. (Brown University, 1962) as recorded by S. Katsura, *Adv. Phys.* **12**, 416 (1963); and *Physics* **3**, 255–283 (1967). Note that the 1967 article is the text of the talk presented at the Centennial Conference on Phase Transitions held at the University of Kentucky, 18–20 March 1965 (as illustrated in Figures 7 and 8, above).

[7] The concept of an essential singularity at condensation was independently advanced by A. F. Andreev, *Sov. Phys. JETP* **18**, 1415 (1964). For a lattice gas the result was proved with full mathematical rigor by S. N. Isakov, *Commun. Math. Phys.* **95**, 427 (1984): see also the discussion in M. E. Fisher, *Proc. Gibbs Symp.* (Yale University, 15–17 May 1989; 1990 American Mathematical Society), pp. 47–50.

real axis must influence the thermodynamics at condensation. Recently, however, Ren-Bao Liu,[8] of the Chinese University of Hong Kong, and his coworkers have demonstrated how Lee–Yang zeroes may be investigated by "measuring quantum coherence of a probe spin coupled to a Ising-type spinbath." Indeed, Dr. Ren-Bao Liu presented his results at the Yang–Mills Conference.

Of course, as the saying has it: "What's good for the goose is good for the gander!" Thus, as pointed out some time ago,[9] the complex plane is also valuable for other variables, perhaps most notably the temperature. This leads to what have been called[8] "Fisher zeroes."

For an Ising model on a rectangular lattice one knows from Kaufman[1] that the canonical partition function for an $m \times n$ torus can be expressed as[9]

$$Z_N^{(1)} = \prod_{r=1}^{m}\prod_{s=1}^{n} \left\{ \frac{1+v^2}{1-v^2}\frac{1+v'^2}{1-v'^2} - \frac{2v}{1-v^2}\cos\frac{2\pi r}{m} - \frac{2v'}{1-v'^2}\cos\frac{2\pi s}{n} \right\}, \quad (5.1)$$

together with three other quite similar products. Here the natural thermal variables, vanishing like $1/T$ as $T \to \infty$, are

$$v = \frac{1-x_1}{1+x_1} = \tanh\left(\frac{J_1}{k_B T}\right), \quad v' = \frac{1-x_2}{1+x_2} = \tanh\left(\frac{J_2}{k_B T}\right). \quad (5.2)$$

For the symmetric case $v = v'$ (or $J_1 = J_2$) it is not hard to see, as illustrated in Figure 14, that all the zeros of (5.1) lie on two circles in the complex v plane. Where one of the circles crosses the real positive axis at $\text{Re}(v) < 1$ corresponds to the ferromagnetic critical point. Conversely, where the other circle crosses the negative v axis [at $\text{Re}(v) > -1$] locates the critical point of the antiferromagnet (for which $J < 0$). If θ is an angle describing the circle centered at $\text{Re}(v) = -1$, the density of zeroes varies as

$$g(\theta) = |\sin\theta| F(\theta), \quad (5.3)$$

where $F(\theta)$ is an analytic function periodic in θ. It then readily follows[9] that the specific heat varies simply as $A \ln |T - T_c|$ — just as orginally found by Onsager!

---

[8]X. Ping, H. Zhou, B.-B. Wei, J. Cui, J. Du, and R.-B. Liu, *Phys. Rev. Lett.* **114**, 010601 (2015); see also B.-B. Wei and R.-B. Liu, *Phys. Rev. Lett.* **109**, 185701 (2012).
[9]M. E. Fisher, "The Nature of Critical Points," in *Lectures in Theoretical Physics, Vol. VIIC* (Univ. of Colorado Press, Boulder, Colorado, 1965), Secs. 13 and 19.

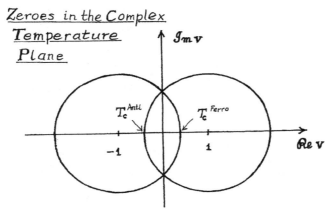

Figure 14. Loci of zeroes of the canonical partition function of the square lattice Ising model in the complex $v = v' = \tanh(J/k_B T)$ plane. The critical points of the ferromagnetic and antiferromagnetic models are indicated.

6. The Universal Yang–Lee Edge Singularity

Let us identify $\mu_\sigma(T)$ as the locus of condensation in the plane of real chemical potential vs. temperature. We may, for convenience, write

$$\Delta\mu(T) \equiv \mu - \mu_\sigma(T) = h'(T) + ih''(T). \tag{6.1}$$

We then know, thanks to the circle theorem, that the Lee–Yang zeroes are confined to the imaginary plane $\text{Re}\{\Delta\mu\} \equiv h'(T) = 0$; hence their density, $g(T; h'')$, is a function only of $h'' \equiv \text{Im}\{\Delta\mu\}$.

Because of the essential singularity at $h'(T) = h''(T) = 0$, always to be expected below the critical temperature, T_c, the density $g(T; h'')$ must be nonvanishing for small h'' when $T < T_c$. On the other hand, when T increases above T_c there must open up a gap, say $h_{\text{YL}}(T) > 0$, that is free of zeroes (in the thermodynamic limit). The gap function, $h_{\text{YL}}(T)$, serves to define the *Yang–Lee edge* as illustrated in Figure 15. On grounds of symmetry we expect two equivalent edges located at $h'' = \pm h_{\text{YL}}(T)$.

Now, as noted by P. J. Kortman and R. B. Griffiths, *Phys. Rev. Lett.* **27**, 1439, (1971) these edges must be branch points of the free energy $F(T, \mu)$. We will call such branch points *Yang–Lee Edge Singularities*[10] and, following Kortman and Griffiths, ask about their nature. It will be argued that they

---
[10] M. E. Fisher, *Phys. Rev. Lett.* **40**, 1610–1613 (1978). See also D. A. Kurtze and M. E. Fisher, *J. Stat. Phys.* **19**, 205–218 (1978) and *Phys. Rev. B* **20**, 2785–2796 (1979).

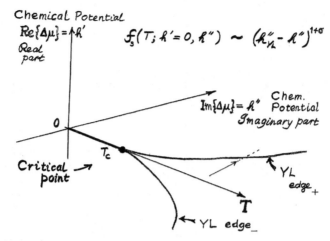

Figure 15. Depiction of the Yang–Lee edges above the critical temperature, T_c, when there is a real condensation locus, $\mu_\sigma(T)$. The real and imaginary parts, (h', h''), of $\Delta\mu \equiv \mu - \mu_\sigma(T)$ are defined in the text; the Lee–Yang zeroes occur only on the plane $h' = 0$.

should be treated as a potentially new class of critical points and analyzed accordingly!

To this end, let us posit that the density of zeroes near the edge varies as

$$g(T, h'') \sim [h_{\text{YL}}(T) - h'']^\sigma, \tag{6.2}$$

when $h'' \to h_{\text{YL}}(T)$ from below and enquire as to the value of the critical exponent σ.

It follows that there is a singular contribution to the total free energy $F(T, h)$, near the Yang–Lee edge which behaves as

$$f_s(T; h' = 0, h'') \approx A_\pm |h_{\text{YL}}(T) - h''|^{1+\sigma}, \tag{6.3}$$

where the amplitudes A_+ and A_- may, in general, depend on the sign of $h'' - h_{\text{YL}}$. This implies that the susceptibility, $\chi \propto (\partial^2 F/\partial h^2)$, diverges with an exponent $\gamma_{\text{YL}} = 1 - \sigma$, where we have anticipated that σ is always less than unity.

Indeed, it is natural to suppose that $\sigma(d)$ depends only on the dimensionality, d, of the system and is, thus, *universal*. This is, in fact, upheld by further investigation.[10] For $d = 1$ the results can be found exactly yielding $\sigma(1) = -\frac{1}{2}$. Furthermore, one can check other aspects of criticality such as *scaling*: specifically, that suggests the correlation function near criticality for an Ising (or more general) system — say with spins s_0 at the origin and s_R

at separation **R** — should behave as

$$G(R; T, h) \equiv \langle s_0 s_\mathbf{R} \rangle - \langle s_0 \rangle \langle s_\mathbf{R} \rangle$$
$$\approx D(Rh^{\nu_c})/R^{d-2+\eta}, \qquad (6.4)$$

when $h, 1/R \to 0$. This can be checked precisely[10] for $d = 1$ yielding the Yang–Lee values $\eta_{\text{YL}}(1) = -1$ and $\nu_c(1) = \frac{1}{2}$. Similarly, hyperscaling relations such as

$$\sigma = \frac{d - 2 + \eta_{\text{YL}}}{d + 2 - \eta_{\text{YL}}} \quad \text{and} \quad \gamma_c = \frac{2}{d + 2 - \eta_{\text{YL}}}, \qquad (6.5)$$

are verified up to a borderline dimension that turns out to be[10] $d_{\text{YL}} = 6$.

Beyond that, consideration of mean-field or Landau-type theories demonstrates that Yang–Lee criticality should, indeed, be universal. However, in place of the standard field theoretic φ^4 analysis, it transpires that an $i\varphi^3$ theory is needed! That being understood, a renormalization-group treatment is possible and leads to

$$\sigma = \tfrac{1}{2} - \tfrac{1}{12}\epsilon \quad \text{and} \quad \eta_{\text{YL}} = -\tfrac{1}{9}\epsilon, \qquad (6.6)$$

to first order in $\epsilon = 6 - d(> 0)$. (In second order the $\frac{1}{9}$ for η is replaced by[10] $\frac{43}{81}$. A third order calculation has been undertaken by Bonfim, Kirkham, and McKane, *J. Phys. A* **14**, 2391 (1981); see also Ref. 11 and Figure 17 below.)

It turns out, however, that this is not the end of the story! In 1984 a physical chemist at Johns Hopkins University, D. Poland, proposed [in *J. Stat. Phys.* **35**, 341 (1984)] that both lattice and continuum *hard-core fluids* are characterized, in the Mayer fugacity expansion of the pressure, by a *universal* dominant singularity on the *negative* fugacity axis, say at $z = z_0 < 0$. On introducing an exponent $\phi(d)$ for this singularity, the pressure displays a contribution of the form

$$p_s(T, z) \sim P(z - z_0)^{\phi(d)}. \qquad (6.7)$$

A subsequent study[11] led to the exponent values $\phi(1) = \frac{1}{2}$, $\phi(2) = \frac{5}{6} = 0.833\ldots$, (derived from Baxter's well known 1980 exact solution of the hard-hexagon model) $\phi(3) \simeq 1.06$, and $\phi(4) \simeq 1.2$. When d became large, it appeared that $\phi(d)$ approached $\frac{3}{2}$.

A natural first question is: "How crucial is the 'hard core' — implying an interaction potential, $u(\mathbf{r})$, which is actually *infinite* over a finite range of particle separation r?" An answer can be found by studying models with a 'soft repulsive core' for which $u(r)$ is always finite. A suitable model, for which low-density series of significant length could be generated for arbitrary d turned out to be the *Gaussian Molecule Mixture* (GMM); on this my student Sheng-Nan Lai labored.[11] This consists of similar A and B particles with interaction potentials

$$u_{AA}(\mathbf{r}) = u_{BB}(\mathbf{r}) = 0 \quad \text{but}$$
$$\exp[-u_{AB}(\mathbf{r})/k_B T] = 1 - e^{-r^2/r_0^2}, \quad (6.8)$$

where r_0 merely sets the scale of the repulsive potential. (Evidently the Mayer f-function is Gaussian.) The singularities arising in the (z_A, z_B) plane for this GMM model are shown in Figure 16 (with $d > 1$).

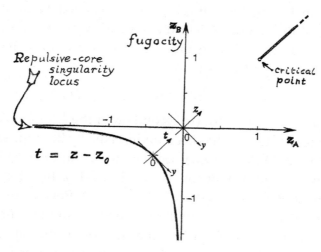

Figure 16. Schematic view of the fugacity plane, (z_A, z_B), for the GMM model showing, in the first quadrant, a first-order phase boundary on the axis of symmetry, $z_A = z_B > 0$, that terminates at a critical point. In the third quadrant is a locus of *repulsive-core singularities* (which, in general, may extend into the second and fourth quadrants).

The numerical values found for $\phi(d)$ as d varied strongly supported the natural hypothesis, namely, that the intrinisic 'hardness' (or otherwise!) did not matter at all! In other words, universality extended to whatever the nature of the particle repulsive cores that generated the singularity. (It is worth remarking that some form of repulsive cores are essential to maintain thermodynamic stability.[12])

---

[11] S.-N. Lai and M. E. Fisher, *J. Chem. Phys.* **103**, 8144–8155 (1995) and references therein.
[12] See, e.g., M. E. Fisher, *Arch. Ratl. Mech. Anal.* **17**, 377–410 (1964).

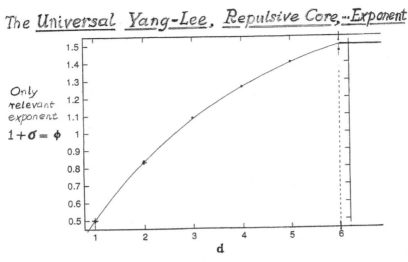

Figure 17. Variation with dimensionality, d, of the universal repulsive-core singularity exponent, $\phi(d)$, now identified with the Yang–Lee edge exponent, $\sigma(d) = \phi(d) - 1$. The borderline dimensionality is $d = 6$; beyond that $\phi = \frac{3}{2}$ and $\sigma = \frac{1}{2}$. The continuous plot embodies an $O(\epsilon^3)$ expansion where $\epsilon = 6 - d > 0$.

In the same way, the similarity of the repulsive core exponents as a function of d to the Yang–Lee exponents led[11] to the explicit proposal

$$\sigma(d) = \phi(d) - 1. \tag{6.9}$$

A theoretical justification for this identification was originally lacking. However, with the aid of a field-theoretic approach which employed separate representations for the repulsive and attractive parts of the pairwise interactions, such a demonstration was eventually constructed.[13] (Note that particle–particle *attractions* are needed to yield condensation, criticality, and a Yang–Lee gap, $h''_{YL}(T)$, above T_c.) Accepting (6.9), enables us to present, in Figure 17 a combined plot of the dimensional dependence of both the Yang–Lee edge exponent, σ, and the repulsive core exponent ϕ.

Is that then the end of the story? Not really! Indeed, separate investigations by many authors[11] have shown not only that $\phi(d)$ describes assorted square and triangular lattice models of hard squares, hard hexagons (as mentioned), and dimers (with holes) but have also established the relations

$$\phi_D(d) = \phi(d-1) \quad \text{and} \quad \phi_I(d) = 1 + \phi(d-2) \tag{6.10}$$

---

[13] Y. Park and M. E. Fisher, *Phys. Rev. E* **60**, 6323–6328 (1999).

for *Directed Site* and *Bond Animals*[14,15] and *Isotropic Site and Bond Animals* or *Branched Polymers*.[16] Note that in the case of animals, etc., of size n one expects the numbers of animals to vary asymptotically as

$$a_n \approx A\lambda^n n^{-\phi_D}, \qquad (6.11)$$

(and similarly for isotropic animals, etc.) where A and λ are lattice-dependent, nonuniversal constants. The corresponding singularities now lie on the *positive* real fugacity axis.[11]

7. The Quantal Many-Body Problem at low T

In Fall 1955 Kerson Huang became a member of the Institute for Advanced Study at Princeton and introduced C. N. Yang to Fermi's *pseudopotential* approach. This soon led to joint work on the quantal many-body problem for Bose and Fermi systems. Ultimately Yang published some 15 or so papers on this topic, many involving a close collaboration with T. D. Lee. The first of these, together with four other early articles are shown in Figure 18. However, unanticipated issues arose — as related in Yang's much later commentary[3] — and, in retrospect, these prove instructive illustrating the vicissitudes to which, even for the most talented scientist, research can lead in the real world!

In the original paper ([A] in Figure 18) the results for the ground-state energy, E_0, of a Fermi system proved satisfactory; but for a Bose system of dilute hard spheres the pseudopotential approach resulted in

$$E_0 = N\frac{4\pi a(N-1)}{L^3}\left\{1 + 2.37\frac{a}{L} + \frac{a^2}{L^2}\left[(2.37)^2 + \frac{\xi}{\pi^2}(2N-5)\right]\right\} + \cdots, \quad (7.1)$$

where a is the hard-sphere diameter, N is the number of spheres, $L \times L \times L$ is the periodic box containing the system, while ξ is a convergent sum over three integers, (l,m,n). For fixed $\rho = N/L^3$, as $L \to \infty$ in the thermodynamic limit, however, the last term in (7.1) does not make sense: it behaves as $\xi N a^2/L^2 = \xi L a^2 \rho$ and so diverges! This observation disturbed the authors but they eventually decided to publish anyway.

Of course, the pressure was then on to find an alternative approach: this led to the *binary collision expansion* method which was presented (under the

---

[14] J. L. Cardy, *J. Phys. A* **15**, L593 (1982).

[15] See D. Dhar, *Phys. Rev. Lett.* **51**, 853 (1983), *ibid.* **49**, 959 (1982), and H. E. Stanley, S. Redner, and Z.-R. Yang, *J. Phys. A* **15**, L569 (1982) (and references therein).

[16] See D. S. Gaunt, *J. Phys. A* **13**, L97 (1980) and G. Parisi and N. Sourlas, *Phys. Rev. Lett.* **46**, 871 (1981).

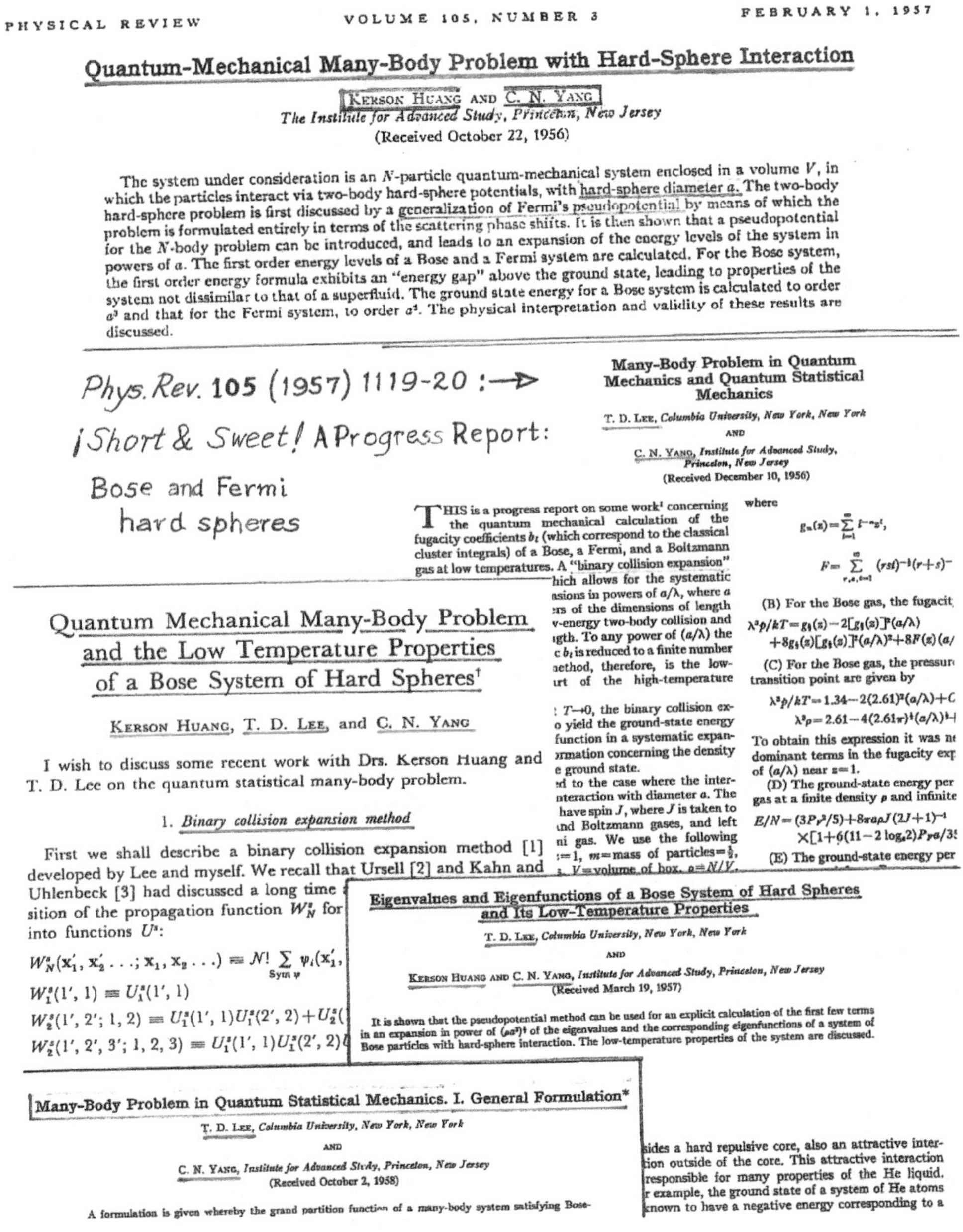

PHYSICAL REVIEW VOLUME 105, NUMBER 3 FEBRUARY 1, 1957

Quantum-Mechanical Many-Body Problem with Hard-Sphere Interaction

KERSON HUANG AND C. N. YANG
The Institute for Advanced Study, Princeton, New Jersey
(Received October 22, 1956)

The system under consideration is an N-particle quantum-mechanical system enclosed in a volume V, in which the particles interact via two-body hard-sphere potentials, with hard-sphere diameter a. The two-body hard-sphere problem is first discussed by a generalization of Fermi's pseudopotential by means of which the problem is formulated entirely in terms of the scattering phase shifts. It is then shown that a pseudopotential for the N-body problem can be introduced, and leads to an expansion of the energy levels of the system in powers of a. The first order energy levels of a Bose and a Fermi system are calculated. For the Bose system, the first order energy formula exhibits an "energy gap" above the ground state, leading to properties of the system not dissimilar to that of a superfluid. The ground state energy for a Bose system is calculated to order a^3 and that for the Fermi system, to order a^2. The physical interpretation and validity of these results are discussed.

Phys. Rev. 105 (1957) 1119-20 :→
¡Short & Sweet! A Progress Report:
Bose and Fermi
hard spheres

Many-Body Problem in Quantum Mechanics and Quantum Statistical Mechanics

T. D. LEE, *Columbia University, New York, New York*
AND
C. N. YANG, *Institute for Advanced Study, Princeton, New Jersey*
(Received December 10, 1956)

THIS is a progress report on some work[1] concerning the quantum mechanical calculation of the fugacity coefficients b_l (which correspond to the classical cluster integrals) of a Bose, a Fermi, and a Boltzmann gas at low temperatures. A "binary collision expansion" ...hich allows for the systematic ...asions in powers of a/λ, where a ...rs of the dimensions of length ...v-energy two-body collision and ...igth. To any power of (a/λ) the ...c b_l is reduced to a finite number ...ethod, therefore, is the low- ...rt of the high-temperature

... $T\to 0$, the binary collision ex- ...o yield the ground-state energy function in a systematic expan- ...rmation concerning the density ...e ground state.
...d to the case where the inter- ...nteraction with diameter a. The ...have spin J, where J is taken to ...nd Boltzmann gases, and left ...ni gas. We use the following ...$=1$, $m=$mass of particles$=\frac{1}{2}$, ... $V=$volume of box, $\rho=N/V$.

where

$$g_n(z)=\sum_{l=1}^{\infty} l^{-n}z^l,$$

$$F=\sum_{r,s,t=1}^{\infty}(rst)^{-\frac{1}{2}}(r+s)^{-}$$

(B) For the Bose gas, the fugacit...

$$\lambda^3 p/kT=g_{\frac{5}{2}}(z)-2[g_{\frac{3}{2}}(z)]^2(a/\lambda)+8g_{\frac{1}{2}}(z)[g_{\frac{3}{2}}(z)]^2(a/\lambda)^2+8F(z)(a/$$

(C) For the Bose gas, the pressur... transition point are given by

$$\lambda^3 p/kT=1.34-2(2.61)^2(a/\lambda)+C$$
$$\lambda^3\rho=2.61-4(2.61\pi)^{\frac{1}{2}}(a/\lambda)^{\frac{3}{2}}$$

To obtain this expression it was n... dominant terms in the fugacity exp... of (a/λ) near $z=1$.

(D) The ground-state energy per ... gas at a finite density ρ and infinite...

$$E/N=(3P_F^2/5)+8\pi a\rho J(2J+1)^{-1}\times[1+6(11-2\log_e 2)P_F a/35$$

(E) The ground-state energy per

Quantum Mechanical Many-Body Problem and the Low Temperature Properties of a Bose System of Hard Spheres†

KERSON HUANG, T. D. LEE, and C. N. YANG

I wish to discuss some recent work with Drs. Kerson Huang and T. D. Lee on the quantum statistical many-body problem.

1. *Binary collision expansion method*

First we shall describe a binary collision expansion method [1] developed by Lee and myself. We recall that Ursell [2] and Kahn and Uhlenbeck [3] had discussed a long time ... sition of the propagation function W_N^s for ... into functions U^s:

$$W_N^s(\mathbf{x}_1', \mathbf{x}_2' \ldots; \mathbf{x}_1, \mathbf{x}_2 \ldots) = \mathcal{N}! \sum_{\text{Sym }\psi} \psi_i(\mathbf{x}_1',$$
$$W_1^s(1', 1) \equiv U_1^s(1', 1)$$
$$W_2^s(1', 2'; 1, 2) \equiv U_1^s(1', 1)U_1^s(2', 2)+U_2^s($$
$$W_2^s(1', 2', 3'; 1, 2, 3) \equiv U_1^s(1', 1)U_1^s(2', 2)$$

Eigenvalues and Eigenfunctions of a Bose System of Hard Spheres and Its Low-Temperature Properties

T. D. LEE, *Columbia University, New York, New York*
AND
KERSON HUANG AND C. N. YANG, *Institute for Advanced Study, Princeton, New Jersey*
(Received March 19, 1957)

It is shown that the pseudopotential method can be used for an explicit calculation of the first few terms in an expansion in power of $(\rho a^3)^{\frac{1}{2}}$ of the eigenvalues and the corresponding eigenfunctions of a system of Bose particles with hard-sphere interaction. The low-temperature properties of the system are discussed.

Many-Body Problem in Quantum Statistical Mechanics. I. General Formulation*

T. D. LEE, *Columbia University, New York, New York*
AND
C. N. YANG, *Institute for Advanced Study, Princeton, New Jersey*
(Received October 2, 1958)

A formulation is given whereby the grand partition function of a many-body system satisfying Bose-

...sides a hard repulsive core, also an attractive inter- ...tion outside of the core. This attractive interaction ...responsible for many properties of the He liquid. ...r example, the ground state of a system of He atoms ...known to have a negative energy corresponding to a

Figure 18. Papers published by Yang on the quantal many-body problem, initially with Kerson Huang and later with T. D. Lee. From the first, at the top, and downwards, the specific references are: [A] *Phys. Rev.* **105**, 767 (1957); [B] *Phys. Rev.* **105**, 1119 (1957); [C] pp. 165–175 in *The Many-Body Problem*, ed. J. K. Percus (Wiley-Interscience, New York, 1963); [D] *Phys. Rev.* **106**, 1135 (1957); and [E] *Phys. Rev.* **113**, 1165 (1959).

names of all three authors) as a lecture by Yang at the *Steven Conference on the Many-Body Problem in* January 1957. This is displayed as [C] in Figure 18; but publication was unhappily delayed for six years. The technique employs a summation of the most divergent terms: applying this Lee and Yang found that the troublesome term identified in (7.1) dropped out after summation leading to

$$\frac{E_0}{N} = 4\pi\rho a \left[1 + \frac{128}{15}\sqrt{\frac{\rho a^3}{\pi}} + \cdots\right]. \tag{7.2}$$

The correction term of order $(\rho a^3)^{1/2}$ can be found merely by summing the most divergent terms (as shown in [C] of Figure 18). The next term, of order $\rho a^3 \ln(\rho a^3)$, was published first by T. T. Wu, *Phys. Rev.* **115**, 1390 (1959) but soon confirmed by others. But T. T. Wu also found that in some cases summation of the most divergent terms — always a risky enterprise! — actually fails: see *Phys. Rev.* **149**, 380 (1960).

Meanwhile a defect had been found in the original application of the pseudopotential method; that invalidated (7.1). Once it was understood that a careful technique of subtractions was needed, it turned out that the pseudopotential method could, in fact, yield the now well-established result (7.2). The details were published in [D], the penultimate article exhibited in Figure 18. While the various calculations by different methods required thorough-exposition, once the results were clear, it was wisely decided that a very short — less than two pages — "*Progress Report*" was called for! That is the second article, [B], in Figure 18. As more-or-less visible in the figure, this article essentially consists of a series, (A) to (E), of explicit formulae for Fermi, Bose, and for the ground state, Boltzmann systems. It was noted that by a classic argument — which I first learned from the writings of Freeman Dyson — all the low-temperature expansions are likely to be of asymptotic character [allowing contributions like $\exp(1/\rho a^3)$].

Finally, in Figure 18 the last article [E], by Lee and Yang, is the first in a series of five papers, published in 1959–60. The last two articles represented the hope[3] of revealing the true nature of the lambda transition to a superfluid in helium-4. But that was unsuccessful; progress in that direction had to await more general findings regarding critical phenomena, their universality and scaling, the renormalization group and the $\epsilon = 4 - d$ expansion, and so on.

8. Off-Diagonal Long-Range Order

Ever since Einstein accepted the ideas sent to him in 1924 by S. N. Bose and applied them to an ideal gas, it has been hard to avoid the picture presented in Figure 19. But how might it relate to the actual phenomenon of superfluidity in liquid helium? And, perhaps, also have some connection to superconductivity in metallic systems? Quantum mechanics clearly matters; but in helium as in real metals, the interactions between atoms are crucial and an ideal gas cannot be an acceptable model.

Personally, my first encounter with this central issue for many-body quantum mechanics was in 1954. Having rejoined King's College, London (now as a doctoral student after a couple of years National Service in Her Majesty's Royal Air Force), I was sent to Cambridge by Cyril Domb, newly arrived as Professor of Theoretical Physics. The reason for the visit was, I believe, related to my duties as Student Librarian in King's for the Physics Department. At the end of the afternoon in Cambridge I looked up Oliver Penrose, a close friend of my brother-in-law-to-be, David Castillejo (the younger brother of Leonardo Castillejo of some fame in high-energy physics arising from the "CDD" paper by Castillejo, Dalitz and Dyson). To my fascination Oliver Penrose told me of his thoughts, (already published three years previously) regarding the intrinsically "wave-mechanical" nature of the 'order' underlying superfluidity in real helium! Not long afterwards Oliver went, as a postdoctoral scholar, to work with Lars Onsager in

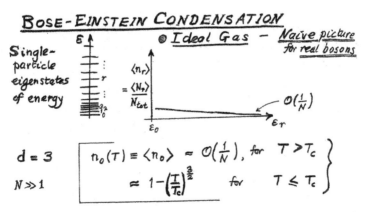

Figure 19. The theoretical picture of condensation in an ideal gas of N bosons. Above T_c the occupancy of the lowest eigenstate of energy remains negligible; but below T_c a finite fraction of the bosons are "condensed" into the ground state, the fraction approaching 100% as $T \to 0$.

> REVIEWS OF MODERN PHYSICS   VOLUME 34 NUMBER 4   OCTOBER 1962
>
> ## Concept of Off-Diagonal Long-Range Order, and the Quantum Phases of Liquid He and of Superconductors
>
> C. N. YANG
>
> *Institute for Advanced Study, Princeton, New Jersey*
>
> **References:**
>
> [1] O. Penrose, Phil. Mag. **42**, 1373 (1951).
>
> [2] O. Penrose and L. Onsager, Phys. Rev. **104**, 576 (1956).
>
> [3] F. London, *Superfluids*, Vol. I (1950), Vol. II (1954).
>
> [4] M. R. Schafroth, Phys. Rev. **96**, 1442 (1954); M.R.S. ··· (1957).
>
> [5] J. Bardeen, L. N. Cooper, and J. R. Schrieffer, Phys. Rev. **108**, 1175 (1957).
>
> [6] N. N. Bogoliubov, Nuovo cimento **7**, 794 (1958).

Figure 20. Yang's fundamental paper on ODLRO, [*Rev. Mod. Phys.* **34**, 694–704 (1962)], showing the first six references visible as footnotes on the title page.

Yale. Together in 1956 they published a basic paper in which, to quote C. N. Yang,[3] they "were the first to give a precise definition of Bose condensation in the case of interacting Bosons."

Yang then introduced the name "*Off-Diagonal Long-Range Order*" (or ODLRO) for the essential concept and, in the article shown in Figure 20, he presented resulting general features for both Bose–Einstein and Fermi–Dirac particles. True to his customary meticulous regard for proper credit, the first six references, reproduced in Figure 20, refer to the previous studies of Penrose and Onsager as well as the less precise ideas of London for superfluids and, for superconductors, of Schafroth and others, including "BCS" and Bogoliubov.

The essential step in defining ODLRO is to recognize that a quantum mechanical system of N particles must be described by a wave function,

$$\Psi_N = \Psi_N(\mathbf{r}_1, \ldots, \mathbf{r}_N; \mathbf{q}) \cong \hat{\psi}(\mathbf{r}), \qquad (8.1)$$

of appropriate symmetry in the positional co-ordinates $\mathbf{r}_1, \ldots, \mathbf{r}_N$, while $\mathbf{q}$ denotes the remaining co-ordinates of the closed system. Following Lev

Landau the overall density matrix ρ is then

$$\rho(\mathbf{r}_i; \mathbf{r}'_j) = \int d\mathbf{q} \Psi_N^*(\mathbf{r}_i; \mathbf{q}) \Psi_N(\mathbf{r}'_j; \mathbf{q}), \tag{8.2}$$

and reduced single-particle, two-particle, ... density matrices can be defined by integrating over the co-ordinates of $(N-1), (N-2), \ldots$ particles.

In terms of the density operator, $\hat{\psi}(\mathbf{r})$ in (8.1), the off-diagonal matrix element of the single-particle density matrix behaves as

$$\rho_1 \equiv \langle \hat{\psi}^\dagger(\mathbf{r}) \hat{\psi}(\mathbf{r}') \rangle \to 0 \quad \text{for} \quad T > T_c,$$
$$\approx \Psi_0^*(T) \Psi_0(T) \neq 0 \quad \text{for} \quad T < T_c, \tag{8.3}$$

when $|\mathbf{r}' - \mathbf{r}| \to \infty$. Clearly T_c locates the transition at which ODLRO, embodied in $n_0(T) = |\Psi_0(T)|^2$, and Bose–Einstein condensation sets in. For liquid helium it is natural to identify $T_c \equiv T_\lambda$ as the lambda point at which superfluidity appears and the specific heat exhibits its extremely sharp spike.

More generally, to include superconductivity one should, following Yang, consider at least ρ_2 the two-particle reduced density matrix. Then large eigenvalues, $\geq O(N)$, and corresponding ODLRO, originate from pairs of fermions, e.g., Cooper pairs of electrons, engaging in Bose–Einstein degeneracy. Thus, the smallest value of n for which ODLRO appears in a reduced density matrix $\rho_m (m \geq n)$ identifies the "collection of n particles that, in a sense, forms the *basic group* exhibiting the long-range correlation." In fact, the article notes that although it is customary to regard liquid helium as a collection of helium atoms obeying Bose statistics, a "much better description is a collection of electrons and He nuclei."

In the 1980's Yang expressed[3] his fondness for this paper but remarked that "it is clearly unfinished." Indeed, there is little doubt that in Yang's mind were also the questions: "How best to define the observable *superfluid density*, $\rho_s(T)$, in microscopic terms?" and, ultimately related: "What is the value of the corresponding critical exponent?" As regards ODLRO it became clear that, in general, this is analogous to $[M_0(T)]^2$ for a ferromagnet; thus the appropriate exponent is 2β. It remained, however, for Josephson[17,18] to convincingly argue that for $\rho_s(T)$ the exponent is slightly

---

[17] B. D. Josephson, *Phys. Lett.* **21**, 608 (1966).
[18] See also M. E. Fisher, M. N. Barber, and D. Jasnow, *Phys. Rev.* A **8**, 1111–1124 (1973), where $\rho_s(T)$ is identified as proportional to the so-called *helicity modulus*, $\Upsilon(T)$, whose definition requires the difference of free energies in *finite-size* systems with *antiperiodic* and *periodic* boundary conditions on the *phase* of the wavefunction (or its operator).

different, namely,

$$2\beta - \eta\nu = 2 - \alpha - 2\nu = (d-2)\nu, \qquad (8.4)$$

where the last equality is restricted to dimensionalities $d \leq 4$. In Eq. (8.4) the exponents α and ν describe, respectively, the specific heat singularity and the divergence of the correlation length, while $\eta < 0.05$ is defined via the critical point decay of the pair correlation function: see Eq. (6.4) above. And, finally, Yang himself drew attention, 35 years ago, to his conjectured formula for the *penetration depth* for superconductors in magnetic fields.

9. Yang–Yang Thermodynamics and the Scaling Axes

When C. N. Yang visited Stanford in 1963 William Little told him of the striking experiments by A. V. Voronel and workers in Russia. This led to the article with his brother, Chen Ping Yang, shown in Figure 21. Voronel's observations on argon and oxygen made it clear that the specific heats of ordinary fluids exhibit very sharp peaks at criticality well described by the form

$$C_V(T;\rho_c) \approx A^{\pm}/|t|^{\alpha} + B, \quad t = (T - T_c)/T_c. \qquad (9.1)$$

This was in direct contradiction to van der Waals (or mean field) predictions. Rather, as Yang and Yang noted, the data were strongly reminscent of the close-to logarithmically divergent specific heat peak for the superfluid transition in helium found by Fairbank, Buckingham, and Kellers in 1957. We now know, however, that in argon and similar fluids, the exponent α introduced in (9.1) is close to 0.109 whereas in superfluid helium it is actually slightly negative with $\alpha \simeq -0.013$. (Note that a logarithmic divergence corresponds, via a simple limit, to $\alpha = 0$.)

Yang and Yang then made a series of remarks of which the first — just visible in Eq. (3) in Figure 21 — was a misleadingly simple thermodynamic formula, which we rewrite as

$$C_V^{\text{total}} = VT\left(\frac{\partial^2 p}{\partial T^2}\right)_V - NT\left(\frac{\partial^2 \mu}{\partial T^2}\right)_V. \qquad (9.2)$$

As they emphasized, this relation, based simply on $SdT = Vdp - Nd\mu$, must hold in the two-phase region below T_c. Hence it must apply to the observations of Voronel and coworkers in (9.1). Consequently, p and μ in (9.2) can be replaced by the condensation loci, $p_\sigma(T)$ and $\mu_\sigma(T)$; thus one

PHYSICAL REVIEW LETTERS

VOLUME 13   31 AUGUST 1964   NUMBER 9

CRITICAL POINT IN LIQUID-GAS TRANSITIONS

C. N. Yang
Institute for Advanced Study, Princeton, New Jersey

and

C. P. Yang
Ohio State University, Columbus, Ohio
(Received 31 July 1964)

In two recent experiments,[1] the specific heats $C_V(T)$ of Ar and of O_2 near the critical temperature T_C have been measured at the critical volume. The measured specific heats display sharp peaks at T_C, suggesting logarithmic infinities. Rough fits of the experimental data yield, for Ar, that

$$C_V = -NT(d^2\mu/dT^2)_V + VT(d^2p/dT^2)_V. \quad (3)$$

Now (3) is applicable to the two-phase region well as the one-phase region. In the two-phase region $d^2\mu/dT^2$ and d^2p/dT^2 are both functions one variable only, T.

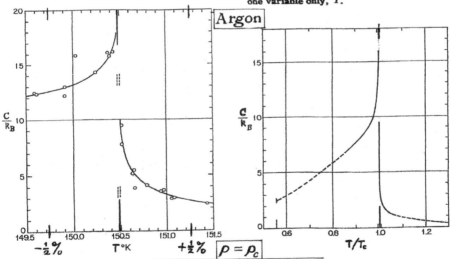

Figure 21. Title and introductory paragraphs of a paper by C. N. Yang and C. P. Yang stimulated by specific heat measurements by Voronel and coworkers [M. I. Bagatskii, A. V. Voronel, and V. G. Gusak, Zh. Eksperim. i Teor. Fiz. **43**, 728 (1962)] on argon, shown in the graphs, and on oxygen a year later.

reaches the question posed pictorially in Figure 22 (and stated verbally[19] in the caption).

Now, on the one hand, it is known that $\mu''_\sigma(T)$ always remains finite in the usual lattice gases, in simple droplet models,[20] and, in other exactly

---

[19] M. E. Fisher and G. Orkoulas, *Phys. Rev. Lett.* **85**, 696–699 (2000); G. Orkoulas, M. E. Fisher, and C. Üstün, *J. Chem. Phys.* **113**, 7530–7545 (2000).

[20] M. E. Fisher, *Physics* **3**, 255–283 (1967). More elaborate cluster-interaction models are in M. E. Fisher

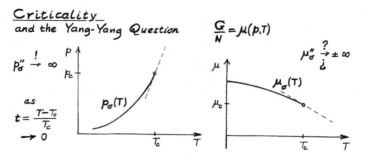

Figure 22. The Yang–Yang Question posed graphically! In words: "Given that the well-defined pressure locus of condensation $p_\sigma(T)$, exhibits a divergent curvature (i.e., second derivative) at the critical point, is the same true (or not?) for the corresponding chemical potential locus, $\mu_\sigma(T)$? And, if true, what is the appropriate[19] dimensionless Yang–Yang ratio, R_μ, measuring the relative strength of the divergence of $\mu_\sigma''(T)$? Might R_μ even be negative? [The dashed lines at criticality indicate a well-defined slope, i.e., first derivative.]

soluble systems.[21] On the other hand in all these cases $p_\sigma''(T)$ becomes singular at T_c! Experimentally, this is also true for H_2O, CO_2 and propane, C_3H_8.[19] For the hard-core square-well fluid this is likewise confirmed by simulation. But as Yang and Yang remark: "For real gases, it is more reasonable to expect that both $(d^2\mu_\sigma/dT^2)$ and (d^2p_σ/dT^2) become ∞" — supposing $\alpha \geq 0$.

But why might this "detail" be significant? The answer is: "Because the orientations of the basic *scaling axes* depend crucially on the behavior of $(d^2\mu_\sigma/dT^2)$." To understand this we define the deviations from critically in the *"space of fields"* (p, μ, T), as

$$\check{p} = \frac{p - p_c}{\rho_c k_B T_c}, \quad \check{\mu} = \frac{\mu - \mu_c}{k_B T_c}, \quad t = \frac{T - T_c}{T_c}. \tag{9.3}$$

Next notice that the full thermodynamics[22] follows from the functional relation between p, μ and T. Consequently, the presence of asymptotic scaling when $\check{p}, \check{\mu}$, and $t \to 0$, can be expressed as

$$\Phi\left(\frac{\check{p}}{t^{2-\alpha}}, \frac{\check{\mu}}{t^{\beta+\gamma}}\right) = 0, \tag{9.4}$$

and B. U. Felderhof, *Ann. Phys. (N.Y.)* **58**, 176–216, 217–267 (1970); B. U. Felderhof and M. E. Fisher, *ibid.* pp. 268–280. None-the-less, some of the models described in Secs. 7 and 8 of these articles do yield a divergence of $\mu_\sigma''(T_c)$.

[21] It is worth noting here, that for ferromagnets (and other systems with a corresponding symmetry) the vanishing of $(d^2\mu_\sigma/dT^2)$ cannot be questioned. Thus the chemical potential is replaced by the magnetic field, H, subject to the symmetry, $H \to -H$, and the phase boundary, $\mu_\sigma(T)$, becomes simply $H = 0$.

[22] See, e.g., E. A. Guggenheim, *Thermodynamics* (2nd Edn., North Holland Publishing, Co., Amsterdam, 1950), Chap. II and, especially, p. 58; R. A. Sack, *Mol. Phys.* **2**, 8 (1959). The corresponding partition function might be described as "great grand canonical."

where, up to second-order corrections, the three *scaling fields* have the form

$$\tilde{p} = \check{p} - k_0 t - l_0 \check{\mu} + O_2(\cdots), \tag{9.5}$$

$$\tilde{\mu} = \check{\mu} - k_1 t, \tag{9.6}$$

$$\tilde{t} = t. \tag{9.7}$$

Note that the dimensionless constants, k_0, k_1, and l_0 (and, to come below, l_1, j_0 and j_1) serve to specify the *nonuniversal orientation* of the scaling axes for $(\tilde{p}, \tilde{\mu}, \tilde{t})$ in the space of fields, (9.2). Recall, however, that the exponents α, β, and γ — for the specific heat [as in (9.1) and (8.4)], for the magnetization or coexistence curve [as in (2.1), (2.2), and (8.4)], and for the susceptibility or compressibility — should be universal.

The coefficient k_1 in (9.6) clearly serves to specify the slope $(d\mu_\sigma/dT_c)$: see dashed line Figure 22. Studies of models, e.g., by Mermin,[23] established the need for a term $-l_1 \check{\mu}$ in (9.7); but, until the question raised by Yang and Yang was considered, that seemed adequate. However, a careful analysis[19] of extensive data for propane provided by Abdulagatov *et al.*[24] — but not available until the late '90s — revealed the existence of a clear "Yang–Yang anomaly," i.e., a divergence of $\mu''_\sigma(T)$ at T_c! Indeed, the Yang–Yang ratio was estimated[19] as $R_\mu = 0.56 \pm 0.04$ (propane). Examination of restricted data[24] suggests a *negative* ratio for CO_2 (with $R_\mu \simeq -0.4$).

Either way it is clear that "pressure mixing" is needed.[19] Thus the previously accepted equations (9.6) and (9.7) must be supplemented to read

$$\tilde{\mu} = \check{\mu} - k_1 t - j_2 \check{p}, \tag{9.8}$$

$$\tilde{t} = t - l_1 \check{\mu} - j_1 \check{p}, \tag{9.9}$$

with two further amplitudes j_1 and j_2 [as well as higher order terms as in (9.5)].

It is then straightforward to see that the condensation boundaries have the form

$$\mu_\sigma(T) = \mu_c + \mu_1 t + A_\mu |t|^{2-\alpha} + \mu_2 t^2 + \cdots, \tag{9.10}$$

$$p_\sigma(T) = p_c + p_1 t + A_p |t|^{2-\alpha} + p_2 t^2 + \cdots. \tag{9.11}$$

On returning to the Yang–Yang expression (9.2), one sees that both amplitudes A_μ and A_p specify the divergence (for $\alpha > 0$) of the specific

---
[23] N. D. Mermin, *Phys. Rev. Lett.* **26**, 169 (1971).
[24] I. M. Abdulagatov *et al.*, *Fluid Phase Equilib.* **127**, 205 (1997); data for CO_2 by I. M. Abdulagatov *et al.*, *J. Chem. Thermodyn.* **26**, 1031 (1994).

heat at constant volume. The Yang–Yang ratio then follows from $R_\mu = -j_2/(1-j_2)$.

But a further surprise remains! Sufficiently accurate measurements of the liquid and vapor phase boundaries, $\rho_{\text{liq}}(T)$ and $\rho_{\text{vap}}(T)$, can determine the coexistence curve diameter. Scaling now predicts

$$\rho_{\text{diam}}(T) = \rho_c[1 + c_{1-\alpha}l_1|t|^{1-\alpha} + c_{2\beta}j_2|t|^{2\beta} + c_1 t + \cdots], \qquad (9.12)$$

with unsuspected power-law correction terms, with amplitudes $c_1, c_{2\beta}$, and $c_{1-\alpha}$. The correction proportional to l_1 is of order $|t|^{0.89}$; the surprise is that pressure mixing induces a new term, of order $j_2|t|^{0.65}$, which dominates that previously known! Clearly, both power-law corrections dominate the anticipated term $c_1 t$, embodied in the *Law of the Rectilinear Diameter*.

10. Yang–Yang Anomaly and "Compressible" Fluid Models

It is natural to ask if there are models that might reveal something about the origins of pressure mixing in the scaling fields as is demanded by Yang–Yang anomalies both observed[19] and subsequently found in precise simulations.[25] To this end consider, first, the standard nearest-neighbor lattice gas (equivalent to a basic Ising model). As illustrated in Figure 23(a), this can be regarded as representing a fluid of hard spheres of diameter a interacting via an attractive square well of range b. Provided the lattice spacing is less than b, which, in turn, must be less than the next-nearest-neighbor lattice distance, the only approximation is the restriction of the particles to lattice sites.

But the lattice gas can be viewed instead[26] as a "cell gas" [illustrated in Figure 23(b)] in which one identifies adjacent lattice *cells* of volume v_0. A cell may be empty or (due to the hard cores) occupied by no more than a single particle that can move freely in its cell. This motion contributes a Boltzmann factor $\exp(-pv_0/k_B T)$ for each particle. The main approximation is that the attractive well comes into play only between particles in nearest-neighbor cells and does not depend on their actual positions within the cells.

---

[25] Y. C. Kim, M. E. Fisher, and E. Luijten, *Phys. Rev. Lett.* **91**, 065701 (2003). This work led to $R_\mu \simeq -0.044$ for the hard-core square-well fluid (with $b/a = 1.50$ in Figure 23) and to $R_\mu \simeq 0.26$ for the restricted primitive model of an electrolyte. See also: G. Orkoulas, M. E. Fisher, and A. Z. Panagiotopoulos, *Phys. Rev. E* **63**, 051507 (2001); Y. C. Kim, M. E. Fisher, and G. Orkoulas, *Phys. Rev. E* **67**, 061506 (2003).

[26] This way of regarding a lattice gas was learnt from Benjamin Widom (private communication).

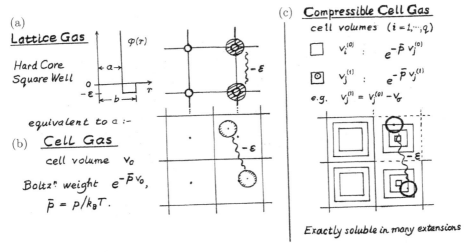

Figure 23. Illustration of (a) the standard *lattice gas* and (b) its reinterpretation as a *cell gas* which, in turn, suggests (c) the *compressible cell gas* model which, in general, yields a Yang–Yang anomaly.

How might this model be improved? Can one relax the "rigidity" imposed by the lattice structure? The alternative (c) in Figure 23 presents one such extension: Even in the absence of particles the lattice volume may fluctuate and respond to pressure: thus individual cells are allowed to assume distinct volumes, $v_i^{(0)}$ ($i = 1, 2, \ldots, q$), that lead to corresponding (but independent) Boltzmann factors $\exp(-pv_i^{(0)}/k_B T)$. The same idea may be used for singly-occupied cells: thus the model supposes that particle motion is restricted to a (possibly) different set of cell volumes, $v_i^{(1)}$. It is natural, but not necessary, to suppose that each $v_i^{(1)}$ is smaller than the associated $v_i^{(0)}$; simplest is to take $v_i^{(1)} = v_i^{(0)} - v_\sigma$ for fixed $v_\sigma > 0$.

Now when the original lattice gas can be analyzed (say, numerically) this *Compressible Cell Gas* (or CCG) proves equally soluble! And the same applies to many of its extensions and different versions.[27] As an illustration Figure 24 presents the simple case where $q = 2$ and $v_2^{(0)} = 2v_1^{(0)}$ while $v_\sigma = v_1^{(0)}$ (the underlying lattice being simple cubic). Evidently the Yang–Yang ratio, R_μ, is now unmistakably *negative*[27] (as, one may recall,[19] seems so for carbon dioxide). However, positive ratios, R_μ, are found when the cell

---

[27] R. T. Willis and M. E. Fisher [poster], *Thermo 2005* Meeting, University of Maryland, 29–30 April 2005; C. A. Cerdeiriña, G. Orkoulas, and M. E. Fisher [abstract and poster], *XV Congress of Stat. Phys.*, Royal Spanish Phys. Soc., Salamanca, 27–29 March 2008; [abstract] *Proc. 7th Liquid Matter Conf.*, Lund, June 2008; *Phys. Rev. Lett.* **116**, 040601 (2016).

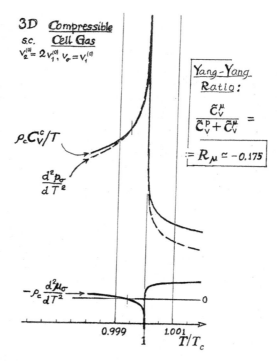

Figure 24. The components of the Yang–Yang relation (9.2) for a simple compressible cell gas on the simple cubic lattice yielding a negative ratio R_μ.

volumes, $v_i^{(1)}$, are coupled with distinct interaction energies, in particular when smaller cells are associated with larger energies.[27]

11. The Richness of Statistical Mechanics

Following his first paper in 1964, C. N. Yang wrote half a dozen further articles with his brother Chen Ping Yang, in the next five years. Many addressed anisotropic Heisenberg spin systems, particularly in one-dimensional chains. Three papers, in particular; *Phys. Rev.* **150**, 321, 327; **151**, 258 (1966), discussed, extended, and exploited the mathematics originally introduced for Heisenberg spins by Hans Bethe over three decades earlier in 1931.[28] A notable by-product, published with B. Sutherland as first author, reported an: "Exact Solution of a Model of a Two-Dimensional Ferroelectric in an Arbitrary External Electric Field," *Phys. Rev. Lett.* **19**, 588 (1967).

---

[28]H. A. Bethe, *Zeit. Physik.* **71**, 205–226 (1931).

From a more mathematical point of view the analysis also led specifically to the so-called *Yang–Baxter equations*.[29] The first publication where these equations appeared was by C. N. Yang in: "Some Exact Results for the Many Body Problem in One Dimension with Repulsive Delta-Function Interaction," *Phys. Rev. Lett.* **19**, 1312 (1967). But space did not allow proofs of some of the crucial equations; complete details, however, were supplied by[30] M. K. Fung in *J. Math. Phys.* **22**, 2017 (1981). A fuller presentation of the 1D model had to wait over twenty years until the appearance in *Commun. Math. Phys.* **122**, 105 (1989) with first author Gu Chao-Hao, who visited the State University of New York at Stony Brook from the Chinese University of Science and Technology in Hefei. Gu and Yang presented a 1D Fermionic model with an explicitly factorized S matrix. The Yang–Baxter equations for the related matrices X_{ij}^{ab}, namely,

$$X_{ij}^{ab} X_{kj}^{cb} X_{ki}^{ca} = X_{ki}^{ca} X_{kj}^{cb} X_{ij}^{ab}, \tag{11.1}$$

appeared as consistency conditions for the validity of Bethe's hypothesis.

An exposition of the essential steps in using the hypothesis had been presented in 1970 at a Winter School of Physics in Kapacz in Poland.[3] Much later, on reviewing his "Journey through Statistical Mechanics" in 1987 at the Nankai Institute of Mathematics,[30] Yang pointed out the similarity of the Yang–Baxter equations to the braid group relation $ABA = BAB$ used in the classification of knots. The reader was invited to conclude that perhaps "these cubic equations embody some fundamental structure still to be explored."

After 33 years at SUNY in Stony Brook, Chen Ning Yang retired in 1999. Four years later, in December 2003, he moved to Tsinghua University in Beijing. There he had the opportunity of developing his passion for the history of physics, especially as regards its philosophical and conceptual roots. Recall that, years earlier in 1982, while still in the United States, he[4] had drawn attention to Joseph Mayer's 1937 attack on the statistical mechanics of condensation in fluids. But still the stimulus of new experiments and the challenge of understanding them led once again to work in statistical physics or, essentially indistinguishable at the theoretical level, "condensed

---

[29] R. J. Baxter, *Phys. Rev. Lett.* **26**, 832–833 (1971), solved exactly the so-called eight-vertex model: see also *Exactly Solved Models in Statistical Mechanics* (Academic Press, London, New York, 1982), Chaps. 9 and 10; and "Some Academic and Personal Reminiscences of Rodney James Baxter", *J. Phys. A Math. Theor.* **48**, 254001 (2015).
[30] See C. N. Yang, *Selected Papers II, With Commentaries* (World Scientific Publishing Co., Singapore, 2013).

matter physics." Note in particular, a paper alone in 2009 and further articles in 2009–2010 concerning fermions in one-dimensional traps with Ma Zhong-Qi and, concerning multicomponent fermions and bosons, with You Yi-Zhuang.[30]

The main focus of the career of Chen Ning Yang — as also true of this meeting — has clearly been on quantum field theory for the understanding of fundamental particles or "high-energy physics." It may, hence, be appropriate to draw attention to the proceedings[31] of a meeting held in Boston University on 1–3 March 1996 concerning the foundations of quantum field theory. The two-day symposium, followed by a workshop, was sponsored by the Center for Philosophy and History of Science at Boston University. Among the speakers were David J. Gross, who featured here in Singapore, and I myself. There were lively discussions and, as the *Preface*[31] put it: "interesting material about the tension between two groups of scholars."

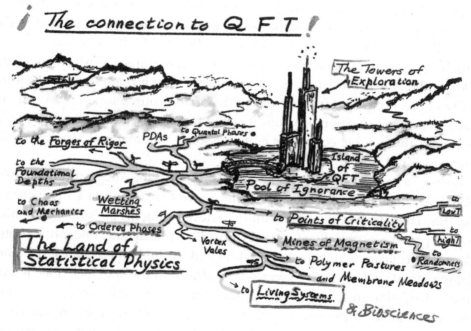

Figure 25. The wide "Land of Statistical Physics" connected by a now sturdy bridge to the "Island of Quantum Field Theory".

---

[31]T. Y. Cao (editor), *Conceptual Foundations of Quantum Field Theory* (Cambridge University Press, Cambridge, 1999), Presentations 1 to 26 in parts I to X.

C. N. Yang is almost unique in having contributed fundamentally to both the broader aspects of statistical mechanics and to central issues in quantum field theory. In concluding my present contribution, I would like to draw attention to the article by David Nelson entitled: *"What is Fundamental Physics? A Renormalization Group Perspective"* and the subsequent discussion (Sec. 18, pages 264–267, in the proceedings[31]). Finally, I offer Figure 25 as a pictorial representation of the richness of statistical physics and the connections to quantum field theory.

CPSIA information can be obtained
at www.ICGtesting.com
Printed in the USA
BVOW07s0726281217
503781BV00029B/114/P